D1511756

Biomathematics

Volume 6

David Smith · Nathan Keyfitz

Mathematical Demography

Selected Papers

With 31 Figures

Springer-Verlag Berlin Heidelberg New York 1977

David P. Smith
Nathan Keyfitz

Harvard University, Center for Population Studies, 9 Bow Street,
Cambridge, Mass. 02138 (USA)

AMS Subject Classifications (1970)
92-01, 92-03, 92A15, 01A55, 01A60, 01A75

ISBN 3-540-07899-1 Springer-Verlag Berlin · Heidelberg · New York
ISBN 0-387-07899-1 Springer-Verlag New York · Heidelberg · Berlin

Library of Congress Cataloging in Publication Data Main entry under title: Mathematical demography. (Bio-mathematics; v. 6) Bibliography: p. Includes index. 1. Demography—Mathematical models—Addresses, essays, lectures. I. Smith, David P., 1944— II. Keyfitz, Nathan, 1913—   HB885.M33   519.5   77-21464

© by Springer-Verlag Berlin · Heidelberg 1977.
Printed in Germany.
Typesetting and printing: Beltz, binding: Konrad Triltsch, Würzburg
2145/3120-543210

216504

For John and Harriet

Preface

This volume is an effort to bring together important contributions to the mathematical development of demography and to suggest briefly their historical context. We have tried to find who first thought of the several concepts and devices commonly used by demographers, what sort of problem he was facing to which the device or concept seemed the solution, and how his invention developed subsequently in the hands of others.

Historically, the book starts with a Roman table of life expectancies from the third century a.d. about which we know little, and with John Graunt's explorations in an area that was still popularly suspect when he wrote in 1662. These are followed by the astronomer Halley, who looked into the field long enough to invent the life table and to notice that Their Majesties would take a sizeable loss on the annuity scheme they had just launched; and by Euler, who was first to devise the formulas of stable population theory and to apply them to filling gaps in data. To these we add the handful of further contributions in the 19th century and many pieces from the explosion of contributions that began in this century with Lotka. We doubt that we have managed to trace everything back to its ultimate beginning, and suspect that our nominees in some cases have been anticipated by predecessors who will be turned up by other students.

The works we include form a living heritage in demography: Graunt; Halley; Euler; Lotka; Milne, who formalized life table construction; Lexis, who was preoccupied with the way members of a population are situated simultaneously in age and in time, and showed how a plane chart, now known as a Lexis diagram, can help analysis. Much less alive, and largely excluded here, are such notions as that of George King, that graduation of data for a life table was more accurate from pivotal death rates calculated at five-year intervals; John Graunt's belief that the right way to describe the dynamics of a population was as the *ratio* of births to deaths, without considering age; and devices that once reduced the labor of numerical calculation but are obsolete in a computer age. These and many other ideas that have proved to be dead ends and are now of merely antiquarian interest we tried to distinguish from those that were part of a chain of development that is still advancing. As far as we could discriminate our excerpts are confined to the latter.

To determine which works most deserve attention among the large number written has not been easy, and we have undoubtedly made mistakes both of inclusion and of omission. We were far from insisting on subtle mathematical ideas, but did look for the effective uses of mathematics that have come to be assimilated into population work. Articles that profess to deal with population but whose main interest was mathematics we tried to avoid, and we avoided them doubly if they were a mere import from some other subject that seemed unlikely ever to be naturalized in population analysis. Some ideas and techniques have a kind of *droit de la cité* in contemporary population study, and we hope these are the ones that predominate in our selections.

To find passages that were self-contained and suitable for contemporary reading was occasionally difficult. Writers often used symbols well known to their place and time, and their immediate readership had no need for definitions we would now miss. To this the earlier works add key formulas with no hint as to how they are derived. Where we expect readers to have trouble as a result, because we did, we include a brief explanation of what is being done.

The choice of excerpts from the classic articles and books rather than complete reprints in all cases was dictated partly by economy of publication, but this was not the only constraint. Benjamin Gompertz fairly compactly introduces his Law of mortality, but spends above fifty pages fitting it to life tables and working out its implications for annuity payments. Harro Bernardelli published the first article on the use of matrices in population projection in the *Journal of the Burma Research Society*, which is not a source that most of us would come across in our ordinary reading. He has top priority for inclusion, but he deals partly with problems of the Burmese economy under British colonial rule and with speculations on cyclic events that do not carry much interest for readers today. Leslie, whose reading in a sickbed had taken him deep into the mathematical properties of matrices, went into cogredient and contragredient transformations that are unlikely to have demographic application. We saw no need to burden the reader with these only to have him discover at the end that he would never need them.

In editing we did not strip down our authors to the point of losing the context of their contribution to our subject. We learned much of an incidental character in our reading and have tried to retain that richness. Where substantive omissions are made we note these for the reader's benefit.

Several topics that fall in the province of demography are not included, among them treatments of human spatial ecology, urbanization, and migration. Omission is partly due to space limitations, and partly to lack of confidence in our ability to decide what is basic in fields whose mathematical explorations are recent and expanding rapidly.

We expect from the reader at least some background in calculus and matrix algebra, and several papers will require an understanding of stochastic processes. The reader lacking a background in elementary mathematics will find the greater part of the book difficult.

Secondary accounts of much of what we present can be found in Keyfitz (1968), and stochastic processes are well handled in Feller (1968) and Chiang (1968). Our chief sources for the early histories given here are Hendriks (1852, 1853), Westergaard (1969), and Lorimer (1959).

Table of Contents

X

The Life Table

Mathematical demography has its modern beginnings in the gradual development of correct procedures for forming life tables, and in a single remarkable paper by Leonard Euler (1760, paper 11 below) that introduced stable age distributions. One far older work of at least some quality survives: a third century a.d. table of life expectancies attributed to Ulpian (paper 1) that remained in use in northern Italy through the 18th century. The table and an accompanying discussion have been taken from C.F. Trenerry (1926).

The mathematician Girolamo Cardano took up the problem briefly in 1570, but without substantive results. Cardano made the assumption that a man who took great care in all things would have a certain life expectancy α, so that $\mathring{e}_x = \alpha - x$ for all ages x, and then asked what part of this might be forfeited by a relaxation of prudence. He proposed letting life expectancy fall by $\frac{1}{40}$ of its value during each year in which a man was reasonably careful but not fastidious: by the nature of the life expectancies, a man might be born with the prospect of living say 260 years and yet die at age 80, having every year thrown away by inattention a part of what remained to him (Cardano 1570, pp. 204—211). A modern interpretation would be that a cohort carries its life table with it. The result was not generalized to populations.

John Graunt's *Natural and Political Observations Upon the Bills of Mortality* (1662) is the first substantive demographic work to have been written. The book is occasionally curious but most often impressive, even from a perspective of three hundred years: Graunt culled a remarkable amount of information from the christening and death lists begun in the later plague period and usually understood its implications. Parts of the treatise are included here as paper 2.

A second work of great importance followed upon Graunt's, Edmund Halley's (1693) presentation of the Breslau (Wroclaw Poland) life table (paper 3). Halley had made an effort to obtain the Breslau lists in order to see what might be done with them, after learning of their apparent quality.

The methods of calculation Halley used in his life table were partly informal, as in his remarks on stationarity and in his unorthodox subtraction (where l_x is the number of survivors to exact age x in the life table from among l_0 births and L_x represents the number between ages x and $x+1$)

$$890\,[=l_1] - 198 = 692\,[=L_6],$$

1

explained by the oblique statement: "198 do die in the *Five Years* between 1 and 6 compleat, taken at a *Medium*." [The terminology has created confusion down to the present century. Raymond Pearl (1922, p. 83), apparently reading the L_x terms that make up Halley's table as l_x, calculated life expectancy at birth as 33.5 years by the table instead of the correct 27.5. The mistake is carried over in Dublin, Lotka and Spiegelman (1949, p. 34)].

Johan DeWit (1671) preceded Halley in the correct calculation of annuities, using exact (l_x) as against Halley's approximate (L_x) denominators, and most of Halley's other *Uses* can be answered differently, but the quality of Halley's table and discussion much surpasses the few earlier works and several of the subsequent ones. His table is graphed below, alongside Ulpian's, Graunt's, DeWit's, and as references the middle level table for Crulai c. 1700 (Gautier and Henry 1958, pp. 163, 190) and one of the Coale and Demeny (1966) model life tables.

After Halley, the next impressive contribution was the series of life tables for annuitants and monastic orders by Antoine Deparcieux, printed in 1746. The accuracy of Deparcieux's data was sufficient for him to show that adult life expectancies had been increasing over the previous half century. Deparcieux calculated his $\mathring{e}_x$ values by the simple but adequate formula

$$\mathring{e}_x = \frac{\sum_{i=x}^{\omega}(l_i-l_{i+1})(i+0.5-x)}{l_x} = \frac{\sum_{i=x}^{\omega}l_i}{l_x} - 0.5 .$$

Of their utility he writes (1760, pp. 58—59): "Les vies moyennes [i.e., $\mathring{e}_x$] sont ce qui m'a paru de plus commode pour faire promptement & sans aucun calcul, la comparaison des différent ordres de mortalité qu'on a établis … [Life expectancies are what have appeared to me most convenient for making promptly and without any calculation a comparison of different orders of mortality that one has established]."

Two later efforts merit attention here: Daniel Bernoulli (1766) introduced continuous analysis and suggested the force of mortality $[\mu(x)]$ in an application of differential calculus to the analysis of smallpox rates. Later Émmanuel Étienne Duvillard (1806), in an article that also introduced the T_x column (defined as $T_x = \sum_x^{\omega}L_i$), applied Bernoulli's method to estimate the increase in life expectancy that would follow if smallpox were eliminated by Edward Jenner's vaccine. The calculus, which these and most modern work employ, dates to a seventy year period (1665—1736) between Isaac Newton's first investigations and the publication of his principal works. Westergaard (1969, pp. 92—93) comments however that it was not until the late nineteenth century that continuous analysis was widely enough understood for Bernoulli's work to be appreciated.

Joshua Milne in his excellent *Treatise on the Valuation of Annuities* (1815), which includes a careful analysis of life tables made prior to his, was first to suggest a formula by which l_x values could be calculated for real populations.

His is the well known expression

$$d_x = l_x \left(\frac{D_x}{P_x + \frac{1}{2}D_x} \right)$$

where d_x is the number dying between ages x and $x+1$ in the life table population, D_x represents calendar year deaths between these ages in an observed population, and $P_x + \frac{1}{2}D_x$ constructs an initial population analogous to l_x except for scale by adding half of the yearly deaths to the observed midyear population P_x.

Excerpts from Milne's discussions of the life table and of age-specific fertility rates (due to Henric Nicander (1800, pp. 323—324)) are given in paper 4. Milne's method for graduating data from grouped to single ages, not the best, has been omitted. His footnoted criticism of Thomas Simpson's (1742) work opens an area of discussion that is easily missed: Milne's life table was misread by William Sutton (1884)—whose clarification in 1874 of the construction of Richard Price's 1771 Northampton Table is a more competent work—but was immediately reestablished by George King (1884). Like other fields, demography does not only move forward.

The fifth article in this section excerpts from George King (1902), whose notation is contemporary, his explanations of terms of the life table. From the middle of the 19th century William Farr standardized much of the life table, but he did not put his work in a form at all comparable to King's excellent textbook. [A recent addition to the life table, from C. L. Chiang (1960a), is the term $_na_x$. This is King's unremembered "average amount of existence between ages x and $x+n$," belonging to those who die between these ages," i. e.: $_na_x = \frac{_nL_x - n\, l_{x+n}}{_nd_x}$.]

Out of sequence, the Lexis (1875) diagram is introduced in paper 6. For most of a century it has been a standby of all analysis attempting to relate age and time. Among contemporary works, those of Roland Pressat (1969, 1972) exploit it most fully.

The important contributions to the life table in this century have been competent abridgement techniques for generating tables by five or ten year age groupings in place of single years of age. The Lowell Reed and Margaret Merrell (1939) article included here as paper 7 did much to establish the validity of abridgement techniques by its introduction of an attractive expression:

$$_nq_x = 1 - \exp\left[-n\,_nm_x - a\,n^3\,_nm_x^2\right]$$

for estimating $_nq_x$ from $_nm_x$ values where wide age groupings are used. In the expression, $_nm_x$ is the age-specific death rate in the life table population ages x to $x+n$ (that is, $_nm_x = {_nd_x}/{_nL_x}$) and $_nq_x$ the probability of dying within the interval for a person of exact age $x({_nq_x} = {_nd_x}/{l_x})$. By empirical examination the authors found that the constant a required by the expression could be the same for all ages above infancy. Reed and Merrell examine two other approximations to

3

$_nq_x$, the first of which:

$$_nq_x = \frac{n \, _nm_x}{1 + \dfrac{n}{2} \, _nm_x}$$

can be derived from Milne's formula for d_x; the other due apparently to Farr (1864, pp. xxiii—xxiv), and evident earlier to Gompertz (1825, paper 30 below):

$$_nq_x = 1 - \exp[-n \, _nm_x].$$

Following their article we include derivations for both expressions.

T. N. E. Greville (1943) was able through ingenious expansions to derive each of these equations by working with the definitions of $_nm_x$ and $_nq_x$ and to show that the Reed-Merrell formula incorporates Gompertz' Law that the force of mortality is an exponential function of age. The assumption is appropriate at older ages and inappropriate for infancy, and thus defines the age range over which the Reed-Merrell formula is applicable. In the same article Greville discusses approximations to $_nL_x$ values where, as before, age groupings are wide (paper 8).

The methods used by Greville can be generalized to take advantage of the observed age structure of a population as well as its mortality schedule. In the simplest case this gives rise to the formula, due to Nathan Keyfitz and James Frauenthal (1975),

$$_nq_x = 1 - \exp\left[-n \, _nM_x + \frac{n}{48 \, _nP_x} (_nP_{x+n} - _nP_{x-n})(_nM_{x+n} - _nM_{x-n}) \right]$$

with as before the caveat that infancy requires separate consideration.

The chapter concludes with excerpts from the well known article by Edward Deevey (1947) in which he evaluates efforts that had been made up to that time to develop life tables for animal populations.

Sampling variances of life table terms are taken up in chapter 7.

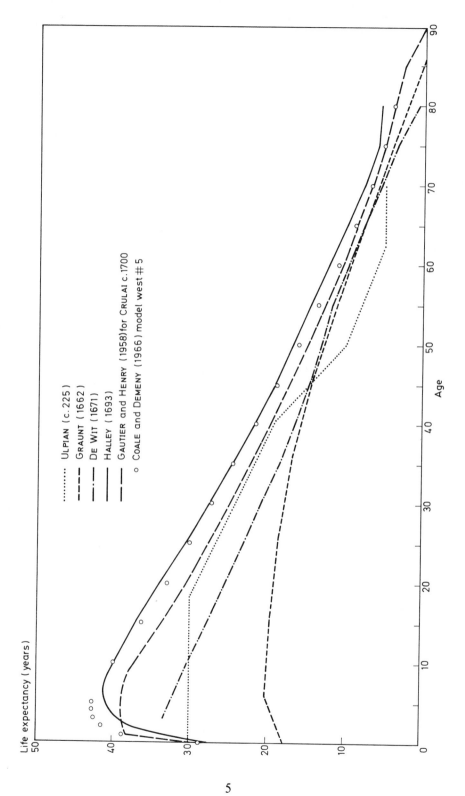

1. Tables of Annuity Values Which Were Sanctioned by the Roman Law for the Purposes of the Lex Falcidia

C. F. TRENERRY (1926)

From *The Origin and Early History of Insurance*, pp. 151—152. London: P.S. King & Son.

On the question of whether Ulpian's table contains an interest factor Hendriks (1852, p. 224) writes: "... as it was the special duty of the jurist to look to a sufficient fund being created for the discharge of the obligations under such annuities, he is not likely in those times to have taken much account of *interest*, which at best could only have been continuously obtained with great difficulty." The table values should be recognized as excessive if an interest rate much above 1% is incorporated, and we believe with Hendriks that none is. The table is reproduced below:

Age	Provision
0—19	30 years
20—24	28
25—29	25
30—34	22
35—39	20
40	19
41	18
42	17
43	16
44	15
45	14
46	13
47	12
48	11
49	10
50—54	9
55—59	7
60+	5

ALTHOUGH there is no trace of any attempt on the part of the Romans to construct a table of mortality or rates of subscriptions payable by members of burial clubs, yet there are records of two tables which were sanctioned by the civil law for the capitalization of annuities. The necessity for such tables arose from the provisions of the *Lex Falcidia*,[1] by which it was ruled that the heir or heirs to an estate should receive not less than one-quarter of the total property left by the testator. As it frequently happened that the will was made when the testator was in better circumstances than at the time of his death, it became necessary in such cases for the values of the various legacies to be proportionately reduced in order that the total value should not exceed three-quarters of the estate. In the case of the ordinary legacy, this reduction presented no difficulty, but when the testator had left a life annuity to one or more legatees, the question arose as to the basis of reduction of the values of such annuities. In order to meet this requirement, it was necessary to capitalize the annuities, and, in order to permit of this being done, a table of annuity values, known subsequently as " Macer's Table," was used. This table was, as will be seen from the quotation below, authorized in a *responsum* of the jurisconsult Macer, who, at the same time, sanctioned the use of another and more correct table, the authorship of which he attributed to Ulpian.[1'] In the case of the first table,

No trace of tables of premiums, etc., used by Romans.

Lex Falcidia.

Two tables of annuity values authorized.

---

[1] Passed 40 B.C. See *Digest*, XXXV, ii, 1.

[1'] *Digest*, XXXV, ii, 68. Æmilius Macer, Lib. 2, ad legem vicesimam hereditatium. "Computationi in alimentis faciendæ hanc formam esse Ulpianus scribit, ut a prima ætate usque ad annum vicesimum quantitas alimentorum triginta annorum computetur, ejusque quantitatis Falcidia præstetur : ab annis vero viginti usque ad annum vicesimum-quintum annorum viginti octo : ab annis vigintiquinque usque ad annos triginta, annorum vigintiquinque : ab annis triginta usque ad annos trigintaquinque, annorum vigintiduo : ab annis trigintaquinque usque ad annos quadraginta, annorum viginti : ab annis quadraginta usque ad annos quinquaginta, (tot) annorum computatio fit, quot ætati ejus ad annum sexagesimum deerit, remisso uno anno : ab anno vero quinquagesimo usque ad annum quinquagesimum quintum, annorum novem : ab annis quinquagentaquinque usque ad annum sexagesimum annorum septem : ab annis sexaginta cujuscunque ætatis sit, annorum quinque. . . . Solitum est tamen a prima ætate usque ad annum trigesimum computationem annorum triginta fieri, ab annis vero triginta, tot annorum computationem inire, quot ad annum sexagesimum deesse videntur, nunquam ergo amplius quam triginta annorum computatio initur."

Construction
of Macer's
Table quite
arbitrary.
Ulpian's
Table based
on life
experience.

it is clear on inspection that the values are arbitrary, and in no way based on a mortality experience. Ulpian's Table, however, shows distinct signs of an attempt to represent more nearly the actual values of annuities.

Table of
Ulpian
probably
table of
values of
a_z, not of e_z

It has been held by various writers [2] that the values given by Ulpian are those of the expectation of life rather than those of annuities, and the reason put forward for this assumption is that the Romans were unlikely to have used the factor of interest when compiling the table. There seems, however, to be no particular justification for this assumption. It is generally admitted that the table of Ulpian shows signs of having been calculated with direct reference to either a mortality or a subscription experience, but, so far, no satisfactory suggestion has been made of the way in which the table was constructed.

[2] Cf. Hendriks, *J.I.A.*, VI, p. 313.

2. Natural and Political Observations Mentioned in a Following Index, and Made Upon the Bills of Mortality

JOHN GRAUNT (1964 (1662))

From *Journal of the Institute of Actuaries* 90. Excerpts are from pages 15, 19—23, 35—38, 44—47.

In extracting from Graunt's observations we have tried to give the reader a feel both for Graunt's imaginativeness and tenacity and for the materials he had to work with. We include both of his estimates for London's population: the first (460,000) he found by attributing to the city one-fourteenth of a national population estimate of 6,440,000 not well derived. The second (384,000), from his more careful knowledge about London, is probably of the correct order of magnitude, and suggests a reasonable total population figure (5 to 5.5 million) (cf. Wrigley 1967, pp. 44—45). Graunt's life table entries and his comments on London's doubling time are not of this quality.

We have mostly avoided Graunt's discussion of specific diseases, and have omitted his remarks on the expansion of London beyond the old city walls, on the healthfulness of the city and countryside, and on the country bills.

CHAPTER I

OF THE BILLS OF MORTALITY, THEIR BEGINNING, AND PROGRESS

THE first of the continued weekly Bills of Mortality extant at the Parish-clerks Hall, begins the 29th of December, 1603, being the first year of King James his reign; since when, a weekly account hath been kept there of burials and christenings. It is true, there were Bills before, viz. for the years 1592,–93,–94, but so interrupted since, that I could not depend upon the sufficiency of them, rather relying upon those accounts which have been kept since, in order, as to all the uses I shall make of them.

2. I believe that the rise of keeping these accounts, was taken from the Plague: for the said Bills (for aught appears) first began in the said year 1592, being a time of great mortality; and after some disuse, were resumed again in the year 1603, after the great Plague then happening likewise. ...

10. We have hitherto described the several steps, whereby the Bills of Mortality are come up to their present state; we come next to shew how they are made and composed, which is in this manner, viz. when anyone dies, then, either by tolling or ringing of a bell, or by bespeaking of a grave of the Sexton, the same is known to the searchers, corresponding with the said Sexton.

11. The Searchers hereupon (who are ancient matrons, sworn to their Office) repair to the place where the dead corpse lies, and by view of the same, and by other enquiries, they examine by what disease or casualty the corpse died. Hereupon they make their Report to the Parish-clerk and he, every Tuesday night, carries in an account of all the burials and christenings, happening that week, to the clerk of the Hall. On Wednesday the general account is made up and printed, and on Thursdays published, and dispersed to the several families, who will pay four shillings per annum for them.

12. *Memorandum.* That although the general yearly Bills have been set out in the several varieties afore-mentioned, yet the original entries in the Hall-books were as exact in the very first year as to all particulars, as now; and the specifying of casualties and diseases, was probably more.

CHAPTER II

GENERAL OBSERVATIONS UPON THE CASUALTIES

IN my discourses upon these Bills I shall first speak of the casualties, then give my observations with reference to the places and parishes comprehended in the Bills; and next of the years, and seasons.

1. There seems to be good reason why the Magistrate should himself take notice of the numbers of burials and christenings, viz. to see whether the City increase or decrease in people; whether it increase proportionately with the rest of the Nation; whether it be grown big enough, or too big, etc. But why the same should be made known to the People, otherwise than to please them as with a curiosity, I see not.

2. Nor could I ever yet learn (from the many I have asked, and those not of the least sagacity) to what purpose the distinction between males and females is inserted, or at all taken notice of; or why that of marriages was not equally given in? Nor is it obvious to everybody, why the account of casualties (whereof we are now speaking) is made? The reason, which seems most obvious for this latter, is, that the state of health in the City may at all times appear.

3. Now it may be objected that the same depends most upon the accounts of epidemic diseases, and upon the chief of them all, the *Plague*; wherefore the mention of the rest seems only a matter of curiosity.

4. But to this we answer, that the knowledge even of the numbers which die of the *Plague*, is not sufficiently deduced from the mere report of the Searchers, which only the Bills afford; but from other ratiocinations, and comparings of the *Plague* with some other casualties.

5. For we shall make it probable that in years of Plague a quarter part more dies of that disease than are set down; the same we shall also prove by the other casualties. Wherefore, if it be necessary to impart to the world a good account of some few casualties, which since it cannot well be done without giving an account of them all, then is our common practice of so doing very apt, and rational.

6. Now, to make these corrections upon the perhaps, ignorant, and careless Searchers' Reports, I considered first of what authority they were in themselves, that is, whether any credit at all were to be given to their distinguishments: and finding that many of the casualties were but matter of sense, as whether a child were *Abortive*, or *Stillborn*; whether men were *Aged*, that is to say, above sixty years old, or thereabouts, when they died, without any curious determination whether such aged persons died purely of *Age*, as for that the innate heat was quite extinct, or the radical moisture quite dried up (for I have heard some candid physicians complain of the darkness which themselves were in hereupon) I say, that these distinguishments being but matter of sense, I concluded the Searchers' Report might be sufficient in the case.

7. As for *Consumptions*, if the Searchers do but truly report (as they may) whether the dead corpse were very lean and worn away, it matters not to many of our purposes whether the disease were exactly the same as physicians define it in their books. Moreover, in case a man of 75 years old died of a cough (of which had he been free, he might have possibly lived to ninety) I esteem it little error (as to many of our purposes) if this person be in the table of casualties, reckoned among the *Aged* and not placed under the title of *Coughs*.

13

8. In the matter of Infants I would desire but to know clearly, what the Searchers mean by Infants, as whether children that cannot speak, as the word Infant seems to signify, or children under two or three years old, although I should not be satisfied, whether the Infant died of *Wind*, or of *Teeth*, or of the *Convulsion*, etc. or were choked with *Phelgm*, or else of *Teeth*, *Convulsion*, and *Scowring*, apart, or together, which, they say, do often cause one another: for, I say, it is somewhat, to know how many die usually before they can speak, or how many live past any assigned number of years.

9. I say it is enough if we know from the Searchers but the most predominant symptoms; as that one died of the *Head-Ache*, who was sorely tormented with it, though the physicians were of opinion that the disease was in the stomach. Again, if one died suddenly, the matter is not great, whether it be reported in the Bills, *Suddenly*, *Apoplexy*, or *Planet-strucken*, etc.

10. To conclude, in many of these cases the Searchers are able to report the opinion of the physician who was with the patient, as they receive the same from the friends of the defunct, and in very many cases, such as *Drowning*, *Scalding*, *Bleeding*, *Vomiting*, *Making-away themselves*, *Lunaticks*, *Sores*, *Small-Pox*, etc. their own senses are sufficient, and the generality of the world, are able pretty well to distinguish the *Gowt*, *Stone*, *Dropsy*, *Falling-sickness*, *Palsy*, *Agues*, *Pleurisy*, *Rickets*, etc. one from another.

11. But now as for those casualties which are aptest to be confounded, and mistaken, I shall in the ensuing discourse presume to touch upon them so far as the learning of these Bills hath enabled me.

12. Having premised these general advertisements, our first observation upon the casualties shall be, that in twenty years there dying of all diseases and casualties, 229,250, that 71,124 died of the *Thrush*, *Convulsion*, *Rickets*, *Teeth*, and *Worms*; and as *Abortives*, *Chrysoms*, *Infants*, *Liver-grown*, and *Overlaid*; that is to say, that about one-third of the whole died of those diseases, which we guess did all light upon children under four or five years old.

13. There died also of the *Small-Pox*, *Swine-Pox*, and *Measles*, and of *Worms* without *Convulsions*, 12,210, of which number we suppose likewise, that about one-half might be children under six years old. Now, if we consider that 16 of the said 229 thousand died of that extraordinary and grand casualty the *Plague*, we shall find that about 36 per centum of all quick conceptions died before six years old.

14. The second observation is, that of the said 229,250 dying of all diseases, there died of acute diseases (the *Plague* excepted) but about 50,000 or 2/9 parts. The which proportion doth give a measure of the state and disposition of this climate and air, as to health, these acute and epidemic diseases happening suddenly and vehemently, upon the like corruptions and alterations in the air.

15. The third observation is, that of the said 229 thousand about 70 died of chronic diseases, which shews (as I conceive) the state and disposition of the country (including as well its food, as air) in reference to health, or rather to longevity: for as the proportion of acute and epidemic diseases shews the aptness of the air to sudden and vehement impressions, so the chronic diseases shew the ordinary temper of the place, so that upon the proportion of chronic diseases seems to hang the judgment of the fitness of the country for long life. For, I conceive, that in countries subject to great epidemic sweeps men may live very long, but where the proportion of the chronic distempers is great, it is not likely to be so; because men being long sick, and always sickly, cannot live to any great age, as we see in several sorts of Metal-men, who although they are less subject to acute diseases than others, yet seldom live to be old, that is, not to reach unto those years which David says is the age of man. ...

17. In the next place, whereas many persons live in great fear and apprehension of some of the more formidable and notorious diseases following; I shall only set down how many died of each: that the respective numbers, being compared with the total 229,250, those persons may the better understand the hazard they are in.

Table of notorious diseases		Table of casualties	
Apoplexy	1,306	*Bleeding*	69
Cut of the Stone	38	*Burnt*, and *Scalded*	125
Falling Sickness	74	*Drowned*	829
Dead in the streets	243	*Excessive drinking*	2
Gowt	134	*Frighted*	22
Head-Ache	51	*Grief*	279
Jaundice	998	*Hanged themselves*	222
Lethargy	67	*Killed by several*	
Leprosy	6	*accidents*	1,021
Lunatick	158	*Murdered*	86
Overlaid, and *Starved*	529	*Poisoned*	14
Palsy	423	*Smothered*	26
Rupture	201	*Shot*	7
Stone and *Strangury,*	863	*Starved*	51
Sciatica	5	*Vomiting*	136
Sodainly	454		

18. In the foregoing observations we ventured to make a standard of the healthfulness of the air from the proportion of acute and epidemic diseases, and of the wholesomeness of the food from that of the chronic. Yet forasmuch as neither of them alone do shew the longevity of the inhabitants, we shall in the next place come to the more absolute standard, and correction of both, which is the proportion of the aged, viz. 15,757 to the total 229,250. That is of about 1 to 15 or 7 per cent. Only the question is, what

number of years the Searchers call *Aged*, which I conceive must be the same, that David calls so, viz. 70. For no man can be said to die properly of *Age*, who is much less: it follows from hence, that if in any other country more than seven of the 100 live beyond 70 such country is to be esteemed more healthful than this of our City. ...

CHAPTER VII

OF THE DIFFERENCE BETWEEN BURIALS AND CHRISTENINGS

THE next observation is, that in the said Bills there are far more burials than christenings. This is plain, depending only upon arithmetical computation; for, in 40 years, from the year 1603 to the year 1644, exclusive of both years, there have been set down (as happening within the same ground, space, or parishes) although differently numbered, and divided, 363,935 burials, and but 330,747 christenings within the 97, 16, and 10 outparishes, those of Westminster, Lambeth, Newington, Redriff, Stepney, Hackney, and Islington, not being included.

2. From this single observation it will follow that London should have decreased in its people, the contrary whereof we see by its daily increase of buildings upon new foundations, and by the turning of great palacious houses into small tenements. It is therefore certain that London is supplied with people from out of the country, whereby not only to repair the overplus difference of burials above-mentioned, but likewise to increase its inhabitants, according to the said increase of housing. ...

4. But, if we consider what I have upon exact enquiry found true, viz. that in the country, within ninety years, there have been 6,339 christenings and but 5,280 burials, the increase of London will be salved without inferring the decrease of the people in the country; and withal, in case all England have but fourteen times more people than London, it will appear, how the said increase of the country may increase the people, both of London and itself; for if there be in the 97, 16, 10, and 7 parishes, usually comprehended within our Bills, but 460,000 souls as hereafter we shall shew, then there are in all England and Wales, 6,440,000 persons, out of which subtract 460,000, for those in and about London, there remains 5,980,000 in the country, the which increasing about 1/7 part in 40 years, as we shall hereafter prove doth happen in the country, the whole increase of the country will be about 854,000 in the said time, out of which number, if but about 250,000 be sent up to London in the said 40 years, viz. about 6,000 per annum, the said missions will make good the alterations which we find to have been in and about London, between the years 1603 and 1644 above-mentioned. But that 250,000 will do the same, I prove thus, viz. in the 8 years, from 1603 to 1612, the burials in all the parishes, and of

all diseases, the *Plague* included, were at a medium 9,750 per annum. And between 1635 and 1644 were 18,000, the difference whereof is 8,250, which is the total of the increase of the burials in 40 years, that is about 206 per annum. Now, to make the burials increase 206 per annum, there must be added to the City thirty times as many (according to the proportion of 3 dying out of 11 families) viz. 6,180 *advenae*, the which number multiplied again by the 40 years, makes the product 247,200, which is less than the 250,000 above propounded; so as there remains above 600,000 of increase in the country within the said 40 years, either to render it more populous, or send forth into other colonies, or wars. But that England hath fourteen times more people is not improbable, for the reasons following.

1. London is observed to bear about the fifteenth proportion of the whole tax.
2. There is in England and Wales about 39,000 square miles of land, and we have computed that in one of the greatest parishes in Hampshire, being also a market town and containing twelve square miles, there are 220 souls in every square mile, out of which I abate 1/4 for the overplus of people more in that parish, than in other wild counties. So as the 3/4 parts of the said 220, multiplied by the total of square miles, produces 6,400,000 souls in all, London included.
3. There are about 10,000 parishes in England and Wales, the which, although they should not contain the 1/3 part of the land nor the 1/4 of the people of that country parish which we have examined, yet may be supposed to contain about 600 people, one with another, according to which account there will be six millions of people in the Nation. I might add, that there are in England and Wales about five and twenty millions of acres at $16\frac{1}{2}$ foot to the perch; and if there be six millions of people, then there is about four acres for every head, which how well it agrees to the rules of plantation, I leave unto others, not only as a means to examine my assertion, but as an hint to their enquiry concerning the fundamental trade, which is husbandry and plantation. ...

CHAPTER VIII

OF THE DIFFERENCE BETWEEN THE NUMBERS OF MALES AND FEMALES

THE next observation is, that there be more males than females.

There have been buried from the year 1628 to the year 1662, exclusive, 209,436 males and but 190,474 females: but it will be objected, that in London it may indeed be so, though otherwise elsewhere; because London is the great stage and shop of business, wherein the masculine sex bears the greatest part. But we answer, that there have been also christened within the same time, 139,782 males and but 130,866 females, and that the country accounts are consonant enough to those of London upon this matter.

2. What the causes hereof are, we shall not trouble ourselves to conjecture, as in other cases, only we shall desire that travellers would enquire whether it be the same in other countries. ...

CHAPTER XI

OF THE NUMBER OF INHABITANTS

I HAVE been several times in company with men of great experience in this City, and have heard them talk seldom under millions of people to be in London, all which I was apt enough to believe until, on a certain day, one of eminent reputation was upon occasion asserting that there was in the year 1661 two millions of people more than in the year 1625, before the great *Plague*; I must confess that, until this provocation, I had been frighted with that misunderstood example of David, from attempting any computation of the people of this populous place; but hereupon I both examined the lawfulness of making such enquiries and, being satisfied thereof, went about the work itself in this manner: viz.

2. First, I imagined that, if the conjecture of the worthy person aforementioned had any truth in it, there must needs be about six or seven millions of people in London now; but repairing to my Bills I found that not above 15,000 per annum were buried, and consequently, that not above one in four hundred must die per annum if the total were but six millions.

3. Next considering, that it is esteemed an even lay whether any man lives ten years longer, I supposed it was the same, that one of any 10 might die within one year. But when I considered, that of the 15,000 aforementioned about 5,000 were *Abortive*, and *Stillborn*, or died of *Teeth*, *Convulsion*, *Rickets*, or as *Infants*, and *Chrysoms*, and *Aged*. I concluded that of men and women, between ten and sixty, there scarce died 10,000 per annum in London, which number being multiplied by 10, there must be 100,000 in all, that is not the one-sixtieth part of what the Alderman imagined. These were but sudden thoughts on both sides, and both far from truth, I thereupon endeavoured to get a little nearer, thus: viz.

4. I considered, that the number of child-bearing women might be double to the births: forasmuch as such women, one with another, have scarce more than one child in two years. The number of births I found, by those years wherein the registries were well kept, to have been somewhat less than the burials. The burials in these late years at a medium are about

13,000 and consequently the christenings not above 12,000. I therefore esteemed the number of teeming women to be 24,000: then I imagined, that there might be twice as many families as of such women; for that there might be twice as many women aged between 16 and 76, as between 16 and 40, or between 20 and 44; and that there were about eight persons in a family, one with another, viz. the man and his wife, three children and three servants, or lodgers: now 8 times 48,000 makes 384,000.

5. Secondly, I find by telling the number of families in some parishes within the Walls, that 3 out of 11 families per annum have died: wherefore, 13,000 having died in the whole, it should follow there were 48,000 families according to the last mentioned account.

6. Thirdly, the account which I made of the trained bands and auxiliary soldiers, doth enough justify this account.

7. And lastly I took the map of London set out in the year 1658 by Richard Newcourt, drawn by a scale of yards. Now I guessed that in 100 yards square there might be about 54 families, supposing every house to be 20 foot in the front: for on two sides of the said square there will be 100 yards of housing in each, and in the two other sides 80 each; in all 360 yards: that is 54 families in each square, of which there are 220 within the Walls, making in all 11,880 families within the Walls. But forasmuch as there die within the walls about 3,200 per annum, and in the whole about 13,000; it follows that the housing within the Walls is 1/4 part of the whole, and consequently, that there are 47,520 families in and about London, which agrees well enough with all my former computations: the worst whereof doth sufficiently demonstrate that there are no millions of people in London, which nevertheless most men do believe, as they do, that there be three women for one man, whereas there are fourteen men for thirteen women, as elsewhere hath been said.

8. We have (though perhaps too much at random) determined the number of the inhabitants of London to be about 384,000: the which being granted, we assert that 199,112 are males and 184,886 females.

9. Whereas we have found that of 100 quick conceptions about 36 of them die before they be six years old, and that perhaps but one surviveth 76, we, having seven decades between six and 76, we sought six mean proportional numbers between 64, the remainder living at six years, and the one which survives 76, and find that the numbers following are practically near enough to the truth; for men do not die in exact proportions, nor in fractions: from whence arises this Table following:

Viz. of 100 there dies		The fourth	6
within the first six years	36	The next	4
The next ten years, or		The next	3
decade	24	The next	2
The second decade	15	The next	1
The third decade	9		

19

10. From whence it follows, that of the said 100 conceived there remains alive at six years end 64.

At sixteen years end	40	At fifty-six	6
At twenty-six	25	At sixty-six	3
At thirty-six	16	At seventy-six	1
At forty-six	10	At eighty	0

11. It follows also, that of all which have been conceived, there are now alive 40 per cent above sixteen years old, 25 above twenty-six years old, & *sic deniceps*, as in the above Table: there are therefore of aged between 16 and 56, the number of 40, less by six, viz. 34; of between 26 and 66, the number of 25 less by three, viz. 22: & *sic deniceps*.

Wherefore, supposing there be 199,112 males, and the number between 16 and 56 being 34. It follows, there are 34 per cent of all those males fighting men in London, that is 67,694, viz. near 70,000: the truth whereof I leave to examination, only the 1/5 of 67,694, viz. 13,539, is to be added for Westminster, Stepney, Lambeth, and the other distant parishes, making in all 81,233 fighting men.

12. The next enquiry shall be, in how long time the City of London shall, by the ordinary proportion of breeding and dying, double its breeding people. I answer in about seven years, and (Plagues considered) eight. Wherefore since there be 24,000 pair of breeders, that is one-eighth of the whole, it follows that in eight times eight years the whole people of the City shall double without the access of foreigners: the which contradicts not our account of its growing from two to five in 56 years with such accesses.

13 According to this proportion, one couple viz. Adam and Eve, doubling themselves every 64 years of the 5,610 years, which is the age of the world according to the Scriptures, shall produce far more people than are now in it. Wherefore the world is not above 100 thousand years old as some vainly imagine, nor above what the Scripture makes it.

3. An Estimate of the Degrees of the Mortality of Mankind

EDMUND HALLEY (1693)

From *Philosophical Transactions* XVII. Excerpts are from pages 596—604, 610, 655—656.

Halley's text may be clearer if the reader notes that he intends: $l_0 = 1238$; $L_0 = 1000$; $l_1 = 890$; $L_1 = 855$. According to Westergaard (1969, p. 34) the Breslau lists showed 1218 births in 1691, among whom 992 were alive on January 1, 1692. Rounding the figure, Halley set $L_0 = 1000$ for 1238 births.

In his *Use* 5 Halley refers to "years purchase", the price charged for an annuity paying one dollar per year. Parliament granted William and Mary permission in 1691 to borrow money through an annuity scheme; the rate Halley refers to was set the following year.

We have omitted Halley's discussion of joint and group annuities.

The Contemplation of the *Mortality* of *Mankind*, has besides the *Moral*, its *Physical* and *Political* Uses, both which have been some years since most judiciously considered by the curious Sir *William Petty*, in his *Natural* and *Political* Observations on the Bills of *Mortality* of *London*, owned by Captain *John Graunt*. And since in a like Treatise on the Bills of *Mortality* of *Dublin*. But the Deduction from those Bills of *Mortality* seemed even to their Authors to be defective: First, In that the *Number* of the People was wanting. Secondly, That the *Ages* of the People dying was not to be had. And Lastly, That both *London* and *Dublin* by reason of the great and casual Accession of *Strangers* who die therein, (as appeared in both, by the great Excess of the *Funerals* above the *Births*) rendred them incapable of being Standards for this purpose; which requires, if it were possible, that the People we treat of should not at all be changed, but die where they were born, without any Adventitious Increase from Abroad, or Decay by Migration elsewhere.

This *Defect* seems in a great measure to be satisfied by the late curious Tables of the Bills of *Mortality* at the City of *Breslaw*, lately communicated to this Honourable Society by Mr. *Justell*, wherein both the *Ages* and *Sexes* of all that die are monthly delivered, and compared with the number of the *Births*, for Five Years last past, *viz.* 1687, 88, 89, 90, 91, seeming to be done with all the Exactness and Sincerity possible.

This City of *Breslaw* is the Capital City of the Province of *Silesia*; or, as the *Germans* call it, *Schlesia*, and is scituated on the Western Bank of the River *Oder*, anciently called *Viadrus*; near the Consines of *Germany* and *Poland*, and very nigh the Latitude of *London*. It is very far from the Sea, and as much a *Mediterranean* Place as can be desired, whence the Confluence of Strangers is but small, and the Manufacture of Linnen employs chiefly the poor People of the place, as well as of the Country round about; whence comes that sort of Linnen we usually call your *Sclesie Linnen*; which is the chief, if not the only Merchandize of the place. For these Reasons the People of this City seem most proper for a *Standard*; and the rather, for that the *Births* do, a small matter, exceed the *Funerals*. The only thing wanting is the Number of the whole People, which in some measure I have endeavoured to supply by comparison of the *Mortality* of the People of all Ages, which I shall from the said Bills trace out with all the Acuracy possible.

It appears that in the Five Years mentioned, *viz.* from 87 to 91 inclusive, there were *born* 6193 Persons, and *buried* 5869; that is, born *per Annum* 1238, and *buried* 1174; whence an *Encrease* of the People may be argued of 64 *per Annum*, or of about a 20th part, which may perhaps be ballanced by the Levies for the *Emperor*'s Service in his Wars. But this being contingent, and the Births certain, I will suppose the People of *Breslaw* to be encreased by 1238 *Births* annually. Of these it appears by the same Tables, that 348 do die *yearly* in the *first Year* of their *Age*, and that but 890 do arrive at a full *Years Age*; and likewise, that 198 do die in the *Five Years* between 1 and 6 compleat, taken at a *Medium*; so that but 692 of the Persons *born* do survive *Six* whole *Years*. From this *Age* the Infants being arrived at some degree of Firmness, grow less and less *Mortal*; and it appears that of the whole People of *Breslaw* there die *yearly*, as in the following Table, wherein the upper Line shews the *Age*, and the next under it the *Number* of Persons of that Age *dying yearly*.

```
 7  . 8   9  . . 14   . 18 . 21 . 27 . 28 . . 35 .
11  . 11 . 6 . 5½ .2 . 3½  5  6 4½ 6½  9  .  8  . 7 .7 .

36  .    42  .    45        49 54 . 55 . 56    .    63
 8  . 9½  8  . 9  . 7  . 7  . 10 11 .  9  .  9  . 10 . 12

    70 71 . 72    77     81     84  .     90  91  .
9½ 14 9  . 11 9½  6  .7  .3  .4  .2  . 1  . 1  . 1  .

98 . 99 . 100 .
 0 . ⅕ . ⅗
```

And where no *Figure* is placed over, it is to be understood of those that die between the Ages of the preceding and consequent *Column*.

From this Table it is evident, that from the Age of 9 to about 25 there does not die above 6 *per Annum* of each *Age*, which is much about one *per Cent.* of those that are of those *Ages*: And whereas in the 14, 15, 16, 17 *Years* there appear to die much fewer, as 2 and 3½, yet that seems rather to be attributed to Chance, as are the other Irregularities in the Series of Ages, which would rectifie themselves, were the number of Years much more considerable, as 20 instead of 5. And by our own Experience in *Christ-Church Hospital*, I am informed there die of the *Young Lads*, much about one *per Cent. per Annum*, they being of the foresaid *Ages*. From 25 to 50 there seem to die from 7 to 8 and 9 *per Annum* of each Age; and after that to 70, they growing more *crasie*, though the number be much diminished, yet the *Mortality increases*, and there are found to die 10 or 11 of each Age *per Annum*: From thence the number of the *Living* being grown very small, they gradually decline till there be none left to *die*; as may be seen at one View in the Table.

From these Considerations I have formed the *adjoyned Table*, whose Uses are manifold, and give a more just *Idea* of the *State* and *Condition* of *Mankind*, than any thing yet extant that I know of. It exhibits the *Number* of *People* in the City of *Breslaw* of all Ages, from the *Birth* to extream *Old Age*, and thereby shews the Chances of *Mortality* at all *Ages*, and likewise how to make a certain Estimate of the value of *Annuities* for *Lives*, which hitherto has been only done by an imaginary *Valuation*: Also the *Chances* that there are that a *Person* of any *Age* proposed does live to any other *Age* given; with many more, as I shall hereafter shew. This *Table* does shew the *number* of *Persons* that are living in the *Age* current annexed thereto.

Thus it appears, that the whole People of *Breslaw* does consist of 34000 *Souls*, being the Sum *Total* of the Persons of all Ages in the *Table*: The first use hereof is to shew the Proportion of *Men* able to bear *Arms* in any *Multitude*, which are those between 18 and 56, rather than 16 and 60; the one being generally too weak to bear the *Fatigues* of *War* and the Weight of *Arms*, and the other too crasie and infirm from *Age*, notwithstanding particular Instances to the contrary. Under 18 from the *Table*, are found in this City 11997 Persons, and 3950 above 56, which together make 15947. So that the Residue to 34000 being 18053 are Persons between those *Ages*. At least one half thereof are Males, or 9027: So that the whole Force this City can raise of *Fencible Men*, as the *Scotch* call them, is about

Age Curt.	Per-sons	Age Curt.	Per-sons	Age Curt.	Per-sons	Age Curt.	Per-sons	Age Curt.	Per-sons	Age Curt	Per-sons		Age	Per-sons
1	1000	8	680	15	628	22	586	29	539	36	481		7	5547
2	855	9	670	16	622	23	579	30	531	37	472		14	4584
3	798	10	661	17	616	24	573	31	523	38	463		21	4270
4	760	11	653	18	610	25	567	32	515	39	454		28	3964
5	732	12	646	19	604	26	560	33	507	40	445		35	3604
6	710	13	640	20	598	27	553	34	499	41	436		42	3178
7	692	14	634	21	592	28	546	35	490	42	427		49	2709
													56	2194

Age Curt.	Per-sons	Age Curt.	Per-sons	Age Curt.	Per-sons	Age Curt.	Per-sons	Age Curt.	Per-sons	Age Curt.	Per-sons		Age	Per-sons
43	417	50	346	57	272	64	202	71	131	78	58		63	1694
44	407	51	335	58	262	65	192	72	120	79	49		70	1204
45	397	52	324	59	252	66	182	73	109	80	41		77	692
46	387	53	313	60	242	67	172	74	98	81	34		84	253
47	377	54	302	61	232	68	162	75	88	82	28		100	107
48	367	55	292	62	222	69	152	76	78	83	23			34000
49	357	56	282	63	212	70	142	77	68	84	20			Sum Total

9000, or $\frac{9}{34}$, or somewhat more than a quarter of the *Number* of *Souls*, which may perhaps pass for a Rule for all other places.

The *Second Use* of this Table is to shew the differing degrees of *Mortality*, or rather *Vitality* in all *Ages*; for if the number of Persons of any *Age* remaining after one year, be divided by the difference between that and the number of the Age proposed, it shews the *odds* that there is, that a Person of that Age does not die in a *Year*. As for Instance, a Person of 25 *Years* of *Age* has the odds of 560 to 7 or 80 to 1, that he does not *die* in a *Year*: Because that of 567, living of 25 years of Age, there do die no more than 7 in a *Year*, leaving 560 of 26 Years old.

So likewise for the *odds*, that any Person does not die before he attain any proposed *Age*: Take the *number* of the remaining Persons of the Age proposed, and divide it by the difference between it and the number of those of the *Age* of the Party proposed; and that shews the *odds* there is between the Chances of the Party's living or dying. As for Instance; What is the *odds* that a Man of 40 lives 7 Years: Take the number of Persons of 47 years, which in the Table is 377, and substract it from the number of Persons of 40 years, which is 445, and the *difference* is 68: Which shews that the *Persons dying* in that 7 years are 68, and that it is 377 to 68 or $5\frac{1}{2}$ to 1, that a Man of 40 does live 7 Years. And the like for any other *number* of *Years*.

Use III. But if it be enquired at what number of *Years*, it is an even Lay that a Person of any *Age* shall die, this Table readily performs it: For if the *number* of Persons *living* of the *Age* proposed be *halfed*, it will be found by the *Table* at what Year the said *number* is reduced to half by *Mortality*; and that is the Age, to which it is an even Wager, that a Person of the *Age* proposed shall arrive before he *die*. As for Instance; A Person of 30 Years of *Age* is proposed, the number of that Age is 531, the half thereof is 265, which number I find to be between 57 and 58 Years; so that a Man of 30 may reasonably expect to live between 27 and 28 Years.

Use IV. By what has been said, the *Price* of *Insurance* upon *Lives* ought to be regulated, and the difference is discovered between the *price* of ensuring the *Life* of a *Man* of 20 and 50, for Example: it being 100 to 1 that a Man of 20 dies not in a year, and but 38 to 1 for a Man of 50 Years of Age.

Use V. On this depends the Valuation of *Annuities* upon *Lives*; for it is plain that the *Purchaser* ought to pay for only such a part of the value of the *Annuity*, as he has Chances that he is living; and this ought to be computed yearly, and the Sum of all those yearly Values being added together, will amount to the value of the *Annuity* for the *Life* of the Person proposed. Now the present value of Money payable after a term of years, at any given rate of Interest, either may be had from Tables already computed; or almost as compendiously, by the Table of Logarithms: For the Arithmetical Complement of the Logarithm of Unity and its yearly Interest (that is, of 1.06 for Six *per Cent.* being 9.974694.) being multiplied by the number of years proposed, gives the present value of One Pound payable after the end of so many years. Then by the foregoing Proposition, it will be as the number of Persons living after that term of years, to the number dead; so are the Odds that any one Person is Alive or Dead. And by consequence, as the Sum of both or the number of Persons living of the Age first proposed, to the number remaining after so many years, (both given by the Table) so the present value of the yearly Sum payable after the term proposed, to the Sum which ought to be paid for the Chance the person has to enjoy such an Annuity after so many Years. And this being repeated for every year of the persons Life, the Sum of all the present Values of those Chances is the true Value of the Annuity. This will without doubt appear to be a most laborious Calculation, but it being one of the principal Uses of this Speculation, and having found some *Compendia* for the Work, I took the pains to compute the following Table, being the short Result of a not ordinary number of Arithmetical Operations; It shews the Value of Annuities for every Fifth Year of Age, to the Seventieth, as follows.

Age	Years Purchase	Age	Years Purchase	Age	Years Purchase
1	10.28	25	12.27	50	9.21
5	13.40	30	11.72	55	8.51
10	13.44	35	11.12	60	7.60
15	13.33	40	10.57	65	6.54
20	12.78	45	9.91	70	5.32

This shews the great Advantage of putting Money into the present *Fund* lately granted to their Majesties, giving 14 *per Cent.* per *Annum*, or at the rate of 7 years purchase for a Life; when young Lives, at the usual rate of Interest, are worth above 13 years Purchase. It shews likewise the Advantage of young Lives over those in Years; a Life of Ten Years being almost worth $13\frac{1}{2}$ years purchase, whereas one of 36 is worth but 11. . . .

25

It may be objected, that the different *Salubrity* of places does hinder this Proposal from being *universal*; nor can it be denied. But by the number that die, being 1174 *per Annum* in 34000, it does appear that about a 30th part die yearly, as Sir *William Petty* has computed for *London*; and the number that die in Infancy, is a good Argument that the Air is but indifferently salubrious. So that by what I can learn, there cannot perhaps be one better place proposed for a Standard. At least 'tis desired that in imitation hereof the Curious in other Cities would attempt something of the same nature, than which nothing perhaps can be more useful. ...

A second Observation I make upon the said Table, is that the Growth and Encrease of Mankind is not so much stinted by any thing in the Nature of the *Species*, as it is from the cautious difficulty most People make to adventure on the state of *Marriage*, from the prospect of the Trouble and Charge of providing for a Family. Nor are the poorer sort of People herein to be blamed, since their difficulty of subsisting is occasion'd by the unequal Distribution of Possessions, all being necessarily fed from the Earth, of which yet so few are Masters. So that besides themselves and Families, they are yet to work for those who own the Ground that feeds them: And of such does by very much the greater part of Mankind consist; otherwise it is plain, that there might well be four times as many Births as we now find. For by computation from the Table, I find that there are nearly 15000 Persons above 16 and under 45, of which at least 7000 are Women capable to bear Children. Of these notwithstanding there are but 1238 born yearly, which is but little more than a sixth part: So that about one in six of these Women do breed yearly; whereas were they all married, it would not appear strange or unlikely, that four of six should bring a Child every year. The Political Consequences hereof I shall not insist on, only the Strength and Glory of a King being in the multitude of his Subjects, I shall only hint, that above all things, Celibacy ought to be discouraged, as, by extraordinary Taxing and Military Service: And those who have numerous Families of Children to be countenanced and encouraged by such Laws as the *Jus trium Liberorum* among the *Romans*. But especially, by an effectual Care to provide for the Subsistence of the Poor, by finding them Employments, whereby they may earn their Bread, without being chargeable to the Publick.

4. A Treatise on the Valuation of Annuities and Assurances on Lives and Survivors

JOSHUA MILNE (1815)

London. Excerpts are from pages vi—xii, 89—91, 97—100, 487—489, 582.

Milne's notation and principal equations translate into modern form as:

$$^{n}\mathcal{E} = l_{n}$$ = survivors to exact age n from among l_0 births at observed age-specific death rates

$$S^{n}\mathcal{E} = \sum_{x=n}^{\omega} l_{x}$$

$$\overset{n}{L} = P_{n}$$ = observed population at ages n to $n+1$

$$\overset{n}{D} = D_{n}$$ = annual deaths to persons ages n to $n+1$

$$\delta = d_{n}$$ = deaths to persons ages n to $n+1$ in the life table

$$\frac{S^{n}\mathcal{E} - \frac{1}{2}{}^{n}\mathcal{E}}{{}^{n}\mathcal{E}} = \frac{\sum_{x=n}^{\omega} l_{x} - \frac{1}{2} l_{n}}{l_{n}} \doteq \mathring{e}_{n} = \text{life expectancy at age } n$$

$$\frac{{}^{n}\mathcal{E}\,\overset{n}{D}}{\overset{n}{L} + \frac{1}{2}\overset{n}{D}} = l_{n}\left(\frac{D_{n}}{P_{n} + \frac{1}{2}D_{n}}\right) \doteq d_{n}.$$

The general theorems throughout the work are adapted to any law of mortality whatever, and even to different laws, for all the different lives that may be involved; but they cannot be applied to practical purposes, unless one or more tables of mortality, adapted to the lives proposed, be given. The construction of such tables, although one of the most important parts of the subject, has remained the longest imperfect; principally from the want of *data*, and partly from its having occupied but little of the attention of mathematicians.

This is the object of the third chapter.

It is first assumed that the population remains stationary without being affected by migration, the method of constructing a table of mortality from the registers of burials in these circumstances is then shown, and the principal properties of such tables are demonstrated (143—165).

These might with more propriety be called tables of vitality than of mortality, as their principal use is, to show the mean duration of life, and the probability of its continuance; were it not, that the measure of vitality is often materially different from that of the continuance of life. For one individual may, and certainly often does, to every purpose of utility, enjoyment, or suffering, live much more than another during the same time. So that the absolute quantity of vitality depends upon its intensity, as well as its duration.

They are also, when properly constructed according to the general form of article 162, tables of population as well as mortality, since they show the proportion of the people of each sex in every interval of age, as well as the proportions of them that annually enter upon and die in each of these intervals, when the population is stationary, (art. 163 and 840).

Considered in this point of view, they are of importance in the science of Political Economy, and as the subject is treated here, in a manner which, at the same time that it is general and comprehensive, is perfectly elementary, this part of the work may, perhaps, be useful to those who cultivate that science, but have made little or no progress in mathematics.

When the state of the population is variable, and affected by migration, it is very difficult to ascertain the law of mortality, or even to approach near to it by means of the registers of burials alone. Mr. Simpson[1] endeavoured to make such allowances for the influence of the continual influx of new settlers upon the London bills, as to deduce the law of mortality from them, and from his great judgment and accuracy there is reason to believe that his success was considerable, but he did not explain distinctly how he proceeded.

Dr. Price afterwards, in his "Essay on the Method of forming Tables of Observations[2]," showed how the number of the inhabitants of a town, at every age, might be determined from the bills of mortality, when the number at each age was maintained stationary by the influx of new settlers, although the deaths of every year exceeded the births; provided that the annual number of settlers of each age could be ascertained, and remained always the same: But besides that the practical applications of that method are extremely limited by the hypotheses, the requisite *data* can hardly in any instance be obtained.

The principle of the method is here demonstrated in article 168, is extended in article 169 to places wherein the annual births exceed the deaths, while the population is kept stationary by emigration; and in article 170, a general theorem is given, that includes both these. But it is shown in article 172, that such a table could not exhibit the true law of mortality, nor enable us to determine the probabilities of life.

For that purpose, wherever the population is variable, that is, strictly speaking, in all cases, it is necessary that the number of the living, as well as that of the annual deaths at each age, should be known; and the method of constructing an accurate table of mortality from these *data*, is next given (174—183).

Almost all who have hitherto treated the subject of this chapter, have neglected the use of symbols, and have been content to illustrate their reasoning by examples in numbers. But in this, and all other inquiries that admit of the application of mathematical reasoning, the employment of symbols is attended with many advantages, since they enable us to treat of generals with almost the same facility as particulars; they relieve the mind from the fatigue and obscurity that attend circumlocution and ambiguity of language, and increase its powers much more than mechanical engines do those of the body.

It has generally been assumed that the number of the people when not affected by migration, increases in geometrical progression; a few theorems are given which determine the relations between the time elapsed, the rate of augmentation, and its amount, upon that hypothesis (184—187). Then it is shown how, upon the same hypothesis, a table of mortality may be constructed from the registers of deaths alone, when the annual rate of increase is given (188—191).

---

[1] See his excellent little tract, entitled *The Doctrine of Annuities and Reversions*, etc. (8vo. London, 1742), and his supplement to it, printed first in his *Select Exercises for young proficients in the mathematics*, (8vo. London, 1752), but since separately.

In that supplement, the author complained with some bitterness of M. Deparcieux having criticised, with more severity than judgment, the alterations which he had made in Mr. Smart's table of mortality for London.

But Mr. Simpson's information of what M. Deparcieux had advanced in his *Essai sur les Probabilités de la durée de la vie humaine*, was derived from the account of the work given in the History of the Royal Academy of Sciences at Paris (An. 1746, p. 45.). What Mr. Simpson objected to, and was really injurious, appears to have been given by the writer of that account, rather as his own sentiments than those of M. Deparcieux, which he does not appear to have understood, any more than Mr. Simpson's. I consider the remarks which M. Deparcieux did make on Mr. Simpson's table, to be both candid and judicious, and, in justice to his memory, will here insert what is most material in them.

After observing upon Dr. Halley's table for Breslaw, and that of Mr. Smart for London, which Mr. Simpson corrected, he proceeds thus, "Il est bien difficile, pour ne pas dire impossible, qu'on puisse établir un ordre de mortalité approchant du vrai, par le moyen des regîtres d'une ville comme celle de Londres, à cause de la quantité prodigieuse d'étrangers qui vont s'y établir et mourir. Aussi M. Simpson a-t-il jugé à propos d'y faire quelques corrections, sans trop dire comment. On verra dans la suite par la comparaison qu'on fera de cette table, avec quelques autres, si on peut beaucoup y compter, malgré la correction." (p. 38.).

He then shows several of the difficulties that attend the construction of tables of mortality for large towns, from their mortuary registers, and adds, "Il suit de toutes ces raisons, que la table du Docteur Hallei doit être préférée à celle de M. Simpson. Il est vrai que ce dernier semble ne vouloir donner la sienne que pour les habitans de Londres, ce qui pourroit être approchant du vrai s'il n'entendoit parler seulement que de ceux qui naissent dans cette ville; ce qui ne peut servir de regle pour aucun autre endroit qu'on ne l'ait examiné." (p. 41.).

[2] See his valuable work, entitled Observations on Reversionary Payments, vol. ii. p. 73, of the 7th and last edition (2 vols. 8vo. London, 1812), which is that always quoted in this work.

In the Memoirs of the Academy of Sciences at Berlin for the year 1760, there is a paper by the celebrated EULER, entitled *Recherches générales sur la mortalité et la multiplication du genre humain*, wherein the subject is treated algebraically. He assumes that the population is not affected by migration, and that both the annual births and deaths are always as the contemporaneous population; consequently, that the number of the people increases or decreases in geometrical progression. Then he gives several theorems, exhibiting the relations that would obtain between the annual births and deaths, and the population; and determines the law of mortality upon these hypotheses, but does not show how it may be deduced from actual observations independent of hypotheses; neither does he undertake the construction of any table of mortality, but, by way of example, gives that of M. Kerseboom, with the changes in the numbers which became necessary in consequence of his altering the radix from 1400 annual births to 1000^3.

M. Euler was perfectly aware how much the application of his theory was limited by the hypotheses it was derived from, he allows that his conclusions will not apply where migration has place, nor in case of any extraordinary increase of the people, such, he says, as takes place in new colonies; but he appears to have considered the increase in geometrical progression to prevail generally, though I believe it seldom does, either in countries that have been long or very recently settled (192—197). ...

162. In a society, none of the members of which enter it but by birth, nor go out of it but by death, where the number of the members remains always the same, and one uniform law of mortality obtains; if the number of deaths at every age, in any one year be given, a table may be constructed which will exhibit, the numbers of the living in that society at all ages, and the law of mortality, according to which the members continually pass out of it by death, while their places are supplied by others that are continually rising from birth towards the greater ages, and which shall also show the expectation of life at every age—Thus:

Let there be five columns, in the first of which insert the ages 0, 1, 2, 3, ... $(\omega - 1)$; then, against every age, insert in the fifth column the given number that died in the year between that and the next greater age; which being done, the numbers to be inserted in the third and fourth columns may be easily determined, by beginning at the greatest age in the table, and proceeding regularly, year by year, to the least—Thus:

To the number against any age in the fourth column, add that against the next less age in the fifth, and the sum will be the number to be inserted against that next less age in the fourth column (150).

To the sum of the numbers in the third and fourth columns, against any age, add half the number in the fifth column, against the next less age; and the sum last obtained will be the number to be inserted against that next less age in the third column (161).

---

[3] He assumes unity for the number of the born, consequently, the number completing each year of age is expressed by the fraction which measures the probability that a child just born will attain to that age. Perhaps it is from this property that these have, by some writers, been called tables of *the probabilities of life;* but although they show those probabilities directly at birth, when the radix is any power of 10; they only furnish the means of determining them at all ages after that.

Lastly, divide the number against any age in the third column, by the number against the same age in the fourth; the quotient will be the expectation of life at that age (154 and 157) to be inserted in the second column:

And the general form of the table will be this:

No. of Col.	1	2	3	4	5
	Age	Expectation of life at that age	No. of the living at that age and upwards	No. that annually complete that year of their age	No. that die annually in their next succeeding year
	n	$\dfrac{S^{n}e - \frac{1}{2}{}^{n}e}{{}^{n}e}$	$S^{n}e - \frac{1}{2}{}^{n}e$	${}^{n}e$	${}^{n}e - {}^{n+1}e$
Greatest age	$\omega - 1$	$\dfrac{\frac{1}{2}{}^{\omega-1}e}{{}^{\omega-1}e} = \dfrac{1}{2}$	$\frac{1}{2}{}^{\omega-1}e$	${}^{\omega-1}e$	${}^{\omega-1}e$
Limiting age	ω	0	0	0	0

163. From the third and fourth columns of this table, it will be easy to determine the numbers both of the living, and of the annual deaths, between any two ages that may be assigned; for we have only to subtract the number in each column against the greater age, from that against the less; the remainder will, in each case, be the number sought. ...

174. But whether the population be stationary, or increasing, or decreasing; and whether such changes be produced by procreation, mortality, or migration, or by the joint operation of any two or more of those causes; provided that the mode of their operation be uniform, or nearly so, and not by sudden starts, the law of mortality may be approached near enough for any useful purpose, by actual enumeration and the bills of mortality—Thus:

175. Let the number of persons, in each year of their age, that are resident in a place at any one time be taken; and let an accurate register be kept of the number that die annually in each year of their age, during a term of eight or ten years at the least, whereof the first half may precede, and the second follow, the time of the enumeration.

Then, if the number of inhabitants of every age either increase or decrease uniformly during that term, the mean number of annual deaths in every year of age, thus registered, will be the same as if the population of the place had continued throughout that term the same as when the enumeration was made.

176. In consequence of the several causes of change mentioned in art. 174, by which the population we are considering may be affected, the proportion between the numbers that annually complete any two ages therein may differ materially from that which would obtain in a society not affected by migration, but subject to the same law of mortality at every age, and maintaining its population stationary, by a constant equality between the annual numbers of the births and deaths.

31

And therefore the probabilities of life in that place cannot be determined immediately from the proportions furnished by the enumeration.

But from the enumeration, and the register together, the law of mortality, and consequently those probabilities, may be determined in the following manner.

177. Let $\overset{n}{L}$ denote the number of the living in the place, at the time of the enumeration, of the age of n years, that is, in the $(n+1)$th year of their age; and $\overset{n}{D}$, the mean number that died annually in the same year of their age, during the term for which the register was kept (175); while $\overset{n}{\epsilon}$ represents the number, in the society mentioned in the last article, that annually complete the nth year of their age; and $\delta = \overset{n}{\epsilon} - {}^{n+1}\epsilon$, the number of the same society that die annually in their next succeeding year, as in art. 149:

Then will $\overset{n}{L} : \overset{n}{D} :: \frac{1}{2}(\overset{n}{\epsilon} + {}^{n+1}\epsilon) : \overset{n}{\epsilon} - {}^{n+1}\epsilon :: \overset{n}{\epsilon} - \frac{1}{2}\delta : \delta$, and $\overset{n}{\epsilon}\,\overset{n}{D} - \frac{1}{2}\delta\,\overset{n}{D} = \delta\,\overset{n}{L}$;

whence $\dfrac{\overset{n}{\epsilon}\,\overset{n}{D}}{\overset{n}{L} + \frac{1}{2}\overset{n}{D}} = \delta$.

But $\overset{n}{L}$ and $\overset{n}{D}$ are always given by the enumeration and the register; therefore when $\overset{n}{\epsilon}$ is given, δ, and thence ${}^{n+1}\epsilon = \overset{n}{\epsilon} - \delta$, may be easily obtained.

In this manner, assuming at pleasure $\overset{0}{\epsilon}$ or ϵ, the number of annual births, and proceeding year by year, to determine successively the number that die in each year of their age, and, consequently, the number that complete that year, the numbers against every age in the fourth and fifth columns of the table in article 162 may be inserted; from whence the numbers for the second and third columns thereof may also be supplied, as directed in that article.

178. From about the age of seven or eight years, till forty or fifty, $\frac{1}{2}\overset{n}{D}$ is generally so small in comparison with $\overset{n}{L}$, that it might be neglected in the denominator of the fraction in the formula just given, without producing any considerable error in the value of δ, whereby that formula would become

$\dfrac{\overset{n}{\epsilon}\,\overset{n}{D}}{\overset{n}{L}} = \delta$, and this would facilitate the calculation a little.

179. This approximate value of δ would also result from the supposition that all who enter upon any year of their age continue alive throughout that year; and, that such of them as do not enter upon the next expire at its commencement; which hypothesis has generally been assumed in constructing tables of this kind.

But as the accuracy of such tables is of great importance, and the abridgment of the labour of constructing them in this way is but small, it will be better always to use the correct formula given in article 177. ...

788. In K. V. Ac. Handl. 1800, s. 323. Mr. Nicander has given statements of the mean number of women in all Sweden and Finland, with the annual average number of deliveries, and the proportion delivered annually in each of the under-

mentioned intervals of age; during sixteen years ending with 1795, which are presented at one view in the following table[4]:

Between the ages of	Mean Number of Females living	Annual average Number of Deliveries	One Woman of	Proportion of 1000 Deliveries
15 & 20	134,548	3,298	40.8	33
20 & 25	129,748	16,507	7.8	165
25 & 30	121,707	26,329	4.6	263
30 & 35	111,373	25,618	4.3	256
35 & 40	97,543	18,093	5.4	181
40 & 45	90,852	8,518	10.6	85
45 & 50	78,897	1,694	46.5	17
Above 50	69,268	39	1776.0	0.4
Above 15	833,936	100,096	8.3	1000

According to the proportion of legitimate and illegitimate children, Mr. Nicander found, that of 100,096[5] women delivered annually, 96,124 were married, and 3972 unmarried.

Upon comparing the legitimate births with the married women, and the illegitimate with the unmarried above fifteen years of age, he also found that 10 married women out of 54, and 10 unmarried out of 918, were annually brought to bed.

Some other results given by Mr. Nicander are presented under table XI., which table differs but little from that given above. What difference there is, arises from hence, that I have deduced the mean number of women in each interval of age from that gentleman's table marked Q, in K. V. Ac. Handl. 1801,1 Qu., and have calculated the two last columns from that and the column of deliveries.

It should be observed, that as the annual numbers here given are the average of sixteen years' observations, the number of deliveries they were derived from was 1,601,536.

789. The proportion of males born to females, appears to be the result of a uniform law of nature, and is probably always the same under the same circumstances, though it seems to vary with the situation of the parents, and I should rather suspect that variation to depend principally upon their age.

---

[4] It will be observed that this table only shows the actual fecundity, which is always kept below the natural aptitude or physical power, by the obstacles to marriage; as appears by the table itself, for the proportion of women delivered annually between the ages of 30 and 35, is almost twice as great as between 20 and 25; which can only arise from the proportion of the married in the first of these intervals of age being at least so much greater than in the second.

[5] This number is printed 100,098 by Mr. Nicander, and I believe it should be so, as it occurs several times, but if that be right, some other number in the column of deliveries must be wrong, and as I have not the means of determining which it is, I have thought it better to state the amount of them as they stand, because the error bears a less proportion to that than any other, it is indeed of no kind of consequence.

The following table exhibits the proportion with considerable precision, under a few varieties in the circumstances.

Place	Term.		Number of Births of		No. of Male Births for every 10,000 Females
	Years	ending with	Males	Females	
England and Wales	29	1800	3,285,188	3,150,922	10,426
	10	1810	1,468,677	1,410,229	10,415
Sweden and Finland	9	1763	400,086	387,702	10,436
	20	1795	1,006,420	965,000	10,429
France	3	1802	110,312	105,287	10,477
England	29	1800	2,997,842	2,879,011	10,413
	10	1810	1,391,977	1,338,998	10,396
Wales	29	1800	177,401	166,593	10,649
	10	1810	76,700	71,231	10,768
Scotland	29	1800	67,353	62,636	10,753
Carlisle	18	1796	2,400	2,271	10,568
Montpellier	21	1792	12,919	12,145	10,637
Diocess of Borgo in Finland	18	1791	93,701	91,404	10,251
	22	1795	117,928	115,191	10,238
	4	1795	24,227	23,787	10,185
Stockholm	9	1763	12,015	11,706	10,264
Illegitimate					
In Sweden and Finland	20	1795	37,700	37,060	10,173
In Montpellier	21	1792	1,373	1,362	10,081

Table XI.   Showing the Fecundity of Women at the different Periods of Life, in *Sweden and Finland*, from 1780 to 1795, both Years inclusive. (Art. 788).

	Absolute Fecundity		Intensity of Fecundity	
Between the Ages of	Mean Number of Females living	Annual average Number of Deliveries	That is, of 10,000 living	Or one of
15 & 20	132,765	3298	248	40.256
20 & 25	131,377	16,507	1257	7.959
25 & 30	121,650	26,329	2164	4.620
30 & 35	112,250	25,618	2282	4.382
35 & 40	98,710	18,093	1833	5.456
40 & 45	89,259	8518	954	10.479
45 & 50	74,002	1694	229	43.686
50 & 55	69,035	39	5.65	1770.1
20 & 40	463,987	86,547	1865	5.361
15 & 55	829,048	100,096	1207	8.283

Double Births, 1 of   58
Triple,   1 of   3,365.
Quadruple,   1 of 143,000.

One Birth in 104 produced the death of the mother.

And the number of deaths from Childbirth, was to the whole number of deaths of females between the ages of 15 and 55, as 2 to 19, or as 1 to 10 nearly.

5. Statistical Applications of the Mortality Table

GEORGE KING (1902)

From *Institute of Actuaries' Textbook*, Part II, Second Edition, pages 56—58, 63. London: Charles and Edward Layton.

We omit several numerical examples given by King of the uses of the life table, which concern populations at various ages, annuity payments, and the effects of immigration on observed death rates. The article introduces the life table in modern notation.

4. The fundamental column of the Life Table is the column of l_x. The first value in that column, l_0, called the radix, is the number of annual births in the imaginary population; and the succeeding numbers show how many persons, out of l_0 born alive, complete each year of age. In the table the number of annual births is 127,283; and we observe that 100,000 live to complete the tenth year of their age; 89,685 live to complete the thirtieth; and so on. We also see that only 4 live to complete the century; and that, although 1 survives 101 years, all die before reaching the age 102. Age 102 is therefore the *limiting age* of the table, being the year of age on which some lives enter, but which none complete. To the limiting age, the Greek letter ω is assigned for symbol; and therefore $l_\omega = 0$. Also, the difference between the limiting age, and the present age, is called the *complement of life*; so that, at age x, the complement of life is $\omega - x$; in the case of our table, $102 - x$.

5. The column of d_x contains the differences between the numbers in the column of l_x; and shows how many, out of l_0 persons born alive, die in each year of their age. Thus, by the table, out of 127,283 persons born alive, 14,358 die before completing their first year; 691 survive to age 30, but die before reaching 31; and so on. The number, then, in column d opposite any age, x, is the number who complete that year of age, but die before completing the next; that is, the number in column d opposite age x, is the number who die in the $(x+1)$th year of age. As all born must die, it follows that the sum of all the numbers in column d is equal to l_0: also, the sum of the numbers in column d, from age x to the oldest age, is equal to l_x.

6. To find how many die aged between x and $x+n$, we may take the sum of the numbers in column d, from d_x to d_{x+n-1}, inclusive; but unless n be very small, it will be easier to obtain the result by means of the l column; because $l_x - l_{x+n} = d_x + d_{x+1} + \&c. + d_{x+n-1}$. Thus, by the table, the number of persons, out of 127,283 born alive, who die between ages 20 and 30, is $l_{20} - l_{30} = 6376$.

7. Passing now to the column L, we have the population living in a stationary community. Such a community, sustained by l_0 annual births, will, on the supposition of uniform distribution of births and deaths, always contain $\frac{1}{2}(l_0 + l_1) = L_0$ children in the first year of their age; $\frac{1}{2}(l_1+l_2)=L_1$ in the second year of their age; and so on. Thus, by the table, a population sustained by 127,283 annual births, will always contain 95,787 young persons aged between 20 and 21. The total population at all ages will be the sum of all the numbers in column L ; and that is given in column T. By the table, the total population that would be supported by 127,283 annual births, is 6,082,031=T_0. The column T bears exactly the same relation to the column L, that the column l bears to the column d: that is, T_x is the sum of the numbers in column L, from age x to the oldest age: therefore T_x is the total population, aged x and upwards, in the community. In the community of T_0 inhabitants, there must be l_0 deaths annually; because there are l_0 births; and, the population being stationary, the deaths must be equal in number to the births. Similarly, there must be l_x deaths annually of persons aged x and upwards; and l_x-l_{x+n} deaths annually, of persons aged between x and $x+n$. Also, the number of inhabitants aged between x and $x+n$, is T_x-T_{x+n}; and, therefore, the proportion of deaths to population, between ages x and $x+n$, is $\dfrac{l_x-l_{x+n}}{T_x-T_{x+n}}$; and, for the whole community, the proportion of deaths to population is $\dfrac{l_0}{T_0}$. When n

is unity, $\dfrac{l_x-l_{x+n}}{T_x-T_{x+n}}=\dfrac{d_x}{L_x}=m_x$, the central death rate at age x. By the table, the proportion of deaths to the population for the whole community is $\dfrac{127,283}{6,082,031}=\cdot020928$ or about 21 per thousand. The proportion for the population aged less than 50, is $\dfrac{127,283-72,795}{6,082,031-1,475,603}$

$=\dfrac{54,488}{4,606,428}=\cdot011829$, or not quite 12 per thousand, while the pro-

portion for the population aged 50 and upwards, is $\dfrac{72,795}{1,475,603}=\cdot049332$,

or over 49 per thousand. These figures illustrate the remarks made in Chapter iii, Art. 13. They show that if from any cause there is an unusual proportion of young persons in a community, the ratio of deaths to population will be diminished; but, as previously remarked, it does not follow that therefore the members of that community enjoy unusual longevity. ...

We have seen that, of l_x persons who attain the precise age x, l_{x+1} will complete a year of life in the first year; and d_x will live on the average half a year each; therefore the quantity of existence in the first year due to the l_x persons will be $l_{x+1}+\frac{1}{2}d_x=L_x$. Similarly for future years; therefore $\Sigma L_x=T_x$ will be the total future existence due to the l_x persons; giving $\frac{T_x}{l_x}=\overset{\circ}{e}_x$ years to each; and the average age at death of the l_x persons will be $x+\overset{\circ}{e}_x$ years. The existence within the next n years due to the l_x persons, is T_x-T_{x+n}; giving $\frac{T_x-T_{x+n}}{l_x}=|_n\overset{\circ}{e}_x$ years for each. Of these years, $n\times l_{x+n}$ are due to those who complete age $x+n$; leaving $T_x-T_{x+n}-nl_{x+n}$ for those who die between age x and age $x+n$. But l_x-l_{x+n} persons die between these ages; therefore the average amount of existence between ages x and $x+n$, belonging to those who die between these ages, is $\frac{T_x-T_{x+n}-nl_{x+n}}{l_x-l_{x+n}}$, and their average age at death is $x+\frac{T_x-T_{x+n}-nl_{x+n}}{l_x-l_{x+n}}$. For example, if $x=20$ and $n=10$, we find $|_{10}\overset{\circ}{e}_{20}=\frac{T_{20}-T_{30}}{l_{20}}=9\cdot680$. Also, the existence within the 10 years, due to all those who reach age 20, is $T_{20}-T_{30}=929,902$; the existence in the period due to those who survive it is $10l_{30}=896,850$; leaving 33,052 years to the $l_{20}-l_{30}=6,376$ persons who die in the 10 years; or $5\cdot184$ to each. The average age at death of those who die between 20 and 30, is therefore $25\cdot184$. Similarly the average age at death of those who die below 20 is $\frac{T_0-T_{20}-20l_{20}}{l_0-l_{20}}=3\cdot734$.

13. We have seen, Art. 7, that the ratio of deaths to population in a stationary community is $\frac{l_0}{T_0}$: also (Chap. iii, Art. 16), that the complete expectation of life is $\frac{T_0}{l_0}$. It therefore follows that the ratio of deaths to population is equal to the reciprocal of the complete expectation of life.

6. Formal Treatment of Aggregate Mortality Data

WILHELM LEXIS (1875)

From *Einleitung in die Theorie der Bevölkerungs-Statistik*, pp. 5—7. Strasbourg: Trubner. Translated by Nathan Keyfitz.

As an example of a population with a single source of variation, let us consider that of a fixed territory subject only to death. (One could consider birth as the alteration of condition, but it seems more natural to take the births as the absolute starting point of the elements). The proposed treatment is valid also for the presentation of any other kind of alteration of condition of those born.

The development of our example yields the formal theory of mortality statistics, which covers not only mortality, but also its correlate, survivorship.

Provisionally disregarding out- and in-migration, we consider a number of individuals who are distinguished for us only by being born at different times, and dying at different times and ages. Thus the population can only be decomposed according to these variables. The decomposition to reveal the inner structure of the population can be seen most simply in a graphical scheme.

Let (Fig. 1) N be an axis of coordinates, on which time is measured from an arbitrary fixed origin O, and let the birth times of individuals be marked on ON. These birth points occur discontinuously and more or less densely. The number of points that lie on equal intervals of the line ON would serve to compare the density of births at different times.

Two kinds of birth points are to be distinguished, those corresponding to live births, and those corresponding to stillbirths. We consider only the first.

Starting from the birth points, we now measure the ages of individuals vertically to the birth axis, and parallel to the axis of ages $O\Omega$. That age at which the event under consideration, namely death, occurs to an individual is represented by a point.

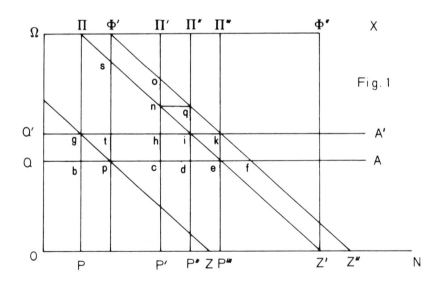

Fig. 1

In that way the plane ΩON is strewn with scattered death points, not shown in the diagram. The diagram is closed by the line ΩX, which runs parallel to ON at a distance equal to $O\Omega = \omega$, which corresponds to the highest age reached. The figure can be thought of as extending indefinitely to the right.

On every life-line there now lies only one death point, but with a high density of births there would tend to be in each small interval of births dying points of all the ages from O to ω.

The mass of points so presented can be divided by parallels to the two axes.

If the birth times of the abcissa are in general expressed by n, the equation $n = OP$ represents the vertical line $P\Pi$. This divides the dying times of the individuals who are born before the time OP from those born after.

If the ages, represented by the ordinates, are designated by a, then $a = OQ$ is the equation of the line QA, which distinguishes the dying points of those who die before reaching the age $O\Omega$ from those who die after.

Finally, there is a third division of the plane, such that the cases of death occurring before a certain absolute time are divided from those that lie beyond this time. If z is the general designation of the time of observation, then the equation

$$n + a = z$$

applies for the points that fall exactly at time z.

Let us now suppose that z takes a fixed value OZ' as measured on the birth axis, corresponding to the age O, and let z go through all values from O to ω, then n goes backwards through all values from z (or OZ') to $z - \omega$. The above equation thus represents the straight line $Z'\Pi$, which on further extension would thus cut the axis $O\Omega$ at a distance z from point O and makes an angle of $45°$ with either axis.

Consider now an age at death that belongs to an arbitrary birth point in the interval $z - \omega$ to z (or PZ'). If it is below that determined by the above equation, (that is, lower than $z - n$) then the death point of the individual concerned lies below line $Z'\Pi$. But if the age at death is higher than would be determined by the n from the above equation, then the death point lies above that line.

The line $Z'\Pi$ divides those who die before the time z from those whose death occurs after this time. For values of n that exceed z it is obvious that the corresponding cases of death can only occur after the time z. It is also plain that persons born before the time $z - \omega$ (or OP) necessarily die before the time z, for they otherwise would exceed the maximum age ω.

7. A Short Method for Constructing an Abridged Life Table

LOWELL J. REED and MARGARET MERRELL (1939)

From *American Journal of Hygiene* 30. Excerpts are from pages 33—38, 51.

For infancy Reed and Merrell revise their q_x formula to correct underenumeration, an approach that is not general. This has been omitted, as have their equations for L_x and T_x values by approximate integration. The article is followed by a note on other $_nq_x$ formulas.

The life table is so valuable a description of the age variation in danger of dying that it is desirable to have rapid methods of obtaining it from the recorded vital statistics. The principal value of the life table lies in the form in which age-specific risks of dying are stated. They are expressed as the probability of dying within a designated interval subsequent to an exact age, or as the probability of surviving from one exact age to another. Such probabilities are not only very descriptive, but they lend themselves readily to the treatment of mortality in a wide variety of problems.

The fundamental element in the construction of a life table for a particular time is thus to convert the observed age-specific death rates for that time into the probabilities of dying within stated age intervals. There is nothing in the definition of the probability of dying which fixes the length of these intervals. They may be chosen to suit the needs of the problem. The term "complete life table" has by custom come to designate a table in which the interval is 1 year, and the probability is stated for every year of age. It should be remembered, however, that this is pure convention, since a table computed for monthly intervals would be more complete and one for weekly intervals still more so.

The term "abridged life table" is less stabilized in its usage, since there are various ways in which a life table may depart from the form called "complete." One form of abridged table is illustrated by Foudray's 1920 Tables for the United States (1923) and by Dublin and Lotka's Appendix Tables in *Length of Life* (1936), in which the probabilities of dying are given for single years of age, but these values are tabulated at intervals of 5 or 10 years rather than for every year. They constitute, therefore, selected probabilities from a complete table, the intermediate values being omitted.

Another form of abridgement, which is the one considered in this paper, increases the interval over which the probabilities are stated, usually to 5 or 10 years. These probabilities cover the interval between the stated ages and therefore form a table which is not incomplete except in the defined technical sense. They differ from the values in a complete life table only in the fact that the interval chosen for study is greater than 1 year. The abridgement in this case, being through a condensation of the complete table rather than through omission, has certain advantages, since the survivorship values can be computed directly from the probabilities of dying, and the probabilities may be used as they stand, without interpolation, in any analysis where an interval greater than 1 year gives sufficient detail.

The method proposed is a short procedure for obtaining these probabilities of dying directly from the observed rates without going through the elaboration of constructing a complete table. The principal function to be derived is therefore $_nq_x$, the probability of dying within n years after age x, and the other life table functions flow directly from these values by procedures which are described.

Direct observation shows that there is a very high degree of association between the observed age-specific death rate over an age interval and the probability of dying within that interval. ...The basis of this relationship may be seen from the two following equations expressing, respectively, the age-specific death rate, $_nm_x$, in a stationary population, and the probability of dying, $_nq_x$, as functions of l_x, the number surviving to age x out of a given number born alive.

---

[1] Paper no. 212 from the Department of Biostatistics, School of Hygiene and Public Health, The Johns Hopkins University, Baltimore, Md.

$$_nm_x = \frac{l_x - l_{x+n}}{\int_x^{x+n} l_x dx}, \qquad (1)$$

$$_nq_x = \frac{l_x - l_{x+n}}{l_x}. \qquad (2)$$

These equations show that $_nq_x$ is expressible in terms of $_nm_x$ and that if the equation for l_x were known, the function relating the two types of rates could be stated explicitly. The assumption that l_x is expressible as a straight line over the interval n leads to the well-known relationship [3]

$$_nq_x = \frac{2n\,_nm_x}{2 + n\,_nm_x}, \qquad (3)$$

which is the equation of a hyperbola, passing through the origin with a slope n, and having an upper asymptote, $_nq_x = 2$. ...

As an alternative, we might assume that l_x is an exponential curve over an interval n, since it is as reasonable to break up the arithlog graph of l_x into linear segments as it is to treat the arithmetic graph in this way. This assumption gives for the relationship between the age-specific rate of the life table and the probability of dying over the same age interval, the catalytic equation [4]

$$_nq_x = 1 - e^{-n\,_nm_x}. \qquad (4)$$

This equation, like the hyperbola (3), passes through the origin with a slope of n, but has an upper limit $_nq_x = 1$, where the hyperbola has a limiting value of 2. Both of these curves have, therefore, a rational position at the origin, the probability of dying being 0 when the death rate is 0. Furthermore, they have a rational slope at the origin since, as the

average annual death rate becomes smaller, the probability of dying within n years approaches n times the value per year. With regard to the upper limit imposed on $_nq_x$, the catalytic with its limit of 1 is more rational than the hyperbola with its limit of 2, since the probability of dying cannot exceed unity. For intervals as small as 1 year, it is hard to say whether the hyperbola (3) or the catalytic (4) is to be preferred as an expression of the relationship between m_x and q_x, but neither proves to be satisfactory for longer intervals.

Other attempts to state an equation describing the l_x curve over longer intervals have led to rather involved equations of association. Furthermore, for practical applications it would be necessary to adjust such equations for the difference between observed and life table values of $_nm_x$. Therefore, the relationship between the observed $_nm_x$ and $_nq_x$ was studied directly without an explicit assumption with regard to the l_x equation. For this purpose Glover's 1910 life tables (1921) were examined as to the relation of the observed 5-year and 10-year death rates to the derived probabilities of dying over the same intervals.

Figure 1 presents for a 5-year interval the association between the observed $_5m_x$ and the derived $_5q_x$ for the 33 tables in Glover's 1910 series, the x values taking all the multiples of 5 through age 80. It is apparent from this graph that the observations fall exceedingly close to a smooth curve. This is especially impressive since the life tables are not at all similar in their mortality schedule. They represent tables for the different sexes and colors, for urban and rural areas, and for various combinations of these factors, with the variation in patterns that comes from such diverse groups. It would thus appear that the relationship of $_5m_x$ and $_5q_x$ is one which need not be stated for a particular age or

[3] For the derivation of equation (3), see appendix 1.

[4] For the derivation of equation (4), see appendix 1.

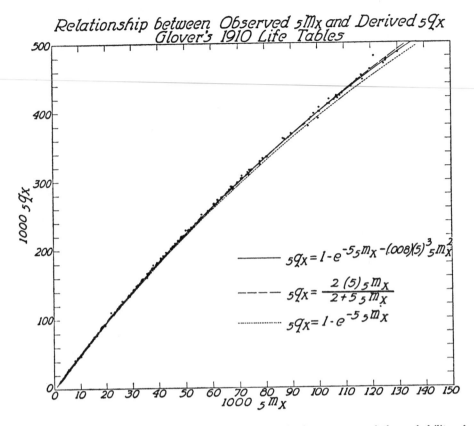

$$_5q_x = 1 - e^{-5 \cdot _5m_x - (.008)(5)^3 \cdot _5m_x^2}$$

$$_5q_x = \frac{2(5) \cdot _5m_x}{2 + 5 \cdot _5m_x}$$

$$_5q_x = 1 - e^{-5 \cdot _5m_x}$$

FIGURE 1.   Correlation between the observed age specific death rate, $_5m_x$, and the probability of dying, $_5q_x$, stated per 1000, from Glover's 1910 life tables.

size of death rate, but is independent of these factors. The statement of this relationship in the form of an equation would clearly permit passing directly from the age-specific death rate to the probability of dying.

The observed relationship is of somewhat the same general form as the hyperbola (3) or the catalytic (4), as may be seen in figure 1. Neither of these theoretically derived equations, however, could be considered an adequate description of the relationship. The catalytic begins to fall below the points when $_5m_x$ is approximately 20 per 1000 (about age 50) and remains below the points throughout the rest of the range. The hyperbola differs from the points less consistently than the catalytic, but does not have the proper curvature. For the ages beyond 80 which are not shown on this graph, it is in general farther above the points than the catalytic is below them, since the l_x curve in extreme old age is better described by an exponential than a linear function. Thus, the catalytic and the hyperbola deviate from the points in different form but neither stays with them. This is due to the fact that the basic assumptions from which they were derived do not hold for intervals much longer than 1 year.

An equation was therefore sought which would describe the entire range of observations and one which would also

46

be suitable for expressing the relationship of the rates for intervals other than 5 years. This requirement proved to be satisfactorily met by the equation

$$_nq_x = 1 - e^{-n\,_nm_x - an^2\,_nm_x{}^2}, \qquad (5)$$

in which a is an arbitrary constant. Its value determined empirically for the method presented in figure 1 is $a = .008$. This equation is seen to be similar in form to (4), passing through the origin with a slope of n and having an upper asymptote $_nq_x = 1$, but it has a corrective term in the exponent which has the effect of lifting the latter part of the curve. Thus it is a rational curve in its position and slope at the origin and in its limiting value, and from figure 1, it is apparent that it fits the points well, rather remarkably so, in view of the fact that it contains only one arbitrary con-- stant. It should be stated that the curve fits the points beyond age 80, which are not shown on this graph, as well as those presented, and the equation may therefore be used from age 5 to the end of life, to derive the life table probabilities from the observed rates.

The complexity of the l_x curve prevents the use of such a simple equation over very broad intervals. The grouping to which it may, with fair accuracy, be extended is indicated by the relationship for 10-year values. Figure 2 shows the scatter of points for all the 10-year rates from Glover's 1910 tables, at 5-year intervals from age 5 to 80. Here again, the points from the various life tables fall into the same pattern, none of the tables showing any marked or consistent divergence from the others. This scatter of points is compared graphically with the three equations just considered. The assumption that l_x could be expressed as a series of linear or exponential equations would naturally be more violent for an interval of 10 than of 5 years. It is not surprising, therefore, that both the hyperbola and

the catalytic depart from the trend of the points in the same way as they did for the 5-year rates, but to a greater degree.

The empirical curve, derived above, with the same value of the arbitrary constant, is seen to give a very satisfactory description of the relationship, and the points hang closely to the central curve. The values beyond age 80 are also well described by the equation and thus for a grouping as broad as 10 years, the equation effects a satisfactory transformation of the observed age-specific death rates into probabilities of dying from age 5 to the end of life.

Although the observations have been presented only for an interval of 5 and of 10 years, it should not be inferred that the equation is limited to these values of n. The equation derived depends implicitly on the area and ordinate relationships in the l_x curve and is sufficiently elaborate to summarize these relationships with a high degree of approximation for an interval as large as 10 years. Consequently it holds equally well for smaller intervals, provided the age-specific rates over these intervals can be obtained, either by direct observation or adjustment of the observations, with the same accuracy as those observed for an n of 5 or 10 years. For an interval as small as 1 year, this curve, the catalytic (3), and the hyperbola (4) give virtually the same results over the entire age range.

The procedure for summarizing the relationship between the observed $_nm_x$ and $_nq_x$ statistically in the form of a regression equation does not take account of the individual variation about the central curve. However, the individual variation is, in general, well within the range of error of observation, and for areas such as states or smaller units, it is even within the variation of simple sampling. Thus the refinement of $_nq_x$ values beyond the value predicted by

47

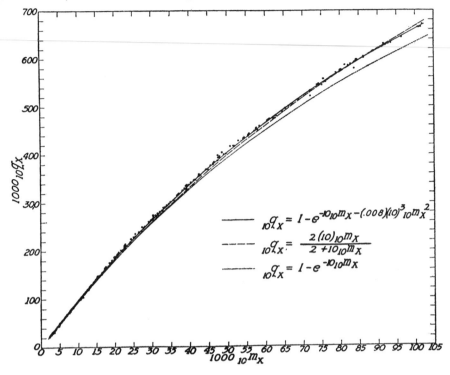

FIGURE 2. Correlation between the observed age specific death rate, $_{10}m_x$, and the probability of dying, $_{10}q_x$, stated per 1000, from Glover's 1910 life tables.

the regression curve from the observed specific rate is, for the usual case, carrying the statistical analysis beyond the point justified by the accuracy of the material. ...

Appendix 1

(b) *Derivation of equation (3) in text:*
Assume that over the interval n,
$$l_x = a + bx.$$
By definition
$$_nm_x = \frac{l_x - l_{x+n}}{\int_x^{x+n} l_x dx} \qquad (1)$$
and
$$_nq_x = \frac{l_x - l_{x+n}}{l_x}. \qquad (2)$$

Then
$$_nm_x = \frac{a + bx - a - b(x + n)}{\int_x^{x+n}(a + bx)dx}$$
$$= \frac{-b}{a + bx + \dfrac{bn}{2}},$$
$$_nq_x = \frac{-bn}{a + bx}.$$

Eliminating $a + bx$ between these two equations gives
$$_nq_x = \frac{2n \; _nm_x}{2 + n \; _nm_x}. \qquad (3)$$

(c) *Derivation of equation (4) in text:*
Assume that over the interval n
$$l_x = e^{a+bx}.$$

48

By substitution in equations (1) and and (2)

$$_nm_x = \frac{e^{a+bx} - e^{a+b(x+n)}}{\displaystyle\int_x^{x+n} e^{a+bx}dx} = -b$$

$$_nq_x = \frac{e^{a+bx}(1 - e^{bn})}{e^{a+bx}} = 1 - e^{bn}.$$

Eliminating b between these two equations gives

$$_nq_x = 1 - e^{-n\,_nm_x}. \qquad (4)$$

REFERENCES

Foudray, E.

1923 United States Life Tables, 1919–20. Washington (Bureau of the Census).

Dublin, L. I. and A. J. Lotka.

1936 Length of Life. New York. Ronald Press Company.

Glover, J. W.

1921 United States Life Tables, 1890, 1901, 1910, and 1901–1910. Washington (Bureau of the Census).

49

Editors' Note: Approximations to $_nq_x$

The equations for $_nq_x$ Reed and Merrell examine apart from their own derive from Milne and Farr. Their *equation (3)*—*equation (1)* in the article by Greville that follows—is found by setting $M_x = m_x$ and generalizing from 1 year to n year intervals Milne's equation (1815, p. 99)

$$\frac{P_x}{D_x} = \frac{\frac{1}{2}(l_x + l_{x+1})}{l_x - l_{x+1}} = \frac{l_x - \frac{1}{2}d_x}{d_x},$$

to yield

$$_nm_x = \frac{_nD_x}{_nP_x} = \frac{l_x - l_{x+n}}{\frac{n}{2}(l_x + l_{x+n})} = \frac{_nd_x}{n(l_x - \frac{1}{2}\,_nd_x)}.$$

On dividing by l_x this becomes

$$_nm_x = \frac{_nq_x}{n(1 - \frac{1}{2}\,_nq_x)} \quad \therefore \quad _nq_x = \frac{n\,_nm_x}{1 + \frac{n}{2}\,_nm_x}.$$

Reed and Merrell arrive at the same result by setting $l_x = a + bx$, which of course Milne assumed. The equation can also be identified as DeMoivre's (1725) hypothesis, discussed in chapter 4.

Reed and Merrell's *equation (4)*—Greville's *equation (2)*—derives from Farr's (1864, pp. xxiii—xxiv) English Life Table No. 3. Farr introduced a better form, which would generalize for abridged tables as

$$_nq_x = 1 - \exp[-\,_nm_x\,r_x^{-n/2}(r_x^n - 1)/\ln r_x],$$

where

$$r_x = (_nm_{x+n}/\,_nm_x)^{1/n}.$$

Here, r is "the rate at which the rate of mortality increases or decreases." The derivation of the formula is explained by Wolfenden (1954, pp. 130—131). Assuming the force of mortality μ_x to be approximately $_nm_{x-n/2}$ and applying Gompertz' Law that the force of mortality increases exponentially with age, we have

$$\mu_{x+t} = _nm_{x+t-n/2} = r_x^{t-n/2}\,_nm_x.$$

This permits the substitution

$$_nq_x = 1 - \exp\left[-\int_0^n \mu(x+t)\,dt \right]$$

$$= 1 - \exp\left[-_nm_x\, r_x^{-n/2} \int_0^n r_x^t\, dt \right]$$

$$= 1 - \exp\left[-_nm_x\, r_x^{-n/2} (r_x^n - 1)/\ln r_x \right].$$

In application Farr used single year rates and dropped the minor term $r_x^{-n/2}$. If the force of mortality is assumed constant (i.e., $r=1$) the equation reduces to

$$_nq_x = 1 - \exp\left[-n\,_nm_x \right].$$

This yields the l_x formula given in the appendix to Reed and Merrell:

$$l_x = \alpha e^{-x\,_nm_x} = e^{a+bx}.$$

Graphed with the simple exponential and the Reed-Merrell formula (see their Figure 2) the Farr equation would be the least accurate for the low $_nm_x$ rates at younger ages and intermediate for the higher rates that come later in life, reflecting its close dependence on Gompertz' Law. Greville's article completes this analysis by uncovering a derivation for the Reed-Merrell formula, which employs Gompertz' Law more effectively as a correction term.

8. Short Methods of Constructing Abridged Life Tables

T. N. E. GREVILLE (1943)

From *Record of the American Institute of Actuaries, Part I* 32. Excerpts are from pages 29, 34—40.

I. INTRODUCTION

In recent years sociologists, public health officers, and students of population problems have been taking an increasing interest in the life table as a description of the age-variation in the chances of death and survival. As a result of this interest in the subject, attention has been focused on rapid methods of constructing abridged life tables. Some of the most interesting and useful methods have been devised by nonactuaries and, probably for that reason, have not received the attention they deserve from the actuarial profession. It is the purpose of this paper to bring some of these methods to the notice of actuaries and also to show that certain formulas which have been put forward by their originators on purely empirical grounds actually have a valid mathematical basis. ...

III. RELATION BETWEEN $_nm_x$ AND $_nq_x$

The basic feature of the type of methods of life-table construction under consideration here is the mathematical relation assumed to hold between $_nm_x$ and $_nq_x$. There are a number of such relations which may be used. Perhaps the simplest is obtained by assuming that l_x can be regarded as a linear function in the age-interval. This leads to the equation

$$_nq_x = \frac{2n \cdot {_nm_x}}{2 + n \cdot {_nm_x}},\tag{1}$$

a formula which, for the particular case $n = 1$, is well known to actuaries [5, p. 5]. Another possible assumption is that l_x can be represented by an exponential function. This leads to the equation

$$\text{colog}_e {_np_x} = n \cdot {_nm_x},\tag{2}$$

a relation which is also familiar to actuaries in the special case of $n = 1$ [5, p. 16].

More precise equations of the same general form as equations (1) and (2) will now be derived. In the life table

$$_nm_x = \frac{l_x - l_{x+n}}{T_x - T_{x+n}} = -\frac{d}{dx} \log_e (T_x - T_{x+n}) = -\frac{d}{dx} \log_e {_nL_x}.\tag{3}$$

Integrating,

$$_nL_x = Ce^{-\int {_nm_x}\,dx}.\tag{4}$$

Now, applying the Euler-Maclaurin summation formula, we have

$$T_x = \sum_{h=0}^{\infty} {}_nL_{x+nh} = C\left\{ \frac{1}{n}\int_x^{\infty} e^{-\int {}_nm_t\,dt}\,dt + \tfrac{1}{2}e^{-\int {}_nm_x\,dx} \right.$$
$$\left. + \frac{n}{12}\, {}_nm_x\, e^{-\int {}_nm_x\,dx} + \cdots \right\}.$$

Differentiating and substituting from equation (4) gives

$$l_x = {}_nL_x\left\{ \frac{1}{n} + \tfrac{1}{2}{}_nm_x + \frac{n}{12}\left({}_nm_x^2 - \frac{d}{dx}{}_nm_x\right) + \cdots \right\}. \quad (5)$$

Making use of equations (3) and (5), we get

$$\left. {}_nq_x = \frac{l_x - l_{x+n}}{l_x} = \frac{{}_nm_x \cdot {}_nL_x}{l_x} \right.$$

$$\left. = \frac{{}_nm_x}{\frac{1}{n} + \tfrac{1}{2}{}_nm_x + \frac{n}{12}\left({}_nm_x^2 - \frac{d}{dx}{}_nm_x\right) + \cdots}, \right\} \quad (6)$$

or,

$${}_nq_x = \frac{2n \cdot {}_nm_x}{2 + n \cdot {}_nm_x + \frac{n^2}{6}\left({}_nm_x^2 - \frac{d}{dx}{}_nm_x\right) + \cdots}, \quad (7)$$

an equation of the same general form as equation (1).

In order to derive an equation of the form of equation (2), we write

$${}_np_x = 1 - {}_nq_x = \frac{2 - n \cdot {}_nm_x + \frac{n^2}{6}\left({}_nm_x^2 - \frac{d}{dx}{}_nm_x\right) + \cdots}{2 + n \cdot {}_nm_x + \frac{n^2}{6}\left({}_nm_x^2 - \frac{d}{dx}{}_nm_x\right) + \cdots},$$

whence

$$\text{colog}_e\, {}_np_x = \log_e\left[1 + \frac{n}{2}{}_nm_x + \frac{n^2}{12}\left({}_nm_x^2 - \frac{d}{dx}{}_nm_x\right) + \cdots\right]$$

$$- \log_e\left[1 - \frac{n}{2}{}_nm_x + \frac{n^2}{12}\left({}_nm_x^2 - \frac{d}{dx}{}_nm_x\right) + \cdots\right].$$

55

This gives, upon expansion and simplification,

$$\operatorname{colog}_e {}_np_x = n \cdot {}_nm_x + \frac{n^3}{12}\, {}_nm_x \frac{d}{dx}\, {}_nm_x + \cdots, \qquad (8)$$

a formula of the same general form as equation (2).

The first question that arises in the actual application of formulas (7) and (8) concerns the evaluation of the derivative of ${}_nm_x$. It turns out that great exactness is not necessary, as the terms into which this derivative enters are in the nature of minor adjustments which do not have a great effect on the resulting ${}_nq_x$ values. If it is assumed that the function ${}_nm_x$ can be represented by a polynomial, it is found that

$$\frac{d}{dx}\, {}_nm_x = \frac{{}_nm_{x+n} - {}_nm_{x-n}}{2n} \qquad (9)$$

or

$$\frac{d}{dx}\, {}_nm_x = \frac{-{}_nm_{x+2n} + 8{}_nm_{x+n} - 8{}_nm_{x-n} + {}_nm_{x-2n}}{12n}, \qquad (10)$$

according to whether a polynomial of the second or the fourth degree is assumed. These formulas, however, are not applicable when neighboring age-groups do not contain the same number of years.

As an alternative, it may be noted that the well-known formula for the derivative of ${}_np_x$ [5, p. 17] gives

$$\frac{d}{dx}\operatorname{colog}_e {}_np_x = \mu_{x+n} - \mu_x .$$

Therefore, differentiating equation (8) gives

$$\frac{d}{dx}\, {}_nm_x = \frac{1}{n}\left(\mu_{x+n} - \mu_x\right)$$

approximately. Substituting this approximation in equation (8) gives, finally,

$$\operatorname{colog}_e {}_np_x = n \cdot {}_nm_x\left[1 + \frac{n}{12}\left(\mu_{x+n} - \mu_x\right)\right].$$

This is the formula given (for the case of $n = 5$) by the editors of *JIA* [1, p. 301; 2, pp. 121–22]. However, in the practical application of this formula it was found necessary to approximate the expression in parentheses in terms of $_n m_x$ values. This makes the formula equivalent to equation (8), used in conjunction with the approximation (9).

Since $_n m_x$ is approximately equal to the force of mortality at the middle of the period, a more logical assumption is that $_n m_x$ is an exponential function, in accordance with Gompertz' Law. If

$$_n m_x = Bc^x \; ;$$

then

$$\frac{d}{dx} \, _n m_x = k \cdot \, _n m_x \, ,$$

where

$$k = \log_c e \, .$$

Substituting this value in equations (6) and (8) gives us

$$_n q_x = \frac{_n m_x}{\dfrac{1}{n} + \, _n m_x \left[\dfrac{1}{2} + \dfrac{n}{12} \left(_n m_x - k \right) \right]} \tag{6a}$$

and

$$\text{colog}_e \, _n p_x = n \cdot \, _n m_x + \frac{k}{12} \, n^3 \cdot \, _n m_x^2 \, . \tag{8a}$$

As a moderate variation in the value of k has little effect on the value of $_n q_x$ except at the older ages, where most mortality tables follow Gompertz' Law fairly closely, and at the very young ages, which are generally dealt with by a different method, k may be safely taken as a constant throughout the table. Henderson states [6, p. 90] that for practically all mortality experiences $\log_{10} c$ lies between .035 and .045. Therefore, $k = \log_e c$ would fall between .080 and .104.

It is interesting to observe that Reed and Merrell, after finding equations (1) and (2) unsatisfactory at the older ages, suggested on empirical grounds the equation

$$_n q_x = 1 - e^{-n \cdot \, _n m_x - an^3 \cdot \, _n m_x^2} , \tag{11}$$

a relation exactly equivalent to equation 8(a) if a is written for $k/12$. By fitting this curve to the thirty-three tables in Glover's 1910 series, they arrived at the value .008 for a. This would correspond to $k = .096$, a value well within the range given by Henderson. Reed and Merrell have published extensive tables of the function (11) with $a = .008$, both for $n = 5$ and for $n = 10$ and also for $n = 3$ over a limited range, intended for use in the age-interval two to five. With the aid of these tables a good computer can construct an entire abridged life table in less than two hours.

Although experiment would indicate that equation (6a) generally gives nearly as good results as equation (8a), the latter has certain theoretical advantages. In both cases $_nq_x = 0$ when $_nm_x = 0$. However, in equation (8a) $_nq_x$ approaches unity as it should do, when $_nm_x$ increases without limit, while in equation (6a) there is a point beyond which $_nq_x$ ceases to increase with $_nm_x$, and its limit is actually zero. ...

Reed and Merrell seem to have considered that, were it not for underenumeration, the expression (11) (with $a = .008$) should be applicable to ages zero and one as well as to the remainder of the life-span [3, p. 690]. Theoretical considerations would indicate that in a period of decreasing mortality $k = \log_e c$ (and therefore a in the Reed-Merrell formula) should be negative, and rough calculations based on recent United States Life Tables indicate for age zero a value of about $-.3$ for a instead of .008. The fact is, however, that for an age-interval of only one year, this adjustment is not of much consequence, and for such an interval equation (11) would be close enough for practical purposes. However, equation (1) or equation (2) would be slightly preferable, as the additional adjustment contained in equation (11) is in the wrong direction.

V. LIFE-TABLE POPULATION

The calculation of the values of l_x and $_nd_x = l_x - l_{x+n}$ from those of $_nq_x$ requires no explanation or comment. However, the computation of the $_nL_x$ column introduces some questions. Two distinct methods of obtaining $_nL_x$ have been suggested. The first [7] is based on the assumption that $_nm_x$ has the same value in the actual population and in the life-table population and employs the relation

$$_nL_x = \frac{_nd_x}{_nm_x} . \tag{12}$$

which means, in practice, using an approximate integration formula, such as

$$_nL_x = \frac{n}{2}\,(l_x + l_{x+n}) + \frac{n}{24}\,(_nd_{x+n} - {_nd_{x-n}}) . \qquad (14)$$

This approximate method, although less direct and in theory less exact, generally produces superior results in actual practice. This is because the values of l_x obtained by the abridged process contain inaccuracies and irregularities, and it turns out that the effect on the value of $_nL_x$ in equation (13) of slight errors in l_x is less than the effect in equation (12) of the corresponding error in $_nd_x$ and $_nm_x$. A mathematical demonstration of this fact may be of interest.

From equation (13) we can write by the Theorem of Mean Value

$$_nL_x = nl_{x+\theta} ,$$

where θ is a number beween o and n. Therefore, if $d(_nL_x)$ denotes the error in the value of $_nL_x$, we have

$$\frac{d(_nL_x)}{_nL_x} = \frac{d(l_{x+\theta})}{l_{x+\theta}} .$$

From this it appears that a given percentage of error in the l_x values will tend to produce about the same percentage of error in $_nL_x$. On the other hand, equation (12) gives

$$\frac{d(_nL_x)}{_nL_x} = \frac{d(_nd_x)}{_nd_x} - \frac{d(_nm_x)}{_nm_x} .$$

This indicates that, when equation (12) is used, a given percentage of error in either $_nd_x$ or $_nm_x$ tends to produce about the same percentage of error in $_nL_x$. However, it is obvious that errors of a given size in the values of l_x would tend to produce errors of the same absolute magnitude (and therefore much greater percentage errors) in $_nd_x$. Likewise, the values of $_nm_x$ are likely to contain greater relative errors than the l_x values. This explains why in practice the approximate integration formula gives better results.

BIBLIOGRAPHY

1. EDITORS of *JIA*. Note appended to the article "On a Short Method of Constructing an Abridged Mortality Table, by GEORGE KING, *JIA*, XLVIII, 301–3.

2. WOLFENDEN, H. H. *Population Statistics and Their Compilation.* ("Actuarial Studies," No. 3.) New York: Actuarial Society of America, 1925.

3. REED, L. J., and MERRELL, M. "A Short Method for Constructing an Abridged Life Table," *American Journal of Hygiene*, Vol. XXX, No. 2. Reprinted by the U.S. Bureau of the Census as *Vital Statistics— Special Reports*, Vol. IX, No. 54. All page references are to the Census publication.

4. DUBLIN, L. I., and LOTKA, A. J. *Length of Life.* New York: Ronald Press Co., 1936.

5. SPURGEON, E. F. *Life Contingencies.* London: Macmillan & Co., Ltd., 1938.

6. HENDERSON, R. *Mathematical Theory of Graduation.* ("Actuarial Studies," No. 4.) New York: Actuarial Society of America, 1938.

7. DOERING, C. R., and FORBES, A. L. "A Skeleton Life Table," *Proceedings of the National Academy of Sciences*, XXIV, 400–405.

8. GLOVER, J. W. *United States Life Tables, 1890, 1900, 1910, and 1901–1910.* Washington: Government Printing Office, 1921.

9. Life Tables for Natural Populations of Animals

EDWARD S. DEEVEY, Jr. (1947)

From *Quarterly Review of Biology* 22. Excerpts are from pages 283—284, 287—292, 294, 296—302, 312.

In selecting from Deevey's extensive review, we have tried to emphasize the different types of problems that arise in working with animal populations; our most serious omission is a detailed study of barnacles that examines crowding effects and mortality. Deevey begins his article with discussions of the life table and of different general survival patterns, which we also omit.

Having gained some idea of the limits circumscribing his own mortality, man has turned to look at the other animals. In 1935 Pearl and Miner, in their discussion of the comparative mortality of lower organisms, attempted to formulate a general theory of mortality. They quickly gave up the attempt upon realizing that the *environmental* determinants of life duration can not, at least as yet, be disentangled from such *biological* determinants as genetic constitution and rate of living. They ended with a plea for "more observational data, carefully and critically collected for different species of animals and plants, that will follow through the life history from birth to death of each individual in a cohort of statistically respectable magnitude." Thus by implication Pearl and Miner appealed to the ecologists, who for the most part have been busy elsewhere. Accounts of the conceptions and methodology of life tables have not yet found their way into textbooks of ecology, and while field naturalists have devoted increasing attention to the dynamics of natural populations most of them have been content to leave the construction of life tables to the statisticians and laboratory ecologists.

This article, which is designed as an introduction to the subject rather than as a formal review, brings together from the ecological literature a mass of information bearing on the survival of animals in nature. This information has not heretofore been considered relevant by biometricians working with human populations, nor has it ever been considered in its context by ecologists. In collecting the material it was immediately obvious that it is still too early to formulate general theories. Serious deficiencies are only too apparent in the data. But the difficulties differ from case to case, and are therefore not insurmountable. Moreover, the bibliography will show that virtually all of this knowledge has been acquired in the twelve years since the appearance of the review by Pearl and Miner. By taking stock now, and by calling attention to gaps in our information, it is hoped that some guidance can be given to ecologists and others in the gathering of new material. ...

Ecological Life Tables

The field ecologist deals with populations which are by no means so elementary as those inside *Drosophila* bottles. Even the total size of the population of a species cannot be easily ascertained for an area large enough to be representative, and calculations of the birth rate and death rate are uncertain at best, largely owing to immigration and emigration. It is seldom indeed that the ecologist knows anything of the age structure of a natural population. In a few cases, growth rings on the scales or otoliths (fish) or horns (ungulates) make it possible to determine the age of an animal. Moore (1935) has shown that annual growth rings occur in the genital plates of sea-urchin tests, as they do in the shells of some molluscs. Moore checked the validity of the age determination by reference to the size-frequency distribution in his catches, and the separation of modal size classes in a population often affords a clue to age, particularly for younger age groups. The age of adult females can be determined in the case of certain mammals (whales, Wheeler, 1934, Laurie, 1937; seals, Bertram, 1940) by counting the corpora lutea in the ovaries. But for most animals it is possible to find out the ages of individuals only by marking them in some way.

Even when the age of a member of a natural population is known, it is not a simple matter to obtain accurate vital statistics. The source of greatest confusion lies in the impracticability of keeping the individuals under continuous observation. Migratory birds, for example, are easy to band as nestlings, but nearly impossible to find between fledging and the time they leave for winter quarters. Often they can not be found at all unless they return to the same area to breed, when they can be trapped in nest boxes. Their mortality between fledging and breeding can be calculated, but the calculation is rendered uncertain by the tendency of young birds not to return to their birthplaces as breeding adults.

As sources of data for the construction of life tables, the ecological information falls into three groups: (1) cases where the age at death (d_x) is directly observed for a large and reasonably random sample of the population; (2) cases where the survival (l_x) of a large cohort (born more or less simultaneously) is followed at fairly close intervals throughout its existence; (3) cases where the age structure is obtained from a sample, assumed to be a random sample of the population, and d_x is inferred from the shrinkage between successive age classes. It should be noticed that only the second sort of information is statistically respectable, since in so far as the breeding can safely be assumed to be simultaneous, it is comparable to that obtained from a *Drosophila* bottle. The first and third types can be used only if one is prepared to assume that the population is stable in time, so that the actual age distribution and the life table age distribution are identical. This assumption would certainly not be true of a human population; it may be approximately true for many natural populations of animals. ...

Age at Death Directly Observed

In the course of his careful investigation of the wolves of Mt. McKinley, Murie (1944) picked up the skulls of 608 Dall mountain sheep (*Ovis d. dalli*) which had died at some time previous to his visit, and an additional 221 skulls of sheep deceased during the four years he spent in the Park. The age of these sheep at death was determinable from the annual rings on the horns. "Time, which antiquates antiquities, and hath an art to make dust of all things, hath yet spared these minor monuments" (Sir Thomas Browne, *Urn Burial*). Most of the deaths presumably occurred directly as a result of predation by wolves. Many skulls showed evidence of a necrotic bone disease, but it is not possible to say whether death was due solely to the disease or whether the disease merely ensured death by predation.

The mean longevity of the later sample is significantly greater (7.83 years) that that of the earlier (7.09 years), but the interpretation of this fact is not clear. The form of the distribution of deaths is sensibly the same in the two samples. As the survival of the members of this population is astonishingly great, it seems best to be conservative, and attention has been focussed on the larger, earlier sample. Except for the "lamb" and "yearling" classes, which are doubtless underrepresented in the data owing to the perishability of their skulls, there is no reason to suppose that either group is anything but a fair sample of the total population, i. e., the probability of finding a skull is not likely to be affected by the age of its owner. A life table for the 608 sheep has accordingly been prepared (Table 1). The survivorship curve, plotted logarithmically in Fig. 2, is remarkably

Table 1. Life table for the Dall Mountain Sheep (Ovis d. dalli) based on the known age at death of 608 sheep dying before 1937 (both sexes combined)*. Mean length of life 7.09 years. Data from Murie (1944)

x	x'	d_x	l_x	1000 q_x	e_x
Age (years)	Age as % Deviation from Mean Length of Life	Number Dying in Age Interval out of 1000 Born	Number Surviving at Beginning of Age Interval out of 1000 Born	Mortality Rate per Thousand Alive at Beginning of Age Interval	Expectation of Life, or Mean Life-Time Remaining to Those Attaining Age Interval (years)
0—0.5	−100	54	1000	54.0	7.06
0.5—1	−93.0	145	946	153.0	—
1—2	−85.9	12	801	15.0	7.7
2—3	−71.8	13	789	16.5	6.8
3—4	−57.7	12	776	15.5	5.9
4—5	−43.5	30	764	39.3	5.0
5—6	−29.5	46	734	62.6	4.2
6—7	−15.4	48	688	69.9	3.4
7—8	− 1.1	69	640	108.0	2.6
8—9	+13.0	132	571	231.0	1.9
9—10	+27.0	187	439	426.0	1.3
10—11	+41.0	156	252	619.0	0.9
11—12	+55.0	90	96	937.0	0.6
12—13	+69.0	3	6	500.0	1.2
13—14	+84.0	3	3	1000	0.7

* A small number of skulls without horns, but judged by their osteology to belong to sheep nine years old or older, have been apportioned *pro rata* among the older age classes.

"human" in showing two periods of relatively heavy mortality, very early and very late, with high and nearly constant survival ratios at intermediate ages.

The adult sheep have two principal methods of defense against wolves, their chief enemies: flight to higher elevations, where wolves can not pursue; and group action or herding. It is clear that these recourses confer a relative immunity to death by predation and that only the very young, which have not learned by experience, and the very old, which are too feeble to escape, suffer heavy losses. ...

The second case to be discussed is that of an aquatic invertebrate, the sessile rotifer *Floscularia conifera*. This species has been studied by Edmondson (1945) under conditions which are fully as natural as those enjoyed by Murie's mountain sheep. *Floscularia* lives attached to water plants, especially *Utricularia*, surrounded by a tube constructed by itself out of pellets of detritus. The tube is added to at the top continuously throughout life, and Edmondson was able to identify all the members of a population living in a pond by dusting the *Utricularia* plant with a suspension of powdered carmine. On subsequent visits the *Floscularia* present at the time of dusting were conspicuously marked by bands of carmine-stained pellets in the walls of their tubes, each band being surmounted by new construction of varying widths. Thus in one operation the stage was set for an analysis of growth, age, birth-plus-immigration, and death in a natural population. Among other spectacular results, Edmondson found that the expectation of life of solitary individuals was only half as great as that of members of colonies of two or more, and he presented separate life tables for each component of the population, calculated from the age at death. To facilitate comparison with other species, however, solitary and colonial individuals have been lumped together (for Edmondson's "Experiment 1") in the life table of Table 2.

Table 2. Life table for the sessile rotifer Floscularia conifera based on the known age at death of 50 rotifers, both solitary and colonial. Mean length of life 4.74 days. From Edmondson (1945), Experiment 1

x	x'	d_x	l_x	$1000\,q_x$	e_x
Age (days)	Age as % Deviation from Mean Length of Life	Number Dying in Age Interval out of 1000 Attaching	Number Surviving at Beginning of Age Interval out of 1000 Attaching	Mortality Rate per Thousand Alive at Beginning of Age Interval	Expectation of Life, or Mean Life Time Remaining to Those Attaining Age Interval (days)
0—1	−100	20	1000	20	4.76
1—2	−78.9	200	980	204	3.78
2—3	−57.8	60	780	77	3.70
3—4	−36.7	0	720	0	2.98
4—5	−15.6	300	720	416	1.97
5—6	+ 5.4	140	420	333	2.02
6—7	+26.7	60	280	214	1.79
7—8	+47.7	140	220	636	1.14
8—9	+68.8	40	80	500	1.25
9—10	+90.0	20	40	500	1.00
10—11	+111.0	20	20	1000	0.50

The survivorship curve (Fig. 2), like that of the Dall sheep, shows unexpectedly good survival. As Edmondson has pointed out, it is not so good as that of other rotifers reared in the laboratory under standard conditions *(Proales decipiens, P. sordida, Lecane inermis)*, but it is only a little less good....

The case of *Floscularia* is almost above reproach as an example of a life table obtained under natural conditions. It is, of course, open to the objection that only the age at death is known, and the age structure of the living animals must be assumed to be constant. Apart from this deficiency, it should also be realized that the origin of the life table is not at birth. The pelagic larval life of the rotifer, like the larval life of barnacles and insects, is omitted from consideration in such a table....

In his delightful book, *The Life of the Robin* (1943a) and in two admirable papers, Lack (1943b, c) has investigated the age at death of certain British birds,

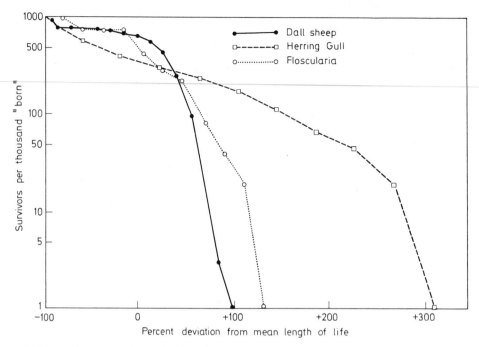

Fig. 2. Survivorship (l_x) Curves for the Dall Mountain Sheep, the Sessile Rotifer Floscularia conifera, and the Herring Gull, Age being Expressed as Percentage Deviation from the Mean Length of Life

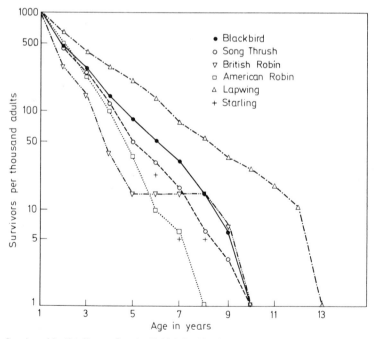

Fig. 3. Survivorship (l_x) Curves for the British Robin, Song Thrush, Blackbird, Starling, Lapwing and American Robin, Age being Expressed in Years

66

as obtained by recoveries of individuals banded as nestlings. Because banded nestlings are likely to be picked up near the banding stations or not at all, it is impossible to estimate the whole of the mortality in the first year of life with any accuracy, and Lack begins his life tables on August 1. The proportion of birds banded which are later recovered is small, ranging from 1.0 per cent for the robin to 18.4 per cent for the cormorant; but after August 1 of the first year it is considered that the ages at death of birds recovered are not likely to differ from the ages at death among the total population. The samples are small and of course become progressively smaller with increasing age, so that Lack does not regard the mortality rates and expectation of life as reliable beyond the fourth or fifth year.

Several of Lack's life tables are reproduced in Fig. 3. ...

The striking feature of these survivorship curves is their diagonal form. The mortality in the first year varies from 380 per thousand for the lapwing to 723 per thousand for the robin, but for a given species the mortality remains approximately constant throughout life, or at least for as long as the data are reliable. ...

Survivorship Directly Observed

The cases now to be discussed differ from the preceding in the character of the original observations. Instead of a fairly large sample of individuals about which little or nothing can be told except their age at death, we have a group of individuals known to have been born at a particular time and to have been present or absent at some later time. Their presence gives their survivorship, their absence implies death in the interval since they were last observed. This is the best sort of information to have, since it does not require the assumption that the age composition of the population is stable in time. Provided only that the season of birth is a small fraction of the age interval between successive observations, so that births can be assumed to be simultaneous, as in a *Drosophila* bottle, a horizontal life table can be directly constructed from the survivorship data. Unfortunately, most of the species which have been studied in this way have short spans of natural life, and when census data are obtained only once a year the number of points on the survivorship curve is too small to be satisfactory.

The best example of such observed survivorship comes from Hatton's work (1938) with the barnacle, *Balanus balanoides*. ... The case is very nearly ideal. The barnacle settles on rocks during a short time (two to six weeks) in early spring. Test areas were scraped clean one winter, and after new populations had settled, the survival of their members was followed at intervals of one to four months for three years. Barnacles which disappeared from the areas between observations were certainly dead, for emigration does not complicate the problem. Immigration, however, does present difficulties, though since it is confined to the attachment seasons of subsequent years it should be possible to control it in subsequent work. There is one further disadvantage in that the life tables necessarily start at metamorphosis, leaving out of account mortality during pelagic larval stages. A life table for a typical population of barnacles is presented in Table 6.

Table 6. Life table for a typical population of Balanus balanoides, based on the observed survival of adult barnacles settling on a cleaned rock surface in the spring of 1930. The population is that at Cité, (St. Malo, France), a moderately sheltered location, at Level III, at half-tide level. The initial settling density (2200 per 100 cm^2) is taken as the maximum density attained on May 15. Mean length of life 12.1 months. Data from Hatton (1938)

x	x'	d_x	l_x	$1000\,q_x$	e_x
*Age (months)	*Age as % Deviation from Mean Length of Life	Number Dying in Age Interval of 1000 Attaching	Number Surviving to Beginning of Age Interval out of 1000 Attaching	Mortality Rate per Thousand Alive at Beginning of Age Interval	Expectation of Further Life (months)
0—2	−100	90	1000	90	12.1
2—4	−83.5	100	910	110	11.3
4—6	−67.0	50	810	62	10.5
6—8	−50.4	60	760	79	9.1
8—10	−33.9	80	700	114	7.8
10—12	−17.4	160	620	258	6.7
12—14	− 0.9	80	460	174	6.7
14—16	+16.0	100	380	263	5.9
16—18	+32.2	50	280	179	5.7
18—20	+49.0	40	230	174	4.7
20—22	+65.4	100	190	526	2.4
22—24	+82.0	60	90	667	1.9
24—26	+98.8	20	30	667	1.8
26—28	+115.0	8	10	800	1.4
28—30	+132.0	2	2	1000	1.0

* Survivorship data given graphically by Hatton were smoothed by eye, and values at every other month were then read from the curve. The original observations were made at irregular intervals during three years.

The remaining examples suffer from more serious defects, and the data do not justify extended treatment. Green and Evans (1940) in their important study of the snowshoe rabbit *(Lepus americanus)* in Minnesota, followed the survival of marked individuals of several year classes, the total population present on the area and the number in each age-class being obtained by the mark-and-recapture method—also known as the "Lincoln index" (Jackson, 1939). Marking was done during most of the winter, and the annual census was made in February. It is perhaps unnecessarily, and certainly uncharitable, to point out two sources of error in this excellent and ingenious work. In the first place, when marked individuals are released into a population and later recaptured, the calculation of the total population from the fraction

$$\frac{\text{size of sample when recapturing}}{\text{number recaptured}} \times \text{number marked}$$

depends on two assumptions, neither of which is likely to be true in this case: that there is no mortality between marking and recapturing; and that the marked individuals disperse at random through the whole population. Secondly, the flow of vital events in this population was so rapid, very few rabbits more than three years old ever having been found, that observations made annually can give only a very rough idea of the life table.

The latter objection applies with equal force to the study of a pheasant population made by Leopold et al. (1943) in Wisconsin. The former objection, though doubtless it could be urged, has less validity here, since the population, as ascertained by trapping, was checked by census drives.

Nice's thoroughgoing work (1937) on the song sparrow *(Melospiza melodia)* included a consideration of the survival of banded birds from year to year. The number of individuals which could be kept under continuous observation was necessarily small, and to find a sample large enough to use as the basis of a life table, it is necessary to take the 144 males banded in the breeding season between 1928 and 1935. Unfortunately, some of these males were of unknown age when first banded. Even if one assumes, (and the assumption is not far from the truth) that all new males appearing are first-year birds born elsewhere, the survival ratios from year to year will be too low if any adult males were still alive but failed to return to the area. Evidently such emigration is of minor importance with adult male song sparrows. With adult females, however, it is so serious that Nice did not think it worth while to publish the data on their return. Clearly, work on the survival of migratory birds is full of uncertainties, though the same may be

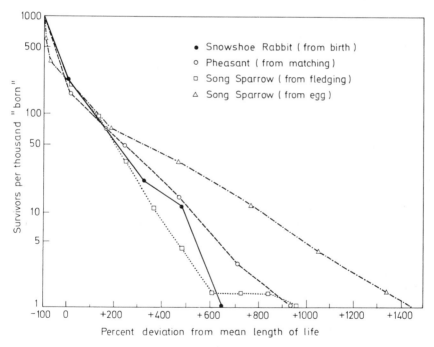

Fig. 5. Synthetic Survivorship (l_x) Curves for the Snowshoe Rabbit, the Pheasant, and the Song Sparrow, the Latter Calculated from Two Different Biological Ages. Data for different parts of the life spans are derived in different ways, as explained in the text.

said of resident species such as the wren-tit (Erickson, 1938) and the robin (Lack, 1943a).

All of these cases, snowshoe rabbit, pheasant, and song sparrow have one defect in common. This is the necessity of calculating the survival between birth and the first year of adult life from other data than those given by banding. For the snowshoe rabbit, the initial strength of the year-class is calculated from the estimated breeding population present and its known fertility. Leopold et al., lacking observations of their own on the pheasant mortality between birth and the first census period, used the estimates given by Errington for pheasants in another state. Nice calculates the survival of fledged young song sparrows to their first breeding season, by assuming a stable population and combining the estimated mean length of life of adults with their average nesting success. These procedures, while perfectly defensible as approaches to the problem, are inadequate substitutes for direct observation.

The three sets of data, with all their uncertainties, have been used as bases for synthetic life tables, and the survivorship curves are presented in Fig. 5. ...

Age Structure Directly Observed

Ecological information of a third sort is available for a number of natural populations, principally of fishes and birds. In these cases the investigator has been able to determine how many individuals of each age are living in the population, and the age at death, though calculable from the shrinkage between successive age classes, has not been directly observed. This kind of information lends itself just as well as either of the others to the computation of life tables. As in the group where only the age at death is known, of course, it is necessary to assume the age composition to be unchanged with time. When this assumption is unreasonable, as it often is for fish populations, with their outrageous fluctuation in strength of year-classes, average age compositions obtained from several years' work can often be used. As it happens, however, all the life tables which fall in this third group are incomplete for one reason or another, and the data do not bear close comparison with such examples of natural life tables as those of the Dall sheep and the barnacle.

Kortlandt (1942) has recently given a very elaborate analysis of the Netherlands population of cormorants (*Phalacrocorax carbo sinensis*). Birds banded as nestlings were later observed in their breeding and playing colonies, the numbers on the bands being read with the aid of a telescope. The age distribution of the banded birds being known in 1940 and 1941, it should be possible to infer the age distribution of the total population and from this to compute the annual mortality suffered by each year-class. A number of complicating conditions are present in this case, however, making direct calculation unreliable and necessitating a more circuitous approach: (1) the size of the Dutch cormorant population is not constant, but has been increasing by about 10 per cent per year, as estimated by counts of nests at the breeding colonies; (2) differences between the observed sex ratio among sexually mature birds and the sex ratio predicted on the basis of estimated mortality by sex and age class show clearly that there is some *band mortality;* that is, some birds either lose bands or die because of the band, making estimates of natural mortality too high by a factor of about 2; (3) it is not possible to infer the complete age structure from observations made at breeding colonies, since the one- and two-year-old birds occupy "colonies" elsewhere, returning to their birthplaces to breed no sooner than their third year.

In view of these difficulties, and others which need not be discussed here, Kortlandt's results must be regarded as schematic and preliminary only, and scarcely warrant the construction of a life table. His computations suggest that cormorants suffer mortality somewhat as follows: 17 per cent between fledging and the first May 1; 8 per cent in the first year; 6 per cent in the second year; and about 4 per cent per year in the third to twelfth years. These are astonishingly low figures for a natural population, but it must be remembered that the population is increasing rapidly. ...

The literature of fisheries biology is full of attempts to estimate the mortality of fishes, to distinguish natural mortality from rate of exploitation, and to determine the rate of exploitation which, given certain mortality rates and certain relationships between age and size, will steadily yield an optimum catch. These complex questions are fully discussed in the important works of Russell (1942), Thompson and Bell (1934), and Ricker (1940, 1944), and by various authors in the *Rapports et Procès-Verbaux of the International Council for the Study of the Sea*, Volume 110, 1939. Little of this information can be directly used for our purpose. The explanation is as simple as it is regrettable: although the age of a caught fish can be ascertained with more or less complete confidence, fishes of all ages can not be caught with equal facility. Inevitably the methods so skilfully developed for catching fish of desirable sizes will fail to catch fish of undesirable sizes. It is true that on various occasions the whole fish population of a lake has been removed by poisoning or drainage. The estimates given by Eschmeyer (1939) for the abundance of large-mouth black bass *(Huro salmoides)* in Howe Lake, Michigan, at the time of its poisoning in 1937, may be cited as a example:

Age	Number
0	18,374
0 (cannibals)	229
I	25
II	10
III	105
IV	7
V and older	9
Total	18,759

The implication of enormously greater mortality in the first year of life is plain from these figures, but such data can not be taken as they stand, partly because of very variable annual recruitment, and partly because young of the year were removed from the lake at various times for hatchery purposes. ...

This section may logically be concluded with a brief reference to the data for the fin whale, in which, as recent investigations have shown, the age of the female can be determined from the number of old corpora lutea in the ovary. By this method Wheeler (1934) arrived at the following as the age structure, observed over five seasons, 1926—1931, of the catch of female fin whales in the Antarctic:

Age	Number caught
3—4 years	130
5—6 years	95
7—8 years	72
9—10 years	53
11—12 years	37
13—14 years	28
15—16 years	10
17—18 years	4
19—20 years	1
21—22 years	1

The data imply (subject to the usual qualifications) a biennial mortality of about 26 per cent, increasing beyond the 15th year to much higher values. The author considers that the increased rate of loss with age is not real, but is due to failure of the older whales to return from their winter quarters in the north. This belief may or may not be well founded, but one suspects it to be predicated on the idea that mortality, at least when it is primarily due to exploitation, is constant among animals with respect to age. Edser (to whom the statistical analysis is credited) assumed, for the purpose of a rough calculation of the necessary rate of replacement, that the mortality between birth and breeding age is also 26 per cent. The improbability of this assumption may be surmised by reference to the life table for the Dall sheep (Table 1). Edser's calculation has the great merit of yielding a minimal estimate of the alarming exploitation being conducted by the whaling industry in the Antarctic. More realistic assumptions would darken the picture even more. In any case the data can not yet be cast into a life table. ...

Comparisons and Conclusions

Both in nature and in the laboratory, animals differ characteristically in their order of dying. When the mortality rate at all ages is constant, the survivorship (l_x) curve is diagonal on semi-logarithmic graph paper. Such a curve is found for many birds from adults stages onward; the mortality of adult birds is about 320 per thousand per hundred centiles of mean life span. If the constant age-specific mortality rate observed for the first few years of adult life is really maintained throughout life, the oldest bird in a cohort of 1000 lives 6.6 times as long as the average bird. Not all animals resemble birds in this respect, however, although many (e. g., fish) are assumed to do so. The Dall sheep, the rotifer, and possibly the barnacle are more like civilized man in that they seem to have evolved a mechanism for stretching the mean life span toward the maximum, so that the survivorship curve is convex. In these cases the maximum life span (among a sample of 1000) is only two or three times the mean. On the other hand there are undoubtedly species in which juvenile mortality is very heavy, but the few survivors to advanced ages die at reduced rates. This J-shaped or concave survivorship line, with the maximum longevity perhaps 15 or more times the mean, is presumed to characterize the oyster and other species, but it has not yet been formally recognized either in the laboratory or in nature. The closest approach to it, so far, is found when the survival of song sparrows is reckoned from the egg stage; but the mackerel will almost certainly provide an even better example.

Detailed comparisons between species cannot yet be made, partly because of the diverse statistical foundations of the life tables and partly because the data begin at different biological ages (birth, hatching, metamorphosis, sexual maturity, etc.) in the different cases. In all cases it is the youngest ages about which we know least, and ecologists should therefore concentrate their efforts on this segment of the life span of animals in nature.

List of Literature

Bertram, G. C. L. 1940. The biology of the Weddell and Crabeater Seals, with a study of the comparative behaviour of the Pinnipedia. *Brit. Mus. (Nat. Hist.)*: *British Graham Land Expedition* 1934—37, *Sci. Rep.*, 1 (1): 139 pp., pls. 1—10.

Edmondson, W. T. 1945. Ecological studies of sessile Rotatoria, Part II. Dynamics of populations and social structures. *Ecol. Mon.*, 15: 141—172.

Erickson, M. M. 1938. Territory, annual cycle, and numbers in a population of wren-tits *(Chamaea fasciata)*. *Univ. Calif. Pub. Zool.*, 42 (5): 247—334, pls. 9—14.

Errington, P. L. 1945. Some contributions of a fifteen-year local study of the northern bobwhite to a knowledge of population phenomena. *Ecol. Mon.*, 15:1—34.

Eschmeyer, R. W. 1939. Analysis of the complete fish population from Howe Lake, Crawford County, Michigan. *Papers Mich. Acad. Sci. Arts Lett.*, 24 (II): 117—137.

Green, R. G., and C. A. Evans. 1940. Studies on a population cycle of snowshoe hares on the Lake Alexander area. I. Gross annual censuses, 1932—1939. *J. Wildl. Man.*, 4: 220—238. II. Mortality according to age groups and seasons. Ibid., 4: 267—278. III. Effect of reproduction and mortality of young hares on the cycle. Ibid., 4: 347—358.

Hatton, H. 1938. Essais de bionomie explicative sur quelques espèces intercotidales d'algues et d'animaux. *Ann. Inst. Océanogr.*, 17: 241—348.

Jackson, C. H. N. 1939. The analysis of an animal population. *J. Anim. Ecol.*, 8: 238—246.

Kortlandt, A. 1942. Levensloop, samenstelling en structuur der Nederlandse aalscholver bevolking. *Ardea*, 31: 175—280.

Lack, D. 1943a. *The life of the robin.* 200 pp. H. F. & G. Witherby, London.

Lack, D. 1943b. The age of the blackbird. *Brit. Birds*, 36: 166—175.

Lack, D. 1943c. The age of some more British birds. *Brit. Birds*, 36: 193—197, 214—221.

Laurie, A. H. 1937. The age of female blue whales and the effect of whaling on the stock. *Discov. Rep.*, 15: 223—284.

Leopold, A., T. M. Sperry, W. S. Feeney, and J. S. Catenhusen. 1943. Population turnover on a Wisconsin pheasant refuge. *J. Wildl. Man.*, 7: 383—394.

Moore, H. B. 1935. A comparison of the biology of *Echinus esculentus* in different habitats. Part II. *J. Mar. Biol. Ass.*, n.s., 20:109—128.

Murie, A. 1944. *The wolves of Mount McKinley.* (Fauna of the National Parks of the U.S., Fauna Series No. 5, xx + 238 pp.) U.S. Dept. Int., Nat. Park Service, Washington.

Nice, M. M. 1937. Studies on the life history of the song sparrow. Vol. I. A population study of the song sparrow. *Trans. Linn. Soc. N. Y.*, 4: vi + 247 pp.

Pearl, R. and J. R. Miner. 1935. Experimental studies on the duration of life. XIV. The comparative mortality of certain lower organisms. *Quart. Rev. Biol.*, 10: 60—79.

Ricker, W. E. 1940. Relation of "catch per unit effort" to abundance and rate of exploitation. *J. Fish. Res. Bd. Canada*, 5: 43—70.

Ricker, W. E. 1944. Further notes on fishing mortality and effort. *Copeia*, 1944: 23—44.

Russell, E. S. 1942. *The overfishing problem*. viii +130 pp. Cambridge Univ. Press, Cambridge.

Thompson, W. F., and F. H. Bell. 1934. Biological statistics of the Pacific halibut fishery. (2) Effect of changes in intensity upon total yield and yield per unit of gear. *Rep. Intern. Fish. Comm.*, 8: 49 pp.

Wheeler, J. F. G. 1934. On the stock of whales at South Georgia. *Discov. Rep.*, 9: 351—372.

Stable Population Theory

In 1760 Leonard Euler introduced the concept of a stable age structure in which proportions in all age categories would remain fixed if mortality were constant and births increased exponentially over time. In his analysis, Euler also included a series of problems to demonstrate the usefulness of stable population theory for filling in gaps in population information. His paper was familiar to Milne (1815: his comments on Euler are included in paper 4 above), who recognized its theoretical interest but emphasized its restricted applicability. Milne's comments were appropriate for his time: the very limited empirical data that existed for Europe did not suggest stability, and did not indicate how abusive of the facts the assumption might be. [Some years later Milne applied Euler's theory in a communication to Robert Malthus (1970 (1830), pp. 228—229) demonstrating that three stable populations, one with a 25 year doubling time, one with a 50 year doubling time, and one stationary, would have substantially different crude death rates if they shared a common mortality schedule. Malthus used the analysis to show that crude death rates could not adequately measure the health of a population.] Euler's article and an earlier note to Johann Peter Süssmilch, who may have brought population mathematics to his attention, are presented here as papers 10 and 11.

The link between stable theory and real populations was very largely uncovered by Alfred Lotka (1907, 1922) and F.R. Sharpe and Lotka (1911), in works that form a singular achievement in demography. The three papers, included here, will indicate how Lotka's insight expanded from his realization that population could be represented as a renewal process displaying some stability, to his discovery that populations would nearly always stabilize, by predictable paths, if fertility and mortality were held constant.

The papers introduce the renewal equation

$$B(t) = G(t) + \int_0^t B(t-x)p(x)m(x)\,dx,$$

where $B(t)$ are births at time t, and are composed of births to the population alive at time zero $[G(t)]$ and births to those born since: $B(t-x)$ being persons born $t-x$ years ago and, subject to their survival probability $p(x)$, currently age x; $m(x)$ being their chance of giving birth in the interval x to $x+dx$. For

$G(t)=0$, substitution of an exponential birth function $B(t)=B_0 e^{rt}$ provides the characteristic equation

$$\psi(r) = \int_0^\infty e^{-rx} p(x) m(x) dx = 1 ,$$

whose single real root is the intrinsic growth rate of the population and whose other roots define the rate of convergence toward the stable age structure. [An examination of high order roots is provided in Coale (1972).]

The conjecture that populations stabilize was attributed by R.R. Kuczynski (1931b, 1935) to L. von Bortkiewicz (1911), but this seems not to be the meaning of the relevant text. Bortkiewicz essentially followed Euler in examining relationships under stability, with the distinctions that he employed density functions corresponding to rectangular sections of the Lexis diagram, and provided applications to resolve current disputes in the literature. Except for a brief note on Euler, the approach to stability is not discussed. Lotka's debate with Kuczynski, which touches on other early contributions to stable theory, is the subject of paper 15 by Paul A. Samuelson.

Lotka's arguments were also subject to mathematical criticism and did not gain universal acceptance until a rigorous mathematical proof that in the main they held for both discrete and continuous cases was given by William Feller (1941), whose article is included here as paper 16. Widespread application of the theory has come more recently, with the introduction of projection matrices and of modern computing equipment to facilitate the extraction of characteristic roots and vectors.

A more direct and considerably simpler method for calculating the intrinsic rate of growth than that developed by Lotka has been given by Ansley Coale (1957b) and is included here as paper 17. It is followed by a selection from R.A. Fisher (1958 (1930)) in which he develops the concept of Reproductive Value, complementing stable theory. It is found in Lotka [1939, equation (214)] as the term P_s and in Feller [1941, equation (3.2)] as $\gamma(s)$, but without the intuitive meaning Fisher gives it.

In paper 19 we present an application of stable population theory by Coale (1957a). In the work he discusses the effects that changes in birth and death rates have on age distributions of real populations, among other points correcting the widespread misconception that a lowering of death rates necessarily results in an older population. The article also introduces the concept of weak ergodicity, an important contribution to stable theory that will be taken up in chapter 3 in conjunction with projection matrices.

We conclude the chapter with two articles that attack a problem commonly overlooked in stable theory: the treatment of reproduction as a discontinuous process in work with very small populations. W.R. Thompson's (1931) pioneering analysis is included here as paper 20. To take into account chance fluctuations, a stochastic approach becomes more relevant to Thompson's small populations than the deterministic solution he gives, and readers should be aware that this limitation exists. It does not detract from Thompson's contribution: as he indicates, both the nature of the problem and of the solution change radically when numbers are small. (We present in the introduction to Thompson's paper a note on its matrix formulation which will relate it to the discussion of matrices in chapter 3. The stochastic treatment of birth processes is taken up in chapters 6 and 7.) The closing article by Lamont Cole (1954) clarifies Thompson and ties his work to stable theory. It includes as well a fascinating note on Fibonacci numbers and their relation to the Thompson problem.

10. An Illustration of Population Growth

JOHANN PETER SÜSSMILCH (1761)

From *Die göttliche Ordnung*, Vol. 1, pp. 291—299. Berlin. Translated by Nathan Keyfitz.

The extract which follows assembles several fragments of Euler's work in stable theory that complement, and apparently precede, his better known 1760 article (paper 11 below). Euler here prepares for Süssmilch a population projection that begins with one couple of age 20 who give birth to an additional couple at ages 22, 24 and 26, and who die at age 40, later generations repeating the same fertility and mortality schedules. Euler demonstrates that after 300 years the population will be growing approximately geometrically and recognizes this as the limit of a recurrent series; i. e., he is working with a very general result.

The recurrent series is in this case a net maternity function expressed as a difference equation. We have:

$$B(t) = B(t - 22) + B(t - 24) + B(t - 26);$$

of which the solution is

$$B_0 \lambda^t = B_0 \lambda^{t-22} + B_0 \lambda^{t-24} + B_0 \lambda^{t-26}$$
$$1 = \lambda^{26} - \lambda^4 - \lambda^2$$
$$\lambda \doteq 1.04696$$

where $B(t)$ represents births at time t, B_0 is a constant and $\lambda = e^r$, the annual intrinsic growth rate. From earlier work (1748) Euler recognized that the ratio between successive terms $B(t)$, $B(t + \alpha)$ in series of this form would approach a constant limit determined by the intrinsic rate of growth as $t \to \infty$; and both he and Joseph Louis Lagrange (1759) had worked with the conversion of the equations to polynomials and extraction of roots, learned in part from Abraham DeMoivre's *Miscellanea Analytica de Seriebus et Quadraturis* (1730, 72—83). A hint of this insight is contained in the closing paragraph of the present article. (Euler's note that the birth series can be generated by dividing out an algebraic fraction is also from DeMoivre. It reappears in Thompson (paper 20 below), in whose notation births at ages 22, 24, and 26 would be representable by the "Generation Law" $G = T^{22} + T^{24} + T^{26}$. The complete birth series is produced by dividing out the fraction $1/(1 - G)$.)

Euler's mathematical contribution to Süssmilch goes no further and is very reminiscent of Fibonacci's rabbits (discussed in Cole, paper 21 below). It is as well the point at which Euler's important 1760 paper begins: from the net ma-

79

216504

ternity function an intrinsic rate of growth has been found; this rate, together with the complete mortality table, provides the stable age distribution and the intrinsic birth and death rates of that paper. E. J. Gumbel (1917) was first to see the full implication of the two papers.

Süssmilch's opening paragraph refers to an earlier table of population doublings (i. e., powers of 2) with imagined doubling times inserted; calculated, with some multiplication errors, by himself. It gives a total population after 300 years of $2^{23} = 8,338,608$ from an initial cohort of 2, a figure twice as great as in the Euler projection and one which corresponds to an annual growth rate of 5.3%. The Euler projection runs to several pages and is shown here only in part.

To show that nothing impossible is contained in the preceding table I present another due to Professor Euler and prepared some years ago at my request. It is too elegant for me to omit. ...

Professor Euler assumes (1) at the outset there exists a married couple aged 20, (2) their descendents also always marry at the age of 20, (3) 6 children are born to each marriage. (This could certainly happen if couples of differing ages were precluded from marrying, if like marry like, and if all are able to marry at the correct time.) Also (4) variations must never occur; therefore twins will always be born, the first pair to each marriage coming in the 22nd, the next in the 24th and the third in the 26th year. It will be assumed (5) that all children survive, marry, and remain living until reaching age 40. (This is the average expectation of life on the whole: lives were of much longer duration and the average much higher before the Flood. If the fertility of the marriages appears too high it can be revised throughout.)

On these assumptions there will be only two people initially, 4 after 2 years, 6 after 4 years, 8 after 8 years. After this time no changes occur until the first two

Table of Growth According to the Conditions Assumed

Year	Number of Births	Births to Date	Deaths to Date	Number Alive
0	0	2	0	2
2	2	4	0	4
4	2	6	0	6
6	2	8	0	8
8	0	8	0	8
10	0	8	0	8
12	0	8	0	8
14	0	8	0	8
16	0	8	0	8
18	0	8	0	8
20	0	8	2	6
22	0	8	2	6
24	2	10	2	8
26	4	14	2	12
28	6	20	2	18
30	4	24	2	22
32	2	26	2	24
34	0	26	2	24
⋮				
280	71 632	1 679 344	280 484	1 398 860
282	122 112	1 791 456	328 610	1 462 846
284	178 036	1 969 492	379 908	1 589 584
286	260 362	2 229 854	428 068	1 801 786
288	342 310	2 572 164	467 934	2 104 230
290	403 268	2 975 432	497 348	2 478 084
292	426 034	3 401 466	517 874	2 883 592
294	404 348	3 805 814	534 572	3 271 242
296	346 570	4 152 384	555 274	3 597 110
298	273 884	4 426 268	589 506	3 836 752
300	214 370	4 640 638	646 684	3 993 954

children reach their 22nd year, which takes place after 24 years, when their first two children come into the world. Two years later this couple will produce 2 more, but the couple born in the 4th year will also produce 2 children; in the 28th year 6 children will arrive; in the 30th again only 4, and so forth. In order to grasp the nature of this initially irregular growth, the attached table shows in its second column the births of even numbered years. The third column shows the sum of all births to date, which would be the same as the number living if no one were to die.

But at age 40 all die, from which are derived the numbers of column 4. If these deaths are subtracted from the number born, the result is the number living each year, found in the 5th column.

One can see from this that after 24 years there is a trebling of the number alive, from which after 1000 or more years an astonishing increase must ensue. After 300 years the number living already runs to 4 million; if the trebling time is set at 25 years there will be 324 million people after 400 years, and after 450 years fully 3000 million, more than now live on the entire earth. The increase before the Flood, while not so great as this, was surely not an insignificant progression; so the earth must at that time have been more heavily populated than now. ...

Notice that although great unorderliness seems to rule in Euler's table, the numbers of births belong to a progression called a recurrent series [einem Geschlecht von Progreßionen, welche man *Series recurrentes* nennet] and which can be produced by dividing out an algebraic fraction. While these progressions initially appear irregular, if they are continued they finally change into a geometric progression; the initially perceived irregularities decrease with time until they finally almost entirely disappear.

11. A General Investigation into the Mortality and Multiplication of the Human Species

LEONARD EULER (1970 (1760))

Theoretical Population Biology 1: 307—314. Translated by Nathan and Beatrice Keyfitz.

Euler's virtually unknown article, published by the Belgian Académie Royale des Sciences et Belles-Lettres in 1760, anticipates important parts of modern stable population theory for a one-sex population closed to migration. Its ideas have been published many times during the subsequent two centuries by writers who independently rediscovered them.

The application to which Euler oriented his argument was inference from incomplete data. Life tables were already in use for working out probabilities and expected values in relation to individuals; Euler saw them as a means of studying populations, provided the assumption of stability was appropriate. He follows a cohort of individuals born "en même tems" (sic) by means of the probability of survival from birth to age x that we know as l_x; he supposed that l_x is available from some source other than the population under consideration, as contemporary users of model life tables do. The equation for the unknown rate of increase λ in paragraph 18, as well as the proposed method of solution, bear a resemblance to procedures now in use for inferring the rate of increase of a population.

If total births and deaths are known (but not by age), then Euler shows how assuming a life table permits the number of the population to be inferred, as well as the age distribution. Euler turned his stable population theory to filling the gaps in information of his own day: Western Europe had baptism and burial records before it had censuses. His model, used in only a slightly different fashion, would enable a present-day underdeveloped country with censuses and no vital statistics to infer its birth rate.

Euler's "hypothesis of mortality" means "life table," and is taken as unchanging in time and operating deterministically. The ratio λ of population increase per year is the same as e^r, if r is Lotka's intrinsic rate of natural increase. Though the cohort with which Euler deals is a group of children born at one moment, and hence projected by the life table function l_x, the alteration needed to work with cohorts evenly spread over the year is slight: We replace the probability of survival l_x by L_x/L_0, where L_x is the integral $L_x = \int_0^1 l_{x+t}\, dt$.

Euler's annual births B occurring at one moment, while his deaths D are spread through the subsequent year, give a simple but unfamiliar expression for the ratio of increase: $\lambda = (P - D)/(P - B)$, where P is the population just after the annual births have occurred, and D the deaths in the succeeding twelve months.

The following list of the problems in stable population theory solved by Euler will serve as a table of contents of the important paragraphs of his work.

Paragraph	Given	Find
18	P, B, l_x	λ
19	λ, l_x	P/B
20	P, B, l_x	P_x
21	P, B, l_x	D
22	B, D, l_x	λ, P
24	B, D, l_x	D_x
26	P, B, D_x	l_x

where P is total population; P_x population by age; D is total deaths; D_x deaths by age; B is births for the given sex; l_x is the probability of surviving to age x; λ is the ratio of annual increase.

Euler uses "hommes" for the population and "enfants" for the births, which could be interpreted as meaning that the sexes are combined in his model. As he nowhere deals with age at childbearing, his argument would apply to the two sexes together, with a consolidated life table. Essentially a one-sex model seems to be intended, with males as the illustration, and this is the rendering of the translation that follows.

N. K.

1. The registrations of births and of deaths at each age which are published in various places every year give rise to many different questions on the mortality and the multiplication of mankind.

2. All of the general questions depend on two hypotheses. I shall call the first the hypothesis or law of mortality (i.e., life table), by which can be determined how many out of a certain number of men born at one time will be still alive at the end of each period of years. Here population increase does not enter the matter at all, and we go on to the second hypothesis, which I shall call that of multiplication: by how many the number of men is increased or diminished in the course of one year. The second hypothesis depends, of course, on the number of marriages and upon fertility, whereas the first is based on the vitality or life potential characteristic of man.

HYPOTHESIS OF MORTALITY

3. For the first hypothesis, let us suppose a number B of children born at the same time; the number of these still alive at the end of one year is Bl_1, at the end of two years Bl_2, of three years Bl_3,.... . These show the decrease of the number of men born at one time; for each region and way of life they will have particular values. The numbers indicated by $l_1, l_2,...$, constitute a decreasing progression of fractions of which the greatest, l_1, is less than unity, and beyond 100 they disappear almost completely. For if of 100 million men not one reaches the age of 125, this means that l_{125} must be less than 1/100,000,000.

4. Having established for a certain place, by a sufficiently large number of observations, the values of the fractions $l_1, l_2,...$, one can resolve several questions which are commonly posed on the probability of human life. First, if the number of children born at the same time is equal to B, the expected number dying each year will be

From Age	To Age	Number of Deaths
0	1	$B - Bl_1$
1	2	$Bl_1 - Bl_2$
2	3	$Bl_2 - Bl_3$
–	–	—
–	–	—
–	–	—

And since of B children born Bl_x will be expected to be still alive after x years, the number of deaths before the end of x years must be equal to $B - Bl_x$.

5. Given a number of men of the same age, find how many of them will be expected to be alive after a certain number of years.

Suppose there are P men, all aged x years, and we want to know how many we can expect to be living after n years. We put $P = Bl_x$ to obtain $B = P/l_x$, where B is the number of male births, of whom P are still alive after x years. Of this number Bl_{x+n} will be the expected number still living $x + n$ years after their birth, and hence n years after the time in question. Then the number sought is equal to $P(l_{x+n}/l_x)$; that is, after n years we can expect there to be that many survivors of P men who are presently all aged x years.

Then of the P men all aged x years, the expected fraction who will die before the end of n years is $1 - l_{x+n}/l_x$.

6–14. (These paragraphs derive life table probabilities for individuals, along with the formula for a life annuity purchasable for the amount of money a. The annuity payable at the end of each year purchasable for a is shown to be

$$\frac{al_x}{l_{x+1}/(1+i) + l_{x+2}/(1+i)^2 + \cdots}$$

per year, where i is the rate of interest. If the annuitant is just born, and the annuity is to be deferred to start at age x, then the annual value purchasable for the amount a is

$$\frac{a}{l_x/(1+i)^x + l_{x+1}/(1+i)^{x+1} + \cdots}.$$

Euler then gives the l_x column by single years of age of a life table due to Keerseboom, and goes on to discuss births.)

15. Just as I assume that the regime of mortality remains ever the same, I shall assume a like constancy of fertility, so that the number of children born each year will always be proportional to the total number living. Let B be the number of children born in the course of one year, and λB the number of children born in the following year. Insofar as the ratio which changes B to λB continues to hold from any year to the next the number of births increases in the ratio 1 to λ. Consequently the births of the third year will be $\lambda^2 B$, of the fourth $\lambda^3 B$,.... . Either the number of annual births will constitute a geometric progression, increasing or decreasing, or it will remain constant, according as $\lambda > 1$, or $\lambda < 1$, or $\lambda = 1$.

16. Suppose then that in a town or province the number of (boy) children born this year is equal to B, and the number born next year is equal to λB, and so on according to this progression:

	Number of Births
This year	B
After 1 year	λB
After 2 years	$\lambda^2 B$
$\vdots$	$\vdots$
After 99 years	$\lambda^{99} B$
After 100 years	$\lambda^{100} B$

If we suppose that after 100 years none of the men alive at this moment will still be living, all the men in existence after 100 years will be the survivors of the above births. Then, bringing the hypothesis of mortality to bear, we could determine the total number of men who will be alive after 100 years. Thus, since the number born in that year will be $\lambda^{100} B$, we will have the ratio of births to the total number living.

17. To make this clearer, let us see how many men will be alive after one hundred years from the births of the preceding years:

	Number of Births	Number Living After 100 Years
At present	B	Bl_{100}
After 1 year	λB	λBl_{99}
After 2 years	$\lambda^2 B$	$\lambda^2 Bl_{98}$
$\vdots$	$\vdots$	$\vdots$
After 99 years	$\lambda^{99} B$	$\lambda^{99} Bl_1$
After 100 years	$\lambda^{100} B$	$\lambda^{100} B$

Thus the total number living after 100 years will be

$$\lambda^{100} B \left(1 + \frac{l_1}{\lambda} + \frac{l_2}{\lambda^2} + \cdots + \frac{l_{100}}{\lambda^{100}}\right).$$

18. The terms of this series disappear finally, by virtue of the hypothesis of mortality, and since the total number living has a certain relationship to the number of births during the course of one year, the multiplication from one year to the next, which we have assumed to be λ, will reveal this relationship. For if (at the end of 100 years) the total number living is equal to P, and the number of children produced in the course of one year is B, we will have

$$P = B \left(1 + \frac{l_1}{\lambda} + \frac{l_2}{\lambda^2} + \cdots + \frac{l_{100}}{\lambda^{100}}\right).$$

Thus, if we know the ratio P/B as well as the hypothesis of mortality, i.e. the values of the fractions l_1, l_2 ,..., l_{100} , this equation will determine the ratio of multiplication λ from any year to the next. For each life table, if we calculate the quantity $1 + l_1/\lambda + l_2/\lambda^2 + \cdots$ for several values of λ, and set up a table of them, it will be easy to ascertain for the given ratio P/B, which expresses fertility, the annual increase λ in the number living.

19. Having this equation, it is indifferent whether we know the fertility P/B, or the multiplication λ, the one being determined by the other through the life table.

20. Given the hypotheses of mortality and fertility and the total number living, find how many there are at each age.
From P, B, and the life table we calculate the ratio of annual multiplication λ. From λ and paragraph 17 we see that there will be among the number P

$$
\begin{array}{ll}
B, & \text{children just born;} \\
Bl_1/\lambda, & \text{aged one year;} \\
Bl_2/\lambda^2, & \text{aged 2 years;} \\
Bl_3/\lambda^3, & \text{aged 3 years;} \\
\vdots & \qquad \vdots
\end{array}
$$

and in general

$$Bl_a/\lambda^a, \qquad \text{aged } a \text{ years.}$$

The sum of all these numbers is equal to P.

21. The same things being given, find the number of persons who will die in one year.

Call P the number of persons living at present, including the number of children B born this year. The quotient P/B will determine the annual increase λ on a given life table. Then next year the number living will be λP, among whom the number just born will be λB; the others, of whom the number is $\lambda P - \lambda B$, are the survivors from the P persons of the previous year; from which it follows by subtraction that $(1 - \lambda) P + \lambda B$ of them have died. Thus if the number now living is P, then $D = (1 - \lambda) P + \lambda B$ of them will die in the course of one year.

22. Knowing both the number of births and of burials occurring in the course of one year, find the total number living, and their annual increase, for a given life table.

The preceding paragraph gave us

$$D = (1 - \lambda) P + \lambda B$$

or

$$P = \frac{\lambda B - D}{\lambda - 1}.$$

From the hypothesis of mortality

$$P = B \left(1 + \frac{l_1}{\lambda} + \frac{l_2}{\lambda^2} + \cdots + \frac{l_{100}}{\lambda^{100}}\right).$$

Substituting $P = (\lambda B - D)/(\lambda - 1)$ (and subtracting B from both sides), we obtain

$$\frac{B - D}{\lambda - 1} = B \left(\frac{l_1}{\lambda} + \frac{l_2}{\lambda^2} + \cdots + \frac{l_{100}}{\lambda^{100}}\right)$$

from which λ may be calculated. (Then from λ, B, and D the total population P can be found.)

24. Given the number of births and of burials in one year, find the number of each age among the deaths.

Call B the number of children born during a year, and D the number of deaths; from the preceding problem we have the number living P, along with the ratio of increase λ. Consider how many persons will be living at each age, this year as well as next year:

	This Year	Next Year
Just born	B	$B\lambda$
Aged 1 year	Bl_1/λ	Bl_1
Aged 2 years	Bl_2/λ^2	Bl_2/λ
Aged 3 years	Bl_3/λ^3	Bl_3/λ^2
$\vdots$	$\vdots$	$\vdots$
Aged 100 years	Bl_{100}/λ^{100}	Bl_{100}/λ^{99}

Thus the deaths in the course of the year must number:

	Number of Deaths
Under one year	$B(1 - l_1)$
From 1 to 2 years	$B(l_1 - l_2)/\lambda$
From 2 to 3 years	$B(l_2 - l_3)/\lambda^2$
$\vdots$	$\vdots$
From 99 to 100 years	$B(l_{99} - l_{100})/\lambda^{99}$
Over 100 years	Bl_{100}/λ^{100}

26. Knowing the total number living P but not their ages, the number of births B, the number of deaths D, and the number of 'deaths at each age during the course of one year, find the law of mortality (i.e., calculate the life table).

We first find the annual multiplication $\lambda = (P - D)/(P - B)$. Next, for this year the number of deaths, from the preceding problem, must be

Under one year	$D_0 = B(1 - l_1)$
From 1 to 2 years	$D_1 = B(l_1 - l_2)/\lambda$
From 2 to 3 years	$D_2 = B(l_2 - l_3)/\lambda^2$
$\vdots$	$\vdots$

and hence we will find the fractions l_1, l_2, l_3,..., which contain the law of mortality:

$$l_1 = 1 - D_0/B,$$

$$l_2 = l_1 - \lambda D_1/B = 1 - \frac{D_0 + \lambda D_1}{B},$$

$$l_3 = l_2 - \lambda^2 D_2/B = 1 - \frac{D_0 + \lambda D_1 + \lambda^2 D_2}{B},$$

$$l_4 = l_3 - \lambda^3 D_3/B = 1 - \frac{D_0 + \lambda D_1 + \lambda^2 D_2 + \lambda^3 D_3}{B},$$

$$\vdots$$

28. I should point out once more that in the calculations which I have developed here I have assumed that the total number living in one place remains the same, or that it increases or decreases uniformly, so that I have had to exclude such extraordinary devastations as plague, wars, famine, as well as extraordinary increases like new colonies. It is well to choose a place where all those born remain in the region, and where strangers do not come in to live or die, which would upset the principles upon which I have based the preceding calculations. For places subject to such irregularities, it would be necessary to keep exact registrations of all those living and dying, and then, by following the principles which I have established here, one would be in a position to apply the same calculation. It always comes back to these two principles, that of mortality and that of fertility, which, once they have been established for a certain place, make it easy to resolve all the questions which one could propose on this subject, of which I am satisfied to have given an account of the principal ones.

12. Relation Between Birth Rates and Death Rates

Alfred J. Lotka (1907)

Science, N.S. 26: 21—22.

A short notice appeared on page 641 of Science, 1907, of a paper read by C. E. Woodruff before the American Association for the Advancement of Science, on the relation between birth rates and death rates, etc.

In this connection, it may be of interest to note that a mathematical expression can be obtained for the relation between the birth rate per head b and the death rate per head d, for the case where the general conditions in the community are constant, and the influence of emigration and immigration is negligible.

Comparison with some figures taken from actual observation shows that these at times approach very nearly the relation deduced on the assumptions indicated above.

I give here the development of the formula, and some figures obtained by calculation by its aid, together with the observed values, for comparison.

Let $c(a)$ be such a coefficient that out of the total number N_t of individuals in the community at time t, the number whose age lies between the values a and $(a+da)$ is given by $N_t c(a) da$.

Now the $N_t c(a) da$ individuals whose age at time t lies between the values a and $(a+da)$, are the survivors of the individuals born in time da at time $(t-a)$.

If we denote by $B_{(t-a)}$ the total birth rate at time $(t-a)$, and by $p(a)$ the probability at its birth, that any individual will reach age a, then the number of the above-mentioned survivors is evidently $B_{(t-a)} p(a) da$.

Hence:

$$N_t c(a) da = B_{(t-a)} p(a) da$$

$$c(a) = \frac{B_{(t-a)}}{N_t} p(a)$$

Now if general conditions in the community are constant, $c(a)$ will tend to assume a fixed form. A little reflection shows that then both N and B will increase in geometric progression with time,[1] at the same rate $r = (b-d)$. We may, therefore, write:

$$B_{(t-a)} = B_t e^{-ra}$$

$$c(a) = \frac{B_t}{N_t} e^{-ra} p(a)$$

$$= b e^{-ra} p(a) \tag{1}$$

Now from the nature of the coefficient $c(a)$ it follows that

$$\int_0^\infty c(a) da = 1$$

Substituting this in (1) we have:

$$\frac{1}{b} = \int_0^\infty e^{-ra} p(a) da \tag{2}$$

---

[1] Compare M. Block, "Traité théorique et pratique de statistique," 1886, p. 209.

Equation (1) then gives the fixed age-distribution, while equation (2) (which may be expanded into a series if desired), gives the relation between b, the birth rate per head, and r, the rate of natural increase per head, and hence between b and d, since $r = b - d$.

Applying these formulae to material furnished by the Reports of the Registrar-General of Births, etc., in England and Wales, the following results were obtained:

England and Wales 1871—80 (Mean)

		Observed[2]	Calculated
Birth-rate per head	b	0.03546	0.0352
Death-rate per head	d	0.02139	0.0211
Excess	$(b-d)=r$	0.01407	(0.0141)

$p(a)$ from Supplement to 45th Ann. Rep. Reg. Gen. Births, etc., England and Wales, pp. vii and viii, assuming ratio:

$$\frac{\text{male births}}{\text{female births}} = 1.04.$$

Age Scale.—1,000 individuals, in age-groups of 5 and 10 years

$a_1 a_2$	$1000 \int_{a_1}^{a_2} c(a)\,da$	
0— 5	136	138
5—10	120	116
10—15	107	106
15—20	97	97
20—25	89	87
25—35	147	148
35—45	113	116
45—55	86	87
55—65	59	59
65—75	33	33
75—∞	13	13

It will be seen that in the above example the values calculated for the age-scale and especially for b and d, show a good agreement with the observed values.

The above development admits of further extension. But this, as well as further numerical tests, must be reserved for a future occasion. In view of the recent note of the work by Major Woodruff, it appeared desirable to the writer to publish this preliminary note.

---

[2] Mean b and d from 46th Ann. Rep. Reg. Gen. Births, etc., England and Wales, p. xxxi.

13. A Problem in Age-Distribution

F. R. SHARPE and ALFRED J. LOTKA (1911)

Philosophical Magazine, Series 6, Volume 21: 435—438.

The age-distribution in a population is more or less variable. Its possible fluctuations are not, however, unlimited. Certain age-distributions will practically never occur; and even if we were by arbitrary interference to impress some extremely unusual form upon the age-distribution of an isolated population, in time the "irregularities" would no doubt become smoothed over. It seems therefore that there must be a limiting "stable" type about which the actual distribution varies, and towards which it tends to return if through any agency disturbed therefrom. It was shown on a former occasion[1] how to calculate the "fixed" age-distribution, which, if once established, will (under constant conditions) maintain itself.

It remains to be determined whether this "fixed" form is also the "stable" distribution: that is to say, whether a given (isolated) population will spontaneously return to this "fixed" age-distribution after a small displacement therefrom.

To answer this question we will proceed first of all to establish the equations for a more general problem, which may be stated as follows:—

"Given the age-distribution in an isolated population at any instant of time, the 'life curve' (life table), the rate of procreation at every age in life, and the ratio of male to female births, to find the age-distribution at any subsequent instant."

1. Let the number of males whose ages at time t lie between the limits a and $a+da$ be $F(a,t)da$, where F is an unknown function of a and t.

Let $p(a)$ denote the probability[2] at birth that a male shall reach the age a, so that $p(0)=1$.

Further, let the male birth-rate (i.e. the total number of males born per unit of time) at time t be $B(t)$.

Now the $F(a,t)da$ males whose age at time t lies between a and $a+da$ are the survivors of the $B(t-a)da$ males born a units of time previously, during an interval of time da. Hence

$$F(a,t)da = B(t-a)p(a)da$$

$$F(a,t) = p(a)B(t-a). \qquad (1)$$

2. Let the number of male births per unit time at time t due to the $F(a,t)da$ males whose age lies between a and $a+da$ be $F(a,t)\beta(a)da$.

If γ is the age at which male reproduction ends, then evidently

$$B(t) = \int_0^{\gamma} F(a,t)\beta(a)da$$

$$= \int_0^{\gamma} B(t-a)p(a)\beta(a)da. \qquad (2)$$

Now in the quite general case $\beta(a)$ will be a function of the age-distribution both of the males and females in the population, and also of the ratio of male births to female births.

---

[1] A.J. Lotka, Am. Journ. Science, 1907, xxiv. pp. 199, 375; 'Science,' 1907, xxvi. p. 21.
[2] As read from the life table.

We are, however, primarily concerned with comparatively small displacements from the "fixed" age-distribution, and for such small displacements we may regard $\beta(a)$ and the ratio of male births to female births as independent of the age-distribution.

The integral equation (2) is then of the type dealt with by Hertz (*Math. Ann.* vol. lxv. p. 86). To solve it we must know the value of $B(t)$ from $t=0$ to $t=\gamma$, or, what is the same thing, the number of males at every age between 0 and γ at time γ. We may leave out of consideration the males above age γ at time γ, as they will soon die out. We then have by Hertz, *loc. cit.*,

$$B(t) = \sum_{h=1}^{h=\infty} \frac{\alpha_h^t \int_0^\gamma \{B(a) - \int_0^a \beta(a_1)p(a_1)B(a-a_1)da_1\} \alpha_h^{-a}da}{\int_0^\gamma a\,\beta(a)p(a)\alpha_h^{-a}da}, \tag{3}$$

where $\alpha_1, \alpha_2, \ldots$ are the roots of the equation for α;

$$1 = \int_0^\gamma \beta(a)p(a)\alpha^{-a}da. \tag{4}$$

The formula (3) gives the value of $B(t)$ for $t > \gamma$, and the age-distribution then follows from

$$F(a,t) = p(a)B(t-a). \tag{1}$$

4. From the nature of the problem $p(a)$ and $\beta(a)$ are never negative. It follows that (4) has one and only one real root r, which is $\gtreqless 1$, according as

$$\int_0^\gamma \beta(a)p(a)da \gtreqless 1. \tag{5}$$

Any other root must have its real part less than r. For if $r_1\,(\cos\theta + i\sin\theta)$ is a root of (4),

$$1 = \int_0^\gamma \frac{\beta(a)p(a)}{r_1^a}\cos a\theta\,da. \tag{6}$$

It follows that for large values of t the term with the real root r outweighs all other terms in (3) and $B(t)$ approaches the value

$$B(t) = A\,r^t. \tag{7}$$

The ultimate age-distribution is therefore given by

$$F(a,t) = A\,p(a)r^{t-a} \tag{8}$$

$$= A\,p(a)e^{r'(t-a)}. \tag{9}$$

99

Formula (9) expresses the "absolute" frequency of the several ages. To find the "relative" frequency $c(a, t)$ we must divide by the total number of male individuals.

$$c(a,t) = \frac{F(a,t)}{\int_0^{\infty} F(a,t)\,da} = \frac{A\,p(a)\,e^{r'(t-a)}}{A\,e^{r't}\int_0^{\infty} e^{-r'a}p(a)\,da} = \frac{p(a)\,e^{-r'a}}{\int_0^{\infty} e^{-r'a}p(a)\,da}$$

$$= b\,e^{-r'a}p(a)\,, \tag{10)^3}$$

where

$$\frac{1}{b} = \int_0^{\infty} e^{r'a}p(a)\,da\,. \tag{11)^3}$$

The expression (10) no longer contains t, showing that the ultimate distribution is of "fixed" form. But it is also "stable;" for if we suppose any small displacement from this "fixed" distribution brought about in any way, say by temporary disturbance *of the otherwise constant conditions*, then we can regard the new distribution as an "initial" distribution to which the above development applies: that is to say, the population will ultimately return to the "fixed" age-distribution.

It may be noted that of course similar considerations apply to the females in the population. The appended table shows the age-distribution calculated according to formula (10) for England and Wales 1871—1880. The requisite data (including the life table) were taken from the Supplement to the 45th Annual Report of the Registrar General of Births, etc. The mean value of r' (mixed sexes) for that period was 0.01401, while the ratio of male births to female births was 1.0382.

It will be seen that at this period the observed age-distribution in England conformed quite closely to the calculated "stable" form.

Table

Age (Years)	Males		Females		Persons	
	Calc.	Obs.	Calc.	Obs.	Calc.	Obs.
0— 5	139	139	136	132	138	136
5—10	118	123	115	117	116	120
10—15	107	110	104	104	106	107
15—20	97	99	95	95	96	97
20—25	88	87	87	91	87	89
25—35	150	144	148	149	149	147
35—45	116	112	116	115	116	113
45—55	86	84	88	87	87	86
55—65	57	59	62	61	59	59
65—75	30	31	35	35	33	33
75—∞	11	12	15	15	13	13

---

[3] Compare Am. Journ. Science, xxiv. 1907, p. 201.

14. The Stability of the Normal Age Distribution

ALFRED J. LOTKA (1922)

Proceedings of the National Academy of Sciences 8: 339—345.

There is a unique age distribution which, in certain circumstances,[2] has the property of perpetuating itself when once set up in a population. This fact is easily established,[3] as is also the analytical form of this unique *fixed* or *normal* age distribution.

More difficult is the demonstration that this age distribution is *stable*, that a population will spontaneously revert to it after displacement therefrom, or will converge to it from an arbitrary initial age distribution. Such a demonstration has hitherto been offered only for the case of small displacements,[4] by a method making use of integral equations. The purpose of the present communication is to offer a proof of stability which employs only elementary analytical operations, and which is readily extended to cover also the case of large displacements. This method presents the further advantage that it is molded in more immediate and clearly recognizable relation to the physical causes that operate to bring about the normal age distribution.

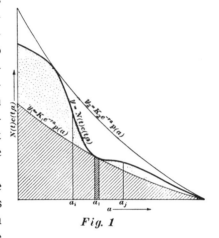

Fig. 1

Consider a population which, at time t has a given age distribution such as that represented by the heavily drawn curve in fig. 1, in which the abscissae represent ages a (in years, say), while the ordinates y are such that the area comprised between two ordinates erected at a_1 and a_2, respectively, represents the number of individuals between the ages a_1 and a_2.

If we denote by $N(t)$ the total population at time t, and if the ordinates of our curve are

$$y = N(t)\, c(t,a) \qquad (1)$$

we have, evidently,

$$\int_{a_1}^{a_2} y\, da = N(t) \int_{a_1}^{a_2} c(t,a) da = N(t,a_1,a_2) \qquad (2)$$

where $N(t,a_1,a_2)$ denotes the number of individuals living at time t and comprised within the age limits a_1 and a_2. We may speak of $c(t,a)$ as the coefficient of age distribution. It is, of course, in general a function of t, only in the special case of the fixed or self-perpetuating age distribution is $c(a)$ independent of the time.

Now, without assuming anything regarding the stability of the self-perpetuating age distribution, it is easy to show[2] that its form must be

102

$$c(a) = \frac{e^{-ra}p(a)}{\displaystyle\int_0^\infty e^{-ra}p(a)da} = be^{-ra}p(a) \tag{3}$$

where r is the real root of the equation

$$1 = \int_{a_i}^{a_j} e^{-ra}p(a)\beta(a)da \tag{4}$$

In this equation r is the natural rate of increase of the population, i.e., the difference $r = b - d$ between the birthrate per head b and the death-rate per head d, and $p(a)$ is the probability, at birth, that a random individual will live to age a (in other words, it is the principal function tabulated in life tables, and there commonly denoted by 1_x). The limits a_i and a_j of the integral are the lower and upper age limits of the reproductive period. The factor $\beta(a)$, which might be termed the procreation factor, or more briefly the birth factor, is the average number of births contributed per annum by a parent of age a. (In a population of mixed sexes it is, of course, immaterial, numerically, to what parent each birth is credited. It will simplify the reasoning, however, if we think of each birth as credited to the female parent only.)

The factor $\beta(a)$ will in general itself depend on the prevailing age distribution. This is most easily seen in the case of extremes, as for example in a population which should consist exclusively of males under one year of age and females over 45. But, except in such extreme cases, $\beta(a)$ will not vary greatly with changes in the age constitution of the population, and we shall first develop our argument on the supposition that $\beta(a)$ is independent of the age distribution. We shall then extend our reasoning to the more general case of $\beta(a)$ variable with $c(t,a)$.

Referring now again to fig. 1, let two auxiliary curves be drawn, a *minor tangent curve* and a *major tangent curve*

$$y_1 = K_1 e^{-ra}p(a) \quad (5) \qquad y_2 = K_2 e^{-ra}p(a) \quad (6)$$

the constants K_1, K_2 being so chosen that the minor tangent curve lies wholly beneath the given arbitrary curve, except where it is tangent thereto, while the major tangent curve lies wholly above the given curve, except where it is tangent thereto.

The given arbitrary curve representing the age constitution of the population at time t then lies wholly within the strip or area enclosed between the minor and the major tangent curves.

Now consider the state of affairs at some subsequent instant t'. Had the population at time t consisted solely of the individuals represented by the lightly shaded area in fig. 1, i.e. the area under the minor tangent curve, then at time t' the population would be represented by the lower curve of fig. 2, whose equation is

$$y'_1 = K'_1 e^{-ra} p(a) \tag{7}$$
$$= K_1 e^{r(t'-t)} e^{-ra} p(a) \tag{8}$$

For the age distribution (5) is of the fixed form (3), and therefore persists in (7); on the other hand, given such fixed age distribution, the population as a whole increases in geometric progression,[3] so that $K'_1 = K_1 e^{r(t'-t)}$. In point of fact, we have left out of reckoning that portion of the population which in fig. 1 is represented by the dotted area. Hence, in addition to the population under the lower curve of fig. 2, there will, at time t', be living a body of population which for our present purposes it is not necessary to determine numerically. We need only know that it is some positive number, so that the curve representing the actual population at time t' must lie wholly above or in contact with the curve (8).

By precisely similar reasoning it is readily shown that at time t' the actual curve lies wholly beneath or in contact with the curve

$$y'_2 = K_2 e^{r(t'-t)} e^{-ra} p(a) \tag{9}$$

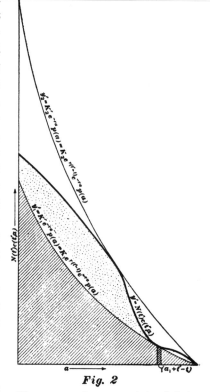

Fig. 2

The curves shown are intended to be interpreted only in a qualitative sense; so, for example, the increase in the ordinates in passing from fig. 1 to fig. 2 is very much exaggerated, to render it obvious to the eye.

Hence at time t' the actual curve lies wholly within the strip comprised between the two curves (8) and (9).

Consider now an elementary strip, of width da, of the original population (shown heavily shaded in fig. 1), which at time t is in contact with the minor tangent curve. Let this contact occur at age a_1, so that $y_t(a_1) = K_1 e^{-ra_1} p(a_1)$. At time t' the survivors of the individuals comprised in this elementary strip will be of age $(a_1 + t' - t)$, so that they will then be represented by a strip of width da and of altitude

$$y_{t'}(a_1 + t' - t) = K'_1 e^{-ra_1} p(a_1 + t' - t)$$
$$= K_1 e^{r(t'-t)} e^{-r(a_1 + t' - t)} p(a_1 + t' - t)$$

From this it is seen that the elementary strip of population which at time t' contacts with the minor tangent curve (8) is built up of the survivors

104

of the strip which at time t contacted with the minor tangent curve (5). In other words, if we follow up, by identity of individuals, the element of the population which at any instant contacts with the minor tangent curve, this element (so long as any part of it survives) continues in contact with that curve. (It must be remembered, however, that the tangent curve itself changes with time according to (5), (8).)

Similarly it follows that the element of population which at any instant contacts with the major tangent curve continues in this contact so long as any part of it survives.

And again, considering any element of the population which does not contact with the minor or the major tangent curves, but has its upper extremity at some point within the area enclosed between these curves, it can be shown by precisely similar reasoning that such element continues in such intermediate position.

Turning now from the consideration of the survivors of the original population, and taking in view the new population added by births since the time $t = t$, we note that if the original population had been that represented by the shaded area in fig. 1, i.e., by the area under the minor tangent curve, then the birthrate would at all times have been such as continuously to reproduce a population represented by the minor tangent curve (5), (8).

In point of fact, provided that contact with the minor tangent curve is not *continuous* over a range of ages equal to or greater than $a_j - a_i$, it is easily seen that the total birthrate

$$B = N \int_{a_i}^{a_j} c(a)\beta(a)da$$

is always greater (equality is here excluded) than that which would result from and in turn reproduce the age distribution represented by the minor tangent curve.

Similarly the total birthrate is always less than that which would result from a population and age distribution represented by the major tangent curve.

From this it is clear that, after the original population has once died out, the representative curve can never again contact with the original minor and major tangent curves, but must henceforth be separated from them everywhere by a finite margin (except, of course, where the several curves terminate upon the axis of a).

We may then begin afresh by drawing a new pair of tangent curves, lying within the original pair, and so on indefinitely, until the minor and major tangent curves coincide, and with them also coincides the actual curve of age distribution, which is then of the form

$$N(t)\, c(a) = K e^{rt} e^{-ra} p(a)$$

105

which we recognize as the fixed age distribution (3), with

$$N(t) = Ke^{rt} \int_0^\infty e^{-ra} p(a) da$$

In this argument we have expressly excluded the case that the original age distribution curve contacts *continuously* with one of its tangent curves over a range greater than the reproductive period $(a_j - a_i) = A$. If such continuous contact occurs in the 0th generation over a range nA, where $a_j/A > n > 1$, then a simple reflection shows that in the next generation (the first) contact will occur over a range $(n-1)A$, in the second generation over $(n-2)A$, etc. A time is therefore reached (in practice very soon), when contact is over a period less than the period of reproduction A. After that the argument set forth above applies.

If the curve representing the original age distribution contacts with one of the tangent curves all the way from $a = 0$ to $a = a_j$, so that $a_j/A \leq n$, then, of course, the fixed age distribution is practically established *ab initio*, or at any rate from the moment the original population above reproducing age a_j has died out.

It remains to consider the effect of variability in the form of $\beta(a)$ with changes in $c(t,a)$. Some such variability undoubtedly exists owing to the influence of the ages of the male and female constituents of the population upon the frequency of matings. We may, nevertheless, in this case also, define a minor and a major tangent curve (5), (6) in terms of the value of r given by equation (4); in order, however, to make this value determinate, it is now necessary to make some definite disposition regarding the form of $\beta(a)$, which is now variable. Merely for purposes of defining r, we shall suppose that the function $\beta(a)$ under the integral sign has that particular form which corresponds to the fixed age distribution.[5]

We cannot, however, now reason, as before, that the portion of the population represented by the shaded area in fig. 1 will, by itself, reproduce its own form of age distribution. For its constitution as to sex will in general differ from that of the "normal" population with self-perpetuating age distribution. We must therefore consider three different possibilities:

1. The shaded area alone will produce a population exceeding at all ages the normal or fixed type continuation (8) of the original shaded area. Should this be the case, then the argument presented with regard to the case of invariable $\beta(a)$ holds *a fortiori*, so far as the minor tangent curve is concerned.

2. The shaded area alone will produce a population deficient at some or all ages, as compared with the normal type continuation (8) of the original shaded area. In that case two alternatives present themselves:

a. The deficiency is more than counterbalanced by the additional population produced by that portion of the original population which is represented by the dotted area in fig. 1. In this case also the original argument, so far as the minor tangent curve is concerned, applies essentially as before.

b. After the contributions from all parts of the population have been taken into account, there remains an unbalanced deficiency short of the population defined by (8), in the population resulting from that originally present. In such case the argument presented on the assumption of invariable $\beta(a)$ fails, and the population may move away from, not towards the age distribution(3). Stability of the fixed age distribution may not extend to such displacements as this.

Similar reflections apply, *mutatis mutandis*, as regards the major tangent curve.

If conditions (1) or (2) prevail with respect to the minor, and corresponding conditions with respect to the major tangent curve, then we can argue as in the case of invariable $\beta(a)$, that after expiration of the existing generation a new pair of tangent curves can be drawn, which will lie within the pair defined by (8), (9). And if conditions (1) or (2) still persist with reference to the new tangent curves, the same process of closing in these curves at the expiration of the current generation can be repeated, and so on, as in the argument first presented. Now conditions (1) or (2) will thus continue to prevail for each new pair of tangents, if, as the minor and major tangents close up, $\beta(a)$ approaches a limiting form. In such case, therefore, the age distribution defined by (3), (4) is stable even for displacements of any magnitude, provided always that conditions (1) or (2) prevail, as indicated.

[1] Papers from the Department of Biometry and Vital Statistics, School of Hygiene and Public Health, Johns Hopkins University, No. 71.

[2] These circumstances are: (*a*) an invariable life curve (life table); (*b*) an invariable ratio of male to female births; (*c*) an invariable rate per head of procreation at each year of age, for any given sex-and-age distribution in the population.

[3] Lotka, *Amer. J. Sci.*, **26**, 1907 (21); *Ibid.*, **24** (199); *J. Wash. Acad. Sci.*, **3**, 1913 (241, 289).

[4] Sharpe, F. R., and Lotka, A. J., *Phil. Mag.*, April 1911, 435.

[5] A certain ambiguity is introduced here by the fact that, with $\beta(a)$ of variable form, equation (4) might have more than one real root for r. In practice, however, in a human population at any rate, probably only one such root exists, or has any effect upon the course of events.

15. Resolving a Historical Confusion in Population Analysis*

PAUL A. SAMUELSON (1976)

This essay was written for this volume, and also appears in *Human Biology* 48, 559—580 (1976).

Introduction

As Robert K. Merton (1973) has so well discussed, creative scientists are not immune to preoccupation with priorities: *their* priorities. Alfred J. Lotka (1880 to 1949), the father of self-renewal models in linear population analysis, was least of all an exception in this regard. A lone pioneer throughout much of his career,[1] with no cadre of graduate students and colleagues, he could naturally be expected to be prickly over failures to recognize and acknowledge his important contributions to demography.

The present brief note attempts to sort out a controversy that arose some 40 years ago between Lotka and another able self-made demographer, R. R. Kuczynski (1876—1947). The matter has an interest at two levels. Substantively, a reader of their polemics has still been left in ignorance of the true merits of the points being argued—whether Bortkiewicz (1911) and other writers had already established the asymptotic approach to a stable exponential equilibrium of a population subject to invariant (one-sex) age-specific mortality *and fertility*, an accomplishment properly attributed today primarily to Lotka. Psychologically, as an exercise in how new science gets itself done, the matter is also of some interest. Once emotions entered in, Lotka ceased to be his own best advocate and came gratuitously to denigrate legitimate theorems in demography that are of interest for their own sake. I shall briefly sort out the truths and misunderstandings of the discussion.

Act I

The story begins with Lotka (1929), a review of Kuczynski (1928), a book with an honorable role in the history of demography for its forceful pointing out an impending decline in European population levels, a decline shown to be concealed by swollen numbers of people of fertile age inherited from earlier generations of higher fertility. Kuczynski could have benefitted from the fundamental findings of Sharpe and Lotka (1911) and Lotka (1913, 1922). Perhaps he did not then know of the 1911 integral-equation finding; and perhaps he had neither the mathematical equipment nor the inclination to master it even were it called to his attention. For the purpose of discerning a shift from a growing to an ultimately declining population, Kuczynski could rely on a common-sense criterion,[2] the *net reproduction rate* (i.e., the expected number of female babies that will be born to a representative female baby who throughout her life will be subject to current age-specific mortality and fertility rates).

Lotka's review is favorable on Kuczynski's substantive thesis, but accuses him of not acknowledging that his analysis is taken from Dublin and Lotka (1925), a work said to receive only cursory acknowledgment by Kuczynski.

Kuczynski (1930) replies sturdily that the finding about swollen fertile-age numbers is in fact an ancient staple in the literature, even having been dealt with by Kuczynski in an 1897 book (when Lotka was only 17 and Kuczynski only 23); moreover, the concept of the net reproduction rate that Kuczynski relied on, Kuczynski points out, was put forth by his mentor, Richard Böckh, as far back as 1886.[3]

In his rebuttal to Kuczynski, Lotka (1930) had to scale down his accusations. But he does assert the incompleteness of the early work Kuczynski cited. And he makes the valid point that, although the net reproductive rate is an accurate indicator of whether the intrinsic rate of population growth is positive or negative, it only gives the rate of growth per ambiguous "generation", and *not* per *year*, so that it falls short of the full analysis of Sharpe and Lotka and of Dublin and Lotka. He concludes with the barb that, if he had known that the discussion of Dublin and Lotka (1925) on the net reproduction rate had been anticipated earlier, *he*, Lotka, would have felt obliged to acknowledge it—whatever Kuczynski's scholarly code might be.

This ends Act I of the drama. The antagonists are now sensitized to each other. In the years between 1930 and 1937, Lotka apparently became increasingly of the opinion that Kuczynski's publications were spreading misleading accounts of the true priorities. This finally culminated in two publications, Lotka (1937a, 1937b), which attempt to set the record straight and clarify the truth. To understand them, I first review some now familiar fundamentals. And then I formulate a number of distinguishable propositions or theorems that are needed to judge the various allegations.

Review of Fundamentals: Mortality Relations

Let $N(a, t)$ be the number of people (females presumably) of age a at time t, with total population number at t given by

$$(1) \qquad N(t) = \int_0^\infty N(a, t)\, da = \int_0^n N(a, t)\, da, \qquad a < n < \infty$$

where n is the maximum length of life. Let $p(a)$ be the fraction of new-born females surviving to age a. Then with births at t given as a prescribed function of time, $N(0, t) = B(t)$,

(2) $\quad N(a,t)=p(a)B(t-a)$

$$=\frac{p(a)}{p(a-\theta)}N(a-\theta,t-\theta),\quad 0<\theta<a\,;$$

$$p(0)=1,\quad p(n)=0,\quad p'(a)\le 0,$$

(3) $\quad \partial N(a,t)/\partial a+\partial N(a,t)/\partial t=[p'(a)/p(a)]N(a,t)\,,$

(4) $\quad N(t)=\int_0^n p(a)B(t-a)\,da\,,$

(5) $\quad N'(t)=B(t)+\int_0^n p'(a)B(t-a)\,da\,.$

A solution to (5), and under suitable specifications to (4), can be given by the "renewal function" so useful in industrial-equipment as well as actuarial analysis, $\Pi(t)$, as discussed in Feller (1941) and Lotka (1933, 1939a), namely by

(6) $\quad B(t)=N'(t)+\int_{-\infty}^{t}\Pi(t-a)N'(a)\,da$

where $\Pi(t-a)$ is the Volterra resolvent kernel to $p'(t-a)$, defined as the solution to

(7) $\quad \Pi(t)=-p'(t)-\int_0^t p'(t-a)\Pi(a)\,da$

$$=-p'(t)-\int_0^t\Pi(t-a)p'(a)\,da\,.$$

One way of solving for $\Pi(t)$ is via its Laplace transform

(8) $\quad \bar{\Pi}(\omega)=\int_0^{\infty}\Pi(t)e^{-\omega t}\,dt$

$$-\bar{p}'(\omega)=\int_0^{\infty}-p'(t)e^{-\omega t}\,dt,$$

$$\bar{\Pi}(\omega)=-\bar{p}'(\omega)-\bar{p}'(\omega)\bar{\pi}(\omega)$$

$$=\frac{-\bar{p}'(\omega)}{1+\bar{p}'(\omega)}\,.$$

Alternatively, writing symbolically,

$$f(t)=-p'(t),\quad f(t)\cdot g(t)=\int_0^t f(t-a)g(a)\,da$$

$$=g(t)\cdot f(t)\,,$$

(9) $\quad \Pi(t)=f(t)+f(t)\cdot f(t)+f(t)\cdot f(t)\cdot f(t)+\cdots$

111

which is a rapidly converging explicit solution, known to have the property

(10) $\lim_{t\to\infty} \Pi(t) = c_0 = 1/\int_0^n -ap'(a)\,da = 1/\int_0^n p(a)\,da > 0$.

Also, with n finite, $\Pi(t)$ can be written as an infinite series of exponentials, of Hertz-Herglotz-Lotka type:

(11) $\Pi(t) = c_0 + \sum_{-\infty}^{\infty} e^{m_j t} c_j$

where m_j are the infinite number of complex roots of the transcendental equation

(12) $\int_0^n -p'(a)e^{-ma}\,da = 1$.

The sole real root is $m_0 = 0$, and the real parts of all the conjugate complex roots are demonstrably negative

$$m_j = \mu_j + iv_j, \qquad m_{-j} = \mu_j - iv_j,$$

(13) $0 = m_0 > \mu_j \qquad (j = 1, 2, \ldots)$.

If any m_j root is multiple, the coefficients c_j and c_{-j} will be polynomials in t rather than constants.

It is worth noting that the non-real roots of (12) are also the roots of

(12′) $\int_0^n p(a)e^{-ma}\,da = 0$

as an integration by parts will verify.

Review: Fertility and Mortality Relations

So far nothing has been said about fertility rates. Writing the number of female births at time t to mothers of age a as $B(a,t)$, we define an invariant age-specific fertility function, $m(a)$, by

(14) $m(a) \equiv B(a,t)/N(a,t) \geq 0$,
 $m(a) \equiv 0, \quad 0 < a < \alpha$,
 $m(a) > 0, \quad \alpha \leq a \leq \beta < n$,
 $m(a) \equiv 0, \quad \beta < a$.

A population self-propelled by invariant age-specific mortality and fertility functions, $p(a)$ and $m(a)$, and starting out from an initial non-negative age distribution, $N(0,a)$, will have births and numbers that forever after satisfy

(15) $\qquad B(t) = \int_{\alpha}^{\beta} m(a)p(a)B(t-a)\,da\,, \qquad t>n>\beta>\alpha>0,$

$\qquad\qquad = \int_{0}^{t} m(t-a)p(t-a)B(a)\,da + \int_{t}^{n} m(a)\frac{p(a)}{p(a-t)}N(a-t,0)\,da, \qquad 0<t<n,$

(16) $\qquad\qquad = \int_{0}^{t} \phi(t-a)B(a)\,da + G(t)\,, \qquad t>0\,,$

where

(17) $\qquad \phi(a) \equiv m(a)p(a)\,,$

$\qquad\qquad G(t) = \int_{t}^{n} m(a)\frac{p(a)}{p(a-t)}N(a-t,0)\,da \geq 0, \qquad 0<t<n$

$\qquad\qquad \equiv 0, \qquad n<t\,.$

Just as $\Pi(t-a)$ was the resolvent kernel of $-p'(t-a)$ in (6)—(9), so is there a useful resolvent kernel to $\phi(t-a)$, which I write as $\phi^*(t-a)$, and which provides a solution to $B(t)$ in (16) and has the other listed verifiable properties:

(18) $\qquad B(t) = G(t) + \int_{0}^{t} \phi^*(t-a)G(a)\,da\,, \qquad t>0\,,$

(19) $\qquad \phi^*(t) = \phi(t) + \int_{0}^{t} \phi(t-a)\phi^*(a)\,da$

$\qquad\qquad = \phi(t) + \int_{0}^{t} \phi^*(t-a)\phi(a)\,da$

$\qquad\qquad = \phi(t) + \phi(t)\cdot\phi(t) + \cdots$

$\qquad\qquad = Q_0 e^{r_0 t} + \sum_{-\infty}^{\infty} Q_j e^{r_j t}, \qquad Q_0>0\,,$

where r_0 is the sole real root and r_j are the complex roots, necessarily infinite in number when β is a finite positive number, of the transcendental equation

(20) $\qquad \psi(r) = \int_{0}^{\infty} \phi(a)e^{-ra}\,da - 1\,,$

(21) $\qquad r_0 > \text{real coefficients of } r_{\pm j} \qquad (j=1,2,\dots)\,.$

The $(Q_{\pm j})$ are constants or polynomials in t, and like Q_0 can be determined from the initial condition $N(a,0)$.

113

To determine whether $r_0 \gtreqless 0$, we can employ the useful Böckh (1886) net reproduction rate as a criterion, namely

(22) $\qquad r_0 \gtreqless 0 \leftrightarrow \int_0^n m(a)p(a)\,da = \int_0^n \phi(a)\,da \gtreqless 1 \; .$

As was done in (8), the Laplace transform may be used to solve for $\phi^*(t)$ in terms of known $\phi(t)$. Lotka (1928) gave the interpretation of $\phi^*(t)$ as the progeny of a Dirac pulse of new births at initial time zero:

(23) $\qquad N(a,0) = \delta(a-0) \; ,$

$\qquad \int_0^n N(a,0)\,da = \int_0^n \delta(a-0)\,da = 1 \; .$

From (19) and (20), or from more general analysis of Feller, one can prove

(24) $\qquad \lim_{t\to\infty} \phi^*(t)e^{-r_0 t} = Q_0 = -1/\psi'(r_0) > 0 \; ,$

(25) $\qquad \lim_{t\to\infty} B(t)e^{-r_0 t} = b_0 > 0 \; ,$

(26) $\qquad \lim_{t\to\infty} N(a,t)e^{-r_0 t} = b_0\, p(a)\, e^{-r_0 a} \; ,$

(27) $\qquad \lim_{t\to\infty} \dfrac{N(a,t)}{N(0,t)} = p(a)\, e^{-r_0 a} \; .$

These two-dozen-odd equations are now standard in the demographic literature, as discussed in Lopez (1961), Keyfitz (1968), Coale (1972), J.H. Pollard (1973), and elsewhere. All of these relations owe much to Lotka; in particular there can be no doubt that equations (15)—(20), (25)—(27) are primarily due to him, so that any account which failed to indicate this fact is open to criticism by an impersonal jury, to say nothing of criticism by the injured scientist himself.

A Bouquet of Theorems

To illuminate the disputed issues, let me write down a number of relevant theorems. The list could be amplified, abbreviated, or arranged differently.

The first three theorems involve essentially nothing more than the actuarial assumption of age-specific mortality.

Theorem 1A. Suppose that, for a finite time interval, $t_0 - \delta < t < t_0 + \delta$

(i) $p(a)$ applies, so that from (3)

$$\partial[\ln N(a,t)]\,\partial a + \partial[\ln N(a,t)]/\partial t \equiv d[\ln p(a)]/da$$

114

(ii) the age distribution is stable, in the sense that

$$N(a,t)/N(0,t) \equiv c(a):$$

Then $N(0,t) = B(t) \equiv B(0)e^{ut}$, $t_0 - \delta < t < t_0 + \delta$, $u \gtreqless 0$

$$N(a,t)/N(0,t) \equiv p(a)e^{-ua},$$

$$b \equiv B(t)/N(t) = 1/\int_0^\infty p(a)e^{-ua}\,da,$$

$$d \equiv D(t)/N(t) = -\left[\int_0^\infty p'(a)e^{-ua}\,da\right]\Big/\int_0^\infty p(a)e^{-ua}\,da,$$

$$u \equiv b - d.$$

Bortkiewicz (1911) essentially states and proves Theorem 1A; however, it would be surprising if a thorough search did not turn it up in the earlier actuarial literature. With charity, and some charity is needed, Lotka (1907a, 1907b), in his maiden demographic papers, can be construed to have glimpsed the truth of 1A.

To prove 1A, using my notation rather than that of Bortkiewicz (1911), combine (i) and (ii) to derive

(28) $$0 \equiv \partial \ln[N(a,t)/N(0,t)]/\partial t$$

$$\equiv d \ln[p(a)]/da - \partial \ln[N(a,t)]/\partial a - B'(t)/B(t),$$

(29) $$c(a) = p(a)e^{-ua},$$

$$u = B'(t)/B(t), \quad \text{a constant of any sign.}$$

No doubt Bortkiewicz (1911) and Lotka (1907a, 1907b) thought of δ as ∞ or as a large number; however, so long as $\delta > 0$, the theorem holds.

Theorem 1B. Suppose, for all positive time, $t \geq 0$,

(i) $p(a)$ applies, and
(ii) $B(t) = B(0)e^{ut}$, $u \gtreqless 0$, $t \geq 0$.

Then

$$N(a,t) \equiv p(a)B(t-a),$$

$$\equiv B(0)[p(a)e^{-ua}]e^{ut}, \quad t \geq n,$$

$$N(t) = B(0)b^{-1}e^{ut}, \quad b^{-1} = \int_0^\infty p(a)e^{-ua}\,da.$$

This almost trivial theorem, whose proof is direct from substitution, is in Euler (1760), paragraph 17; in perusing this celebrated early paper, I was surprised to find that Euler seems to have not gone beyond its fertility analysis, despite his earlier promising remarks that the multiplication of a population "depends, of course, on the number of marriages and upon fertility ...". However, in Euler's private communication to Süssmilch, reported in Süssmilch (1761) and discussed in my footnote 2, Euler did anticipate a case of the Sharpe-Lotka Theorem 2A about to be discussed.

Theorem 1C. Suppose, for *all* $t \geq 0$,

(i) $p(a)$ applies, and
(ii) $N(t)$ is of exponential growth

$$N(t) = N(0) e^{ut}, \quad t \geq 0, \quad u \gtreqless 0:$$

Then

$$\lim_{t \to \infty} \frac{N(a, t)}{N(0, t)} = p(a) e^{-ua},$$

$$\lim_{t \to \infty} B(t) e^{-ut} = b N(0), \quad b^{-1} = \int_0^\infty p(a) e^{-ua} da > 0.$$

If u were zero, and $N(a, 0)$ were Dirac's $\delta(a - 0)$, this would be the renewal equation for $\Pi(t) \equiv B(t)$. However, the theorem is true whatever the admissible specifications of $N(a, 0)$, and it is far from absurd or trivial.

To prove the theorem, utilize (5) and (6). Thus, for $N(t) = e^{ut}, u > \mu_{\pm j}$,

(30) $$N'(t) \equiv u e^{ut}$$

$$\equiv B(t) + \int_0^t p'(t) B(t - a) da - H(t),$$

$$H(t) \equiv \int_t^n - \frac{p'(a)}{p(a)} N(a, t) da \geq 0, \quad 0 < t < n$$

$$\equiv \int_t^n \frac{p'(a)}{p(a)} N(a - t, 0) da,$$

$$H(t) \equiv 0, \quad n < t.$$

Any general solution to (30) is known from the principle of superposition to be the sum of the special exponential solution proportional to e^{ut} plus $Y(t)$, the general solution to the following homogeneous integral equation:

(31) $$Y(t) = \int_0^n p(a) Y(t - a) da$$

$$= \sum_{-\infty}^{\infty} h_j e^{m_j t}, \quad \text{where } m_j \text{ is as in (12').}$$

116

When the h's are tailored to admissible initial $H(t)$, the fact that all μ_j are negative guarantees that as t grows large, the non-exponential $Y(t)$ is damped down to zero. Q.E.D.

Remark: The fact that numbers in any age group can never be negative puts a restriction on how fast total $N(t)$ can fall exponentially: thus, if u in e^{ut} were more negative than the real part of some m_j in (13), μ_j, that would not be an admissible observed situation unless $B(t)$ was *already* exactly proportional to e^{ut}.

This serves as a reminder that one could strengthen the hypothesis in all three theorems to require their postulated conditions to hold for *all* time, $-\infty < t < \infty$. In that case, Theorems 1A and 1B have conclusions that hold for *all* time. And the conclusion of Theorem 1C *holds not merely asymptotically but for all t*, as in the following:

Theorem 1C′: Suppose, for $-\infty < t < \infty$,

(i) $p(a)$ applies,
(ii) $N(t) \equiv N(0) e^{ut}, \quad -\infty < t < \infty$,
(iii) $N(a,t)$ must, of course, *always* be non-negative.

Then

$$N(a,t)/N(0,t) \equiv p(a) e^{-ut},$$

$$B(t) \underset{t}{\equiv} N(0) e^{ut} / \int_0^n p(a) e^{-ua} \, da,$$

$$N(a,0) \equiv \left[N(0) / \int_0^n p(a) e^{-ua} \, da \right] p(a) e^{-ua}.$$

For proof, recall that any *initial* deviation from the stable distribution would, when projected backward in the fashion of Samuelson (1976), generate negative numbers from the backward-anti-damped (!) oscillatory components implied to be present. This shows that there can be no such initial deviation.

The condition of non-negativity, (iii), is important, even if usually left implicit. Thus, Lotka (1931, 1939b) seems a bit casual in assuming that a particular formal solution for $B(t)$ is unique in $\int_0^\infty p(a) B(t-a) \, da =$ a prescribed $N(t)$ function. We can add to such a special solution terms like $\sum k_{\pm j} e^{m_j t}$ and have new *formal* solutions: however, with μ_j all negative in (12′)'s $m_{\pm j} = \mu_j \pm v_j$, these "appendages" would make $B(t)$ negative as $t \to -\infty$; and it is this property that Lotka should utilize in a cogent treatment.

All theorems up to now, 1A, 1B, 1C, or 1C′, have involved only age-specific mortality data as contained in the $p(a)$ survival function. They have not involved age-specific fertility data from an $m(a)$ function, all e^{ut} growth functions having been postulated or deduced from postulates *not* involving $m(a)$. They all belong to the pre-Sharpe-and-Lotka era of Euler, Bortkiewicz, and Lotka (1907a, 1907b).

117

The next set of theorems depend on both $m(a)$ and $p(a)$ belonging to the post-1911 age of Lotka.

The first theorem is the basic one of Sharpe and Lotka (1911), with Bernardelli (1941) and Leslie (1945) equivalences holding for discrete-time, discrete-age models.

Theorem 2A: Suppose, for $t \geq 0$,

(i) $p(a)$ applies,

(ii) age-specific fertility, $m(a)$, applies

$$B(t) = \int_{\alpha}^{\beta} m(a) N(a,t) da, \quad t \geq 0$$

(iii) initial $N(a,0)$ is an arbitrarily given, non-negative, integrable function with some females not yet beyond the fertile ages

$$\int_{0}^{\beta} N(a,0) da > 0:$$

Then

$$\lim_{t \to \infty} B(t) e^{-r_0 t} = c_0 > 0,$$

$$\lim_{t \to \infty} N(t) e^{-r_0 t} = c_0 \int_{0}^{n} p(a) e^{-r_0 a} da,$$

$$\lim_{t \to \infty} \frac{N(a,t)}{N(0,t)} = p(a) e^{-r_0 a},$$

$$\lim_{t \to \infty} N(a,t) e^{-r_0 t} = p(a) e^{-r_0 a} c_0,$$

$$\psi(r_0) = \int_{\alpha}^{\beta} m(a) p(a) e^{-r_0 a} da - 1, \quad r_0 \text{ real},$$

$$c_0 = \frac{\int_{0}^{\beta} e^{-r_0 a} G(a) da}{\int_{\alpha}^{\beta} a \phi(a) e^{-r a} da}; \quad G(t) \text{ as defined in (17)}.$$

Warning: My $\psi(r)$ is often written in the demographic literature as $\psi(r) - 1$.

Two corollaries may be stated.

Corollary 2A. Depending upon whether the net reproduction rate, $\int_{\alpha}^{\beta} m(a) p(a) da$, is greater than, less than, or equal to unity, the population will ultimately grow, decay, or approach a constant level.

Corollary 2AA. The algebraic sign of r_0, the asymptotic or intrinsic rate of natural self-propelled increase, is determined by the algebraic sign of $\int_\alpha^\beta m(a)p(a)\,da - 1 = \psi(0)$, which represents the rate of growth per (ambiguous) "generation".

For $|r_0|$ not too large, good approximations are given to r_0 by r_0', r_0'', r_0''':

$$r_0' = -\psi(0)/\psi'(0) = (R_0 - 1)/R_1 = (R_0 - 1)/\mu R_0,$$

$$R_j = \int_\alpha^\beta a^j \phi(a)\,da, \qquad \mu = R_1/R_0, \qquad \text{average age of becoming a mother,}$$

$$\sigma^2 = (R_2/R_0) - \mu^2 = \text{variance of net fertility,}$$

$$r_0'' = \frac{\mu - \sqrt{\mu^2 - 2\sigma^2 \ln R_0}}{\sigma^2},$$

$$r_0''' = r_0' - \psi(r_0')\psi'(r_0').$$

Corollary 2A is essentially due to Böckh (1886) who intuitively inferred it, and from whom Kuczynski (1928, 1932, 1935) derived his understanding.

Corollary 2AA was popularized by Dublin and Lotka (1925), but its essentials had already been stated in Lotka (1913). The first approximation given above can be derived from various series expansions of $\psi(r)$ and related functions, or as a Newton-Raphson approximation using an initial $r_0 = 0$; the final approximation given above involves a second Newton-Raphson whirl; the intermediate approximation has alternative derivations—from certain ratios of power expansions involving cumulants, or from fitting a Gaussian function to $m(a)p(a)$ (with its bizarre implication of some mothers who are not themselves conceived!). S. D. Wicksell, and others, have given alternative non-Gaussian graduations.

Lotka's formal proofs of 2A were made rigorous by Feller (1941); however, Lopez (1961) showed that the finiteness postulated for β guaranteed that the roots of $\psi(r) = 0$ were infinite in number and sufficed to provide an arbitrary $B(t)$ on the interval $(-\beta, 0)$ with its Fourier-like expansion, $\sum_{-\infty}^{\infty} c_j e^{r_j t}$.

A different theorem from 2A is provided by the following:

Theorem 2B: Suppose that, for *all* time, $-\infty < t < +\infty$,

(i) $p(a)$ applies,
(ii) $m(a)$ applies,
(iii) $N'(t)/N(t) \equiv$ a constant, $-\infty < t < \infty$.

Then, the asymptotic stable state of 2A holds already, and holds all the time

$$N(a,0)/N(0,0) \equiv p(a)e^{-r_0 a}$$

$$\underset{t}{\equiv} N(a,t)/N(0,t), \qquad -\infty < t < \infty,$$

$$B(t) \equiv B(0)e^{r_0 t}, \qquad N(t) \equiv \left[B(0)\int_0^n p(a)e^{-r_0 a}\,da\right]e^{r_0 t},$$

119

$$N(a,t) \equiv [B(0)p(a)e^{-r_0a}]e^{r_0t},$$

$$\psi(r_0) = \int_\alpha^\beta m(a)p(a)e^{-r_0a}da - 1 = 0.$$

Theorem 2B was stated very loosely in Lotka (1937a), and a purported proof of its more careful restatement was offered in Lotka (1937b).

A final theorem may be distinguished from 2A and 2B.

Theorem 2C: Suppose that, for a time period as long as the length of life, namely for $0 < t < \gamma \geq n$,

 (i) $p(a)$ applies,
 (ii) $m(a)$ applies,
(iii) $N'(t)/N(t) \equiv$ a constant.

Then

$$N(a,0) = B(0)p(a)e^{-r_0a}, \quad \psi(r_0) = 0,$$

$$N(a,t) \equiv [B(0)p(a)e^{-r_0a}da]e^{r_0t}, \quad t \geq 0,$$

$$N(a,t)/N(0,t) \equiv p(a)e^{-r_0a}, \quad t \geq 0,$$

$$B(t) \equiv B(0)e^{r_0t}, \quad N(t) \equiv \left[B(0)\int_0^n p(a)e^{-r_0a}da\right]e^{r_0t}, \quad t \geq 0.$$

Remark: If the time interval in the hypothesis of 2C, λ, were permitted to be less than n, the conclusion that a stable age distribution already holds from the beginning could be shown by numerous counter-examples to be definitely false, contrary to what some of the remarks in Lotka (1937a) seem to me to suggest and thereby undermining his contention.

A proof of 2C would show that $N'(t)/N(t)$ would undergo a transient oscillation if any complex-root harmonic e^{r_jt} were present: since this is ruled out by hypothesis, the only admissible initial $N(a,0)$ is *already* in the stable configuration $p(a)e^{-r_0a}B(0)$.

Arbitrating the Quarrels

We are now armed to judge the litigants. Lotka (1937a) begins by alleging that Kuczynski's writings "probably" caused a recent monograph, German Statistical Office (1935), to credit to Bortkiewicz (1911) the first formulation of the Sharpe-Lotka results of my Theorem 2A and my Equation (20). The jury must agree that what Bortkiewicz did accomplish, namely Theorem 1A, cannot possibly be identified with Theorem 2A[4]. So Lotka's complaint, if accurate, is a serious one.

In support of his charge against Kuczynski, Lotka (1937a, p. 104) quotes the following passage from Kuczynski (1935, p. 226):

Bortkiewicz had come to the conclusion that a population constantly subject to the same mortality and with a constant rate of increase must ultimately become stable, that is to say has a stable age composition, a stable birth rate and a stable death rate.

The reader will find this odd, since Kuczynski's quoted words can be construed as attributing to Bortkiewicz not Theorem 2A, but rather my Theorem 1C, which seems to be a perfectly valid theorem, albeit as far as I know not appearing explicitly in the pre-1976 literature.

Lotka goes on to remark that there is no $m(a)$ fertility function in Bortkiewicz, saying[5] in Lotka (1937a, p. 104, n. 2) that even "Kuczynski (1932, p. 43) admits this in his otherwise misleading history of the 'stable' population." Lotka goes on to argue that (1) Kuczynski has not correctly stated what Bortkiewicz (1911) did do—which Lotka correctly states was to *begin* with a stable age distribution and *to deduce from it* a balanced exponential growth rate proportional to my e^{ut} [not to be confused with my $e^{r_0 t}$ since u is *not* deduced from an $m(a)p(a)$ integral], and then to find a numerical estimate for u that made the $p(a)e^{-ua}$ theoretical distribution fit tolerably well the turn-of-the-century German data.

The jury must agree with Lotka's contention under (1), which is that Bortkiewicz (1911) proves my Theorem 1A. Kuczynski, like Homer, nodded. Lotka (1937a, p. 106) however goes a bit far in writing: "All this was old in the literature (which Kuczynski fails to point out) ...", having already been done in Lotka (1907a, 1907b) with data for England and Wales. First, we might deem 1907 not to be long before 1911. Second, Lotka's 1907 utterances of Theorem 1A was, as noted already, loose at best and no semblance of an adequate proof was provided. Third, Bortkiewicz's work is thorough to the point of being tedious, much of it going back to the 1890's, while Lotka's maiden publications are brief and suggestive notes.

(2) Lotka (1937a, pp. 106—7) goes on seemingly to argue that what Kuczynski wrongly attributes to Bortkiewicz is in any case not worth doing, being "inherently absurd" in the light of Lotka's claim that, exponential growth in total self-propelled population *already* implies realization of the stable age distribution—as in Theorem 2B here. It escapes me why the truth of 2B should make it absurd for Bortkiewicz (if only he had done it!) to have formulated Theorem 2A (or, Kuczynski aside, for Bortkiewicz to have formulated Theorem 1C). Indeed, taken literally, Lotka seems to be cutting his own throat and that of Sharpe! Obscurely, Lotka (1937a, p. 106) argues that if Bortkiewicz had done what he is credited with doing, there would have been no need "for Kuczynski to go through the agonizing labor of testing the approach ... to the stable distribution in 15 pages of closely printed figures" of Kuczynski (1932c, p. 65).

This Lotka point seems dubious. The 1911 Sharpe-Lotka proof, or the 1941 proof of Feller of Theorem 2A is admittedly better than Kuczynski's numerical exercise, which (a) starts from an $m(a)p(a)$ such that $\int_0^n m(a)p(a)\,da = 1$ and the population is in *stationary*, stable equilibrium, (b) suddenly lowers $m(a)$ by 10 per cent, and then (c) by laborious numerical projection depicts the 70 years of transient approach to a new exponential equilibrium with negative r_0. Still numerical exercises are frequently performed, and for a non-mathematician like Kuczynski this was a particularly valuable and insightful thing to do.

Lotka (1937a) concludes with his imperfect statement of the truth of Theorem 2B—perhaps not being clear in his own mind that, for the age distribution of a self-propelled system to necessarily *already* be in its stable form, the posited exponential growth for $N(t)$ would have had to have held forever, or at least as in my 2C for $n \approx 100$ years. Lotka (1937b), when it comes to provide a more careful statement of 2B and a proof, evidently assumes that his hypotheses hold *for all* time, $-\infty < t < \infty$; and even then I cannot follow the cogency of his proof.[6]

From my literal account of Lotka's 1937 papers, one can perhaps conclude that the plaintiff has not optimally pressed his case. But the jury cannot conclude from this that the defendant, Kuczynski, is without fault in Kuczynski (1935) or Kuczynski (1932). This requires special investigation, the results of which do not clear Kuczynski of fault in the matter.

Let me audit Kuczynski (1935), a much better-known work than Kuczynski (1932), one which purports to be a basic exposition. Around pp. 6—7, the author concentrates on the qualitative problem of intrinsic growth of population, or decay, properly pointing out that Böckh's net reproduction rate provides an appropriate answer. Fair enough. He goes on to document at length, p. 207, n. 2, how Böckh, Kuczynski, Hirschberg, Rahts had computed in the 1884—1912 period dozens of fertility tables for Germany, Sweden, Denmark, and France—so that Dublin and Lotka (1925) was something of a Johnny-come-lately. Fair enough as documentation of the Kuczynski (1930) reply to the Lotka (1929) review.

But it was quite misleading for Kuczynski to gloss over the difference between the rate of growth per annum, r_0, and the rate of growth per generation: Kuczynski (1935, p. 207) misleadingly says that, from his 1884 calculation of a Berlin fertility table, Böckh "... concluded that the real rate of increase of the Berlin population in 1879 was $\frac{2,172}{2,053} - 1 = 6$ per cent." The growth rate per year or decade is of course not 0.06, and Böckh could not give the correct number.

To conserve time and reduce tedium, I shall reproduce some further misleading passages from Kuczynski (1935, p. 224) with my bracketed editorial comments.

> "Will a population constantly subject to the same fertility and the same mortality ultimately become stable?
>
> The mathematical elements of this problem have for a long time [how long? and when before Sharpe and Lotka of 1911?] attracted the attention of both European [name one!] and American mathematicians.
>
> They [Lotka and who else?] have come to the conclusion that a population with a constant fertility and mortality will in fact ultimately become stable [yes, Lotka, Feller, Leslie, Lopez]. A comparatively easy approach to the computations necessary for ascertaining the age composition and the birth and death rates of the stable popula-

tions is to be found in the report represented by Bortkiewicz to the 1911 congress of the International Statistical Institute [literally correct as written, but Kuczynski (1932, p. 41, n. 1) had already noted that Lotka (1907b) had already done this; and, in any case, this literally true assertion is a *non sequitur* in its seeming implication that Theorem 1A or 1C is Theorem 2A].

There soon follows the p. 226 passage on Bortkiewicz that Lotka quoted in protest, and which I showed to be ambiguous. Kuczynski (1935, p. 226) goes on to say:

> ... But one of his [Bortkiewicz's] assumptions, the stable rate of increase, was not and could not be based on the actual conditions presented by some specific statistical example [being *not* based on any $m(a)$ data, as Lotka pointed out was devastating to Kuczynski's apparent link up of Bortkiewicz with Theorem 2A], his findings, interesting as they were from a theoretical standpoint [misleading in that "theoretical" versus "statistical" is being confused with the Theorem 1A versus Theorem 2A issue], did not attract the attention of demographers.
>
> The attentions of demographers [Kuczynski at least] was indeed only aroused when 14 years later the American mathematician, Lotka, who for a long time had studied the theoretical properties of the stable population [which one? the 1A case? or the 2A case?] published with Dublin ... Their approach is highly mathematical and we shall confine ourselves here to showing how through Lotka's formulae the stable yearly rate of increase (r) may be derived from the net reproduction rate [and the first and second moments of $m(a)$].

A cross that mathematical pioneers in a subject must always bear is to have their pearls dismissed as theoretical and vaguely impractical; later, after capitulation, they receive their revenge.

It is true that in the rarer item, Kuczynski (1932), the author deals at greater length with Lotka's contributions, quoting from him copiously and correcting his numerical errors (but still writing the same misleading sentence about Bortkiewicz!). At one point, Kuczynski (1932, Appendix, pp. 62—3) makes this explicit acknowledgment:

> It goes without saying that we would not have devoted so much space [more than 17 pages!] to the presentation of the trend of Dublin and Lotka's argument and to the translation of their mathematical operations into simple arithmetic if we were not convinced that some of the methods which they apply are of great scientific value and if we did not feel the strong desire that those methods be applied in the future also by such statisticians as are unfamiliar with higher mathematics. We wish even to state expressly that the computation of the exact rate of increase in the stable population, as presented in this study of Dublin and Lotka, in our opinion, marks the only great progress that has been made in the methodology of measuring

net reproduction since Boecke in 1886 puplished his first table of fertility. But just because we so emphatically recommend the application of those methods we feel obliged to show that the manner in which they themselves applied them to statistical data is inadequate.

This is merited if qualified praise, and all the greater the pity that a reader like myself can find it only in the Appendix of a work so rare that, in the end, I had to rely on a xerox from a copy in the Princeton Office of Population library, a copy which I judge from some marginal caligraphy, in comparison with some samples of Lotka's handwriting in my own possession, to have come from Lotka's own copy and which must have been among his books bequeathed to Princeton.

Before concluding, I ought to venture an opinion as to whether Kuczynski was being deliberately ungenerous to Lotka. At the conscious level, I think not. Kuczinski was always a plain spoken scholar, and such people never hesitate to point out the motes that they see in the eyes of others. In this regard he was not extreme: certainly, Bortkiewicz customarily meted out more trenchant criticisms than did Kuczynski, and R.A. Fisher's quill was dipped in stronger acid than Kuczynski's. Moreover, as I read and reread Kuczynski, both a third of a century ago when I developed a mild interest in the matter and recently in the preparation of this article, I sensed that he may really never have fully understood the magic of Lotka's Theorem 2A. He sensed that, if you knew r_0 you could derive the stable age distribution, and then from that you could compute from $p(a)$ alone the death rate; he may also have sensed the element of simultaneity involved, because only if you had happened to guess the right r_0 would your resulting death rate, when subtracted from the birth rate computable from $m(a)$, be consistent with the originally assumed r_0. But he shows signs of being unclear on the essential logic involved.[7] Nevertheless, I ought to point out explicitly that there is a way of reading his train of thought which makes his individual sentences about Bortkiewicz *literally true*, even though misleading in their context. Thus, Kuczynski, in the end seems to have considered the following procedure as optimal: First compute r_0 from the first few moments of the fertility table $m(a)p(a)$ (a result he ought to have clearly excluded Bortkiewicz from in favor of Lotka); second, use Bortkiewicz (or pre-Sharpe Lotka) to get the actuary's stable age distribution and the implied death rate (and, as a residual, the birth rate). This will be faster and digitally more accurate, Kuczynski decided, than computing out a Monte Carlo version of Lotka's Theorem 2A. But why did he not make this clear?

Perhaps at the unconscious level, Kuczynski was a bit grudging in his treatment of Lotka, writing passages, (1932, p. 65), like

> A good mathematician may be a poor statistician; a good statistician may be a poor mathematician. And since the author of this book, if anything, is a poor mathematician ...

A Gestalt psychologist trained in Vienna would expect the reader to complete the *chiasmus* by regarding Lotka as the poor statistician.

A sociologist of science like Merton would not be surprised to observe that controversy sours both contestants. Lotka (1925, p. 112, n. 7) praised Euler's early anticipations, writing, "An exceedingly interesting effort of early date to demonstrate the ultimate approach to geometric increase of the birth rate, independently of the initial conditions (e. g., starting with a single pair of parents) is to be found in L. Euler (1760)." This is generous praise—as I have argued even over-generous, if Lotka did not in 1925 know of Euler's private communication to Süssmilch (1761), and seemingly involving an error like that of Kuczynski's mistake in attributing to Bortkiewicz Theorem 2A rather than 1A—but understandable praise of a great scholar. However, once he has become alarmed for his own property rights, Lotka (1937a, p. 107) takes a shriller tone in defending his own originality, now writing: "... This must be abundantly clear to anyone who takes the trouble to examine the pertinent publications, from the first crude approach by Euler (based on highly unrealistic assumptions, and quite inapplicable to actual statistical data), ...".

As a final word, I ought to emphasize that the reason for now discussing this historical *contretemps* in detail has to do primarily with the need *to clarify the substance of the matter*. By no means was this controversy of unprecedented virulence: neither scholar ever stood in an extreme position with respect to temperament or emotion; both always conducted themselves with honor and dignity.

Moreover, in a sense Lotka has been the ultimate victor. It is he who is accorded full scholarly homage today. The danger is that, if anything, it is Kuczynski's commendable role in the development of the subject that will be lost. Thus, Lotka appears with a full page of references in the excellent Keyfitz (1968, pp. 424—5) bibliography; the works of Kuczynski escape notice. The one notice taken of Kuczynski in J. H. Pollard (1973, p. 82) is only in connection with his 1932 computation of a male NRR for France after World War I that exceeds unity whereas the female NRR falls short of it, a dramatic consequence of war casualties that, ironically, Lotka had adverted to in his original cited review of Kuczynski. Richard Böckh receives no citations in either text. Again, a Merton would understand how the brighter light drowns out the earlier light. By a process somewhat the opposite of Gresham's law the more polished mathematics seems to drive out the less certain experimenting with numerical data.

* I owe thanks to the National Institutes of Health for financial aid on demographic research, NIH Grant #1-R01 HD-09081-01, and to Vicki Elms for help in preparing the manuscript. Wilma Winters, librarian at the Harvard Center for Population Studies, provided me with appreciated help in locating rare items. And I owe gratitude to Professor Ansley Coale for providing me with a Princeton library English translation of Bortkiewicz (1911), and with xerox pages from Kuczynski (1932), which might be Lotka's own copy bequeathed with his other demographic collection to the Princeton Office of Population Research. More than a third of a century ago, I benefitted from some correspondence with Dr. A. J. Lotka on aspects of these questions; however, from so cursory a dialogue, I was not able to sort out then the present formulation of the issues. Professor Nathan Keyfitz's invitation to include this material in the Smith and Keyfitz collection of historical items provided the final stimulus for

the present effort, and I owe thanks to David P. Smith for translation of the Süssmilch (1761) account of Euler's important private communication. All interpretations must, of course, be on my own responsibility.

1. See Spengler (1968) on Lotka in *The International Encyclopedia of the Social Sciences;* also, the obituary notices of Dublin (1950) and Notestein (1950).

2. Edwin Cannan, a no-new-fangled-nonsense economist if there ever was one, showed by arithmetic projections of the absolute number of English births that ultimate U.K. population decline was likely. See Cannan (1895).

3. Cf. Kuczynski (1935, p. 207, n. 2) and his citation of Böckh (1886).

4. On the attainment of stability Bortkiewicz says only (1911, pp. 63, 69—70; the translator is unnamed):

> Three qualifying hypotheses underlie the following statements. It is assumed: 1) that in the population an unchanging order of deaths prevails; 2) that the current age distribution likewise is invariable, and 3) that no immigration or emigration takes place.
>
> Accordingly the "stationary" and the "stable" population appear in a sense, as the terms will be used here, almost as ideal types, to which reality never exactly corresponds, but to which it comes all the closer, the less significant the actual changes of the death order and the age distribution, and, relatively speaking, the smaller the immigration and emigration.
>
> The above hypotheses can be considered as the three characteristics which are held in common with the two concepts of the stationary and stable population. In addition, there comes a fourth characteristic: with the stationary population there is a continual constancy, whereas with the stable population a continual accretion in the total number of persons living.
>
> ... [T]he geometric progression as a standard norm for the growth of the population was established by L. Euler. He proceeded from one human couple, and let it as well as its offspring propagate from generation to generation according to certain invariable conditions. These conditions referred to age of marriage, the number of children begotten by each couple, and the number dying. Hence it resulted that the number of living at the end of every calendar year forms a line which, in its further course, approaches more and more a geometric progression.

The key second paragraph reads in the original German: „Demnach erscheinen die „stationäre" und die „progressive" Bevölkerung in dem Sinne, wie diese Termini hier gebraucht werden, gleichsam als Idealtypen, denen die Wirklichkeit niemals genau entspricht, denen sie aber um so näher kommt, je unerheblicher die tatsächlich vor sich gehenden Wandlungen der Absterbeordnung und der

Altersverteilung und je geringer, relativ genommen, die Zahlen der Ein- und Auswandernden sind."

The statement suggests that real populations may *resemble* the stable ideal type, and not that they *approach* stability. Kuczynski may have given it the second meaning, taking the comment that the geometric is a standard norm as supportive evidence. (The illustration by Euler is from a personal communication to J. P. Süssmilch (1761, Vol. 1, pp. 291—299). Euler's example treats 1 female and 1 male as a "couple", and in effect postulates $B(t) = B(t-22) + B(t-24) + B(t-26)$, $D(t) = B(t-40)$, a Bernardelli-Leslie case that is "cyclic" because of an unfortunate choice of even numbers only: along with dominant $(\lambda^*)^t$ terms go also dominant $(-\lambda^*)^t$ terms and no strict approach to a stable age distribution at both odd and even ages.)

5. I follow Lotka (1937a) in giving the date 1932 to Kuczynski's *Fertility and Reproduction* of 1931; likewise Kuczynski (1928) gives 1886 for Böckh (1884), and I follow him in this (actually, the computation might date from 1886).

6. I am unable to discern the cogency of Lotka (1937b)'s purported proof that if $N(t)$ is observed to be in exponential growth (over *some* consecutive time periods? over all time-periods?—which is another thing), $N(a,t)$ is *already* in the stable configuration. The following discrete-time example meets the only hypotheses he purports to use in his demonstration—yet it is not "already in the stable distribution", and *after* $(t=0,1,2)$ its $N(t)$ ceases to grow like $N(0)(2)^t$! The example is based on $[p(0), p(1), p(2), p(3)] = [1, 2^{-1}, 2^{-2}, 2^{-3}]$, $B(0) = 1,032$, $[m(0), m(1), m(2), m(3)] = [2,0,0,0]$, with initial (and asymptotic!) increase of $N(t)$ like $(1+1)^t$, but lacking such exact exponential growth in the near future.

Period, t	Births, $N(0,t)$	$N(1,t)$	$N(2,t)$	$N(t,3)$	Total $N(t)$
0	1,032	88	192	48	1,360
1	2,064	516	44	96	$2 \times 1,360$
2	4,128	1,032	258	22	$2^2 \times 1,360$
3	8,256	2,064	516	129	$85 + (2^3 \times 1,360)$

The example was fabricated by perturbing an exact exponential solution to the self-propelled Sharpe-Lotka system through adding to a stable initial state, $[N(j,t)] = [1,024, 256, 64, 32]$, a non-exponential solution, $[8, -168, 128, 16]$, of higher harmonics *fitted* to the initial conditions that total $N(t)$ of the add-on *initially* vanish as for $t = (0,1,2)$. Only if Lotka postulates initial exponential growth of $N(t)$ over a long enough initial interval—the whole length of life, $n > \beta$—will it become impossible for me to find such a perturbing add-on. No hint of this appears in his purported proof. The demonstration goes off the tracks because he seems to confuse a functional-equation requirement of the type

$$f(a,t)/g(a,t) \text{ independent of } t \text{ for all } a$$

with his actual type

$$[f_1(t) + f_2(t)]/[g_1(t) + g_2(t)] \text{ independent of } t.$$

127

I must make clear that these last interpretations of mine have to be regarded as only tentative.

7. This surmise is corroborated by Kuczynski (1931, pp. 20, 32, 166): writing apparently simultaneously with his *Fertility and Reproduction*, the author eschews relations like those in (22) and Corollary 2AA, even though he already knows them, in favor of stable-state relations like

$$r \equiv N'(t)/N(t) \equiv B(t)/N(t) - D(t)/N(t),$$

$$r = \frac{\int\limits_{\alpha}^{\beta} m(a)p(a)e^{-ra}\,da}{\int\limits_{0}^{n} p(a)e^{-ra}\,da} - \frac{\int\limits_{0}^{n} -p'(a)e^{-ra}\,da}{\int\limits_{0}^{n} p(a)e^{-ra}\,da}$$

$$= \theta(r) = \frac{\psi(r)+1}{\bar{p}(r)} + \frac{r\,\bar{p}(r)-1}{\bar{p}(r)}$$

$$= \frac{\psi(r)}{\bar{p}(r)} + r$$

where real r is put in the indicated Laplace Transforms and where we solve for the unique r_0 root. Keyfitz (1968, p. 176) calculates r_0 iteratively from

$$r_0'' = \theta(r_0'), \qquad r_0''' = \theta(r_0''), \ldots .$$

Even more rapid convergence would occur for the Newton-Raphson variant

$$r_0'' = r_0' - \theta(r_0')/\theta'(r_0').$$

References

Bernardelli, H.: Population Waves. Journal of the Burma Research Society **31**, Part I, 1—18 (1941).

Böckh, R.: Statistik des Jahres 1884. Statistisches Jahrbuch der Stadt Berlin, Volume 12 (Berlin) (1886).

Bortkiewicz, L. V.: Die Sterbeziffer und der Frauenüberschuß in der stationären und in der progressiven Bevölkerung. Bulletin de l'Institut International de Statistique **19**, 63—183 (1911).

Cannan, E.: The probability of a Cessation of the Growth of Population in England and Wales during the Next Century. Economic Journal **5**, 505—15 (1895).

Coale, A.J.: The Growth and Structure of Human Populations. A Mathematical Investigation. Princeton, N.J.: Princeton University Press 1972.

Dublin, L.I.: Alfred James Lotka, 1880—1949. J. Amer. Statistical Assoc. **45**, 138—9 (1950).

Dublin, L.I., Lotka, A.J.: On the True Rate of Natural Increase of a Population. J. Amer. Statistical Assoc. **20**, 305—39 (1925).

Euler, L.: Recherches générales sur la mortalité et la multiplication. Mémoires de l'Académie Royale des Sciences et Belles Lettres **16**, 144—64 (1760).

Feller, W.: On the Integral Equation of Renewal Theory. The Annals of Mathematical Statistics **12**, 243—67 (1941).

German Statistical Office: Neue Beiträge zum deutschen Bevölkerungsproblem (Berlin) (1935).

Keyfitz, N.: Introduction to the Mathematics of Population. Reading, Mass.: Addison-Wesley Publishing Co. 1968.

Kuczynski, R.R.: The Balance of Births and Deaths, Volume I, Western and Northern Europe. New York: Macmillan Co. for The Brookings Institution 1928.

The Balance of Births and Deaths, Volume II, Eastern and Southern Europe. Washington, D.C.: The Brookings Institution 1931.

Kuczynski, R.R.: A Reply to Dr. Lotka's Review of 'The Balance of Births and Deaths.' J. Amer. Statistical Assoc. **25**, 84—5 (1930).

Kuczynski, R.R.: Fertility and Reproduction. New York: Falcon Press 1931/32.

Kuczynski, R.R.: The Measurement of Population Growth. London: Sedgewick & Jackson, Ltd. 1935; reproduced by Gordon and Breach. New York: Science Publishers 1969.

Leslie, P.H.: On the Use of Matrices in Certain Population Mathematics. Biometrika **33**, 183—212 (1945).

Lopez, A.: Problems in Stable Population Theory. Princeton, N.J.: Office of Population Research 1961.

Lotka, A.J.: Relation between Birth Rates and Death Rates. Science, N.S. **26**, 21—22 (1907a).

Lotka, A.J.: Studies on the Mode of Growth of Material Aggregates. Amer. J. Science **24**, 199—216 (1970b).

Lotka, A.J.: A Natural Population Norm. J. Washington Acad. Sciences **3**, 241—48, 289—93 (1913).

Lotka, A.J.: The Stability of the Normal Age Distribution, Proc. Nat'l Acad. Sciences **7**, 339—45 (1922).

Lotka, A.J.: Elements of Physical Biology. Baltimore: Williams & Wilkins 1925; reproduced in posthumous edition with bibliography as: Elements of Mathematical Biology. New York: Dover Publications, Inc. 1956.

Lotka, A.J.: The Progeny of a Population Element. Amer. J. Hygiene **8**, 875—901 (1928).

Lotka, A.J.: Review of 'The Balance of Births and Deaths, Vol. I, Western and Northern Europe' by R. Kuczynski. New York: Macmillan, 1928; J. Amer. Statistical Assoc. **24**, 332—3 (1929).

Lotka, A.J.: Rejoinder to 'A Reply to Dr. Lotka's Review of The Balance of Births and Deaths by R. Kuczynski.' J. Amer. Statistical Assoc. **25**, 85—6 (1930).

Lotka, A.J.: The Structure of a Growing Population. Human Biology **3**, 459—93 (1931).

Lotka, A.J.: Notes: A Historical Error Corrected. Human Biology **9**, 104—7 (1937a).

Lotka, A.J.: Population Analysis: A Theorem Regarding the Stable Age Distribution. J. Washington Acad. Sciences **27**, 299—303 (1937b).

Lotka, A.J.: A Contribution to the Theory of Self-Renewing Aggregates, with Special Reference to Industrial Replacement. Annals of Mathematical Statistics **10**, 1—25 (1939a).

Merton, R.K.: The Sociology of Science. Theoretical and Empirical Investigations. Chicago: The University of Chicago Press 1973.

Notestein, F.W.: Alfred James Lotka: 1880—1949. Population Index **16**, 22—9 (1950), containing a Lotka bibliography of 114 items.

Pollard, J.H.: Mathematical Models for the Growth of Human Populations. Cambridge: Cambridge University Press 1973.

Samuelson, P.A.: Time Symmetry and Asymmetry in Population and Deterministic Dynamic Systems. J. Theoretical Population Biology **9**, 82—122 (1976).

Sharpe, F.R., Lotka, A.J.: A Problem in Age-Distribution. Philosophical Magazine **21**, 435—8 (1911).

Spengler, J.J.: Alfred J. Lotka. International Encyclopedia of the Social Sciences **9**. New York: Crowell Collier Macmillan 475—6 (1968).

Süssmilch, J.P.: Die göttliche Ordnung, Volume 1, Second Edition. Berlin: 1761.

16. On the Integral Equation of Renewal Theory

WILLIAM FELLER (1941)

From *Annals of Mathematical Statistics* 12. Excerpts are from pages 243—261, 263—266.

Feller's paper is a rigorous treatment of renewal theory, and to assist the reader his principal results are summarized below in demographic form and notation. Equation (1.1) is the renewal equation

(1.1 a)
$$B(t) = G(t) + \int_0^t B(t-a)p(a)m(a)\,da ,$$

(1.1 b)
$$= G(t) + \int_0^t B(t-a)\phi(a)\,da$$

where $B(t)$ is births at time t and is composed of births $G(t)$ to the parent population alive at time 0 plus, inside the integral, births to those born since; $B(t-a)$ being the number born a years ago, $p(a)$ their probability of surviving to age a, and $m(a)$ the probability of their giving birth to offspring in the interval a to $a+da$. In equation (1.1 b), $\phi(a) = p(a)m(a)$. Stieltjes integrals, which translate for discrete distributions as summations, are allowed by a comment following equation (1.3). The equation generalizes from births at time t to births in the interval 0 to t.

For Feller's example ii, equation (2.1) does not apply. We have instead

(2.1 a)
$$G(t) = \begin{cases} \int_0^{\beta-t} k(a)\dfrac{p(t+a)}{p(a)}m(t+a)\,da, & t < \beta \\[2mm] 0 & t \geq \beta \end{cases}$$

where $k(a)$ is the parent population at ages a to $a+da$ at time 0, and other terms are as above. For $t \geq \beta$ the parent population is no longer in the reproductive ages and equation (1.1) takes the homogeneous form

(1.1 c)
$$B(t) = \int_\alpha^\beta B(t-a)p(a)m(a)\,da ,$$

where α is the youngest age at which $m(a) \neq 0$. Equation (2.3) is the net reproduction rate

(2.3 a)
$$\int_0^\infty f(t)\,dt = \int_\alpha^\beta p(a)m(a)\,da = R_0 .$$

131

For Theorems 1 and 2 we have:

$$(3.1\,\text{a}) \qquad F(t) = \int_0^t p(a)m(a)\,da,$$

which for $t \geq \beta$ becomes R_0 as above. Feller's $G(t)$ will be the *integral* of equation (2.1 a), hence it is total births to the initial population over the interval 0 to t. Both are finite and are cumulative functions, hence non-decreasing. The Laplace Transforms (3.2) give their characteristic functions:

$$(3.2\,\text{a}) \qquad \varphi(s) = \phi^*(r) \;\; [\text{or } \psi(r)] = \int_\alpha^\beta e^{-rt} p(t)m(t)\,dt,$$

$$(3.2\,\text{b}) \qquad \gamma(s) = G^*(r) \;\; [\text{or } V(r)] = \int_0^\beta \int_0^{\beta-t} e^{-rt} k(x) \frac{p(t+x)}{p(x)} m(t+x)\,dx\,dt.$$

The first of these is the characteristic equation for r, the intrinsic growth rate; the second is the total reproductive value introduced by R.A. Fisher (1930, paper 18 below). Both integrals are convergent for $t \geq \beta$ since $m(t)$ then becomes 0. Hence, by Theorems 1 and 2 the renewal equation has a unique solution.

Theorem 3 establishes stable properties of mean births $\bar{B}(t)$. We make the substitutions:

(4.2 a)	$a = R_0$,	the net reproduction rate.
(4.2 b)	$b = B$,	total births to the initial population.
(4.4 a)	$m = \mu$,	the mean age at childbearing in the stationary population.
(4.7 a)	$m_1 = -\psi'(r) = A_r$,	the mean age at childbearing in the stable population.

All of these terms are finite, and therefore:

(i) For $R_0 = 1$ the population is stationary and mean births approach the limiting value B/μ;

(ii) For $R_0 < 1$ the *total* number of births accruing to the population after time 0 will be $B/(1 - R_0)$;

(iii) For $R_0 > 1$ mean births when deflated by their intrinsic increase over time approach the limiting value B/A_r.

Theorem 4 establishes conditions under which births $B(t)$ as well as mean births $\bar{B}(t)$ stabilize. The theorem states that for $R_0 = 1$ and B finite as before, $B(t)$ stabilizes if: by equation (5.3) the fertile age distribution has a finite mean and variance; and by equation (5.4) births to the initial population eventually taper out, as by the population dying or ageing out of the reproductive years.

The *Remark* extends Theorem 4 to all R_0, establishing that under the conditions given by the theorem births at time t stabilize toward the limiting value $B(t) = V e^{rt}/A_r$, where

$$m_1 = m'_1 = A_r \quad \text{as above (4.7a)},$$

$$b' = V = \text{total reproductive value} \quad \text{as above (3.2b)}.$$

The result has an intuitive meaning: V/A_r is the number of births at $t = 0$ that would give rise over time to the same population as the parent generation's observed birth distribution. It is found as the backward projection of their births to time 0 at their intrinsic growth rate, adjusted for the delay to childbearing.

[In the *Remark*, note that the characteristic function $\psi(r)$ converges in the interval $0 > r > -\infty$ and that for the case $R_0 < 1$ the dominant root r_1 is also negative; hence the Lemma of Section 4 is applicable. For the case $R_0 = 1$, $V/A_r = B/\mu$. In the discrete case, which Feller does not complete, Theorem 4 holds, to an approximation for finite population size, if the fertility distribution $\phi(a)$ can be expressed as an irreducible and primitive projection matrix. These requirements are discussed in Parlett (1970, paper 29 below).]

Lotka's solution to the renewal equation enters in Theorem 6. He first substituted $B_0 e^{rt}$ for $B(t)$ and $B_0 e^{r(t-a)}$ for $B(t-a)$ in the homogeneous form of the equation, (1.1c), by which it reduces to

$$\int_\alpha^\beta e^{-ra} p(a) m(a) da = 1 .$$

Given an infinite number of simple roots $r_1, r_2, \ldots$ he was able to assert that there existed constants Q such that

(6.2a) $$B(t) = \int_\alpha^\beta B(t-a) p(a) m(a) da = \sum_i Q_i e^{r_i t} ;$$

and to identify the Q_i terms as $Q_i = V(r_i)/-\psi'(r_i)$, where these are the total reproductive value and mean age at childbearing associated with the i'th root of equations (3.2b) and (4.7a) (Lotka 1939, 64—67, 85—87).

Feller shows that the use of the Laplace Transform is both simpler than Lotka's solution and more general since it also applies where the characteristic equation has only a finite number of roots. By taking the transforms of the three terms of equation (1.1b)

$$B(t) = G(t) + \int_0^t B(t-a) \phi(a) da ,$$

and noting that the last term is a convolution whose transform is the product of the transforms of the two functions, Feller comes down to

$$B^*(r) = G^*(r) + B^*(r) \phi^*(r) .$$

The solution becomes simply

$$(6.6\,a) \qquad B^*(r) = \frac{G^*(r)}{1 - \phi^*(r)} = \frac{V(r)}{1 - \psi(r)},$$

where on the right we use the notation of equations (3.2 a) and (3.2 b).
   By Theorem 6, when equation (6.6 a) admits the expansion

$$(6.6\,b) \qquad B^*(r) = \sum_i \frac{V(r_i)}{-\psi'(r_i)} \left[\frac{1}{r - r_i} \right],$$

and $\sum |V(r_i)/ - \psi'(r_i)|$ is convergent, it has the inverse

$$(6.2\,b) \qquad B(t) = \sum_i \frac{V(r_i)}{-\psi'(r_i)} e^{r_i t} = \sum_i Q_i e^{r_i t}$$

given by Lotka for the case of infinite simple roots. In demographic work the conditions of the Theorem are always met.
   Equation (6.10) extends the analysis to multiple roots, which introduce non-geometric components to growth and are not found in demographic work. Section 7 provides further comments on discrete solutions.

1. Introduction. In this paper we consider the behavior of the solutions of the integral equation

$$(1.1) \qquad u(t) = g(t) + \int_0^t u(t - x)f(x)\,dx,$$

where $f(t)$ and $g(t)$ are given non-negative functions.[1] This equation appears, under different forms, in population theory, the theory of industrial replacement and in the general theory of self-renewing aggregates, and a great number of papers have been written on the subject.[2] Unfortunately most of this literature is of a heuristic nature so that the precise conditions for the validity of different methods or statements are seldom known. This literature is, moreover, abundant in controversies and different conjectures which are sometimes supported or disproved by unnecessarily complicated examples. All this renders an orientation exceedingly difficult, and it may therefore be of interest to give a rigorous presentation of the theory. It will be seen that some of the previously announced results need modifications to become correct.

The existence of a solution $u(t)$ of (1.1) could be deduced directly from a well-known result of Paley and Wiener [21] on general integral equations of form (1.1).[3] However, the case of non-negative functions $f(t)$ and $g(t)$, with which we are here concerned, is much too simple to justify the deep methods used by Paley and Wiener in the general case. Under the present conditions, the existence of a solution can be proved in a simple way using properties of completely monotone functions, and this method has also the distinct advantage of showing some properties of the solutions, which otherwise would have to be proved separately. It will be seen in section 3 that the existence proof becomes most natural if equation (1.1) is slightly generalized. Introducing the summatory functions

$$(1.2) \qquad U(t) = \int_0^t u(x)\,dx, \qquad F(t) = \int_0^t f(x)\,dx, \qquad G(t) = \int_0^t g(x)\,dx,$$

equation (1.1) can be rewritten in the form

$$(1.3) \qquad U(t) = G(t) + \int_0^t U(t - x)\,dF(x).$$

However, (1.3) has a meaning even if $F(t)$ and $G(t)$ are not integrals, provided $F(t)$ is of bounded total variation and the integral is interpreted as a Stieltjes integral. Now for many practical applications (and even for numerical calculations) this generalized form of the integral equation seems to be the most

---

[1] For the interpretation of the equation cf. section 2.

[2] Lotka's paper [8] contains a bibliography of 74 papers on our subject published before 1939. Yet it is stated that even this list "is not the result of an exhaustive search." At the end of the present paper the reader will find a list of 16 papers on (1.1) which have appeared during the two years since the publication of Lotka's paper.

[3] This has been remarked also by Hadwiger [3].

appropriate one and, as a matter of fact, it has sometimes been used in a more or less hidden form (e.g., if all individuals of the parent population are of the same age). Our existence theorem refers to this generalized equation.

We then turn to one of the main problems of the theory, namely the asymptotic behavior of $u(t)$ as $t \to \infty$. It is generally supposed that the solution $u(t)$ "in general" either behaves like an exponential function, or that it approaches in an oscillating manner a finite limit q; the latter case should arise if $\int_0^\infty f(t) \, dt = 1$, thus in particular in the cases of a stable population and of industrial replacement. However, special examples have been constructed to show that this is not always so.[4] In order to simplify the problem and to get more general conditions, we shall first (section 4) consider only the question of convergence in mean, that is to say, we shall study the asymptotic behavior not of $u(t)$ itself but of the mean value $u^*(t) = \dfrac{1}{t} \int_0^t u(x) \, dx$. The question can be solved completely using only the simplest Tauberian theorems for Laplace integrals. Of course, if $u(t) \to q$ then also $u^*(t) \to q$, but not conversely. The investigation of the precise asymptotic behavior of $u(t)$ is more delicate and requires more refined tools (section 5).

Most of section 6 is devoted to a study of Lotka's well-known method of expanding $u(t)$ into a series of oscillatory components, and it is hoped that this study will help clarify the true nature of this expansion. It will be seen that Lotka's method can be justified (with some necessary modifications) even in some cases for which it was not intended, e.g., if the characteristic equation has multiple or negative real roots, or if it has only a finite number of roots. On the other hand limitations of the method will also become apparent: thus it can occur in special cases that a formal application of the method will lead to a function $u(t)$ which apparently solves the given equation, whereas in reality it is the solution of quite a different equation.

Of course, most of the difficulties mentioned above arise only when the function $f(t)$ has an infinite tail. However, it is known that even computational considerations sometimes require the use of such curves, and, as matter of fact, exponential and Pearsonian curves have been used most frequently in connection with (1.1). It will be seen that even in these special cases customary methods may lead to incorrect results. Besides, our considerations show how much the solution $u(t)$ is influenced by the values of $f(t)$ for $t \to \infty$, and, accordingly, that extreme caution is needed in practice. The last section contains some simple remarks on the practical computation of the solution.

---

[4] Cf. Hadwiger [2] and also Hadwiger, "Zur Berechnung der Erneuerungsfunktion nach einer Formel von V. A. Kostitzin," *Mitt. Verein. schweizerischer Versich.-Math.*, Vol. 34 (1937), pp. 37–43.

2. Generalities on equations (1.1) and (1.3). This section contains a few remarks on the meaning of our integral equation and on an alternative form under which it is encountered in the literature. A reader interested only in the abstract theory may pass immediately to section 3.

Equation (1.1) can be interpreted in various ways; the most important among them are the following two:

(i) In the theory of industrial replacement (as outlined in particular by Lotka), it is assumed that each individual dropping out is immediately replaced by a new member of zero age. $f(t)$ denotes the density of the probability at the moment of installment that an individual will drop out at age t. The function $g(t)$ is defined by

$$(2.1) \qquad g(t) = \int_0^t \eta(x)f(t-x)\,dx,$$

where $\eta(x)$ represents the age distribution of the population at the moment $t = 0$ (so that the number of individuals of an age between x and $x + \delta x$ is $\eta(x)\delta x + o(\delta x)$). Obviously $g(t)$ then represents the rate of dropping out at time t of individuals belonging to the parent population. Finally, $u(t)$ denotes the rate of dropping out at time t of individuals of the total population. Now each individual dropping out at time t belongs either to the parent population, or it came to the population by the process of replacement at some moment $t - x$ $(0 < x < t)$, and hence $u(t)$ satisfies (1.1). It is worthwhile to note that in this case

$$(2.2) \qquad \int_0^\infty f(t)\,dt = 1,$$

since $f(t)$ represents a density of probability.

(ii) In population theory $u(t)$ measures the rate of female births at time $t > 0$. The function $f(t)$ now represents the reproduction rate of females at age t (that is to say, the average number of female descendants born during $(t, t + \delta t)$ from a female of age t is $f(t)\delta t + o(\delta t)$). If $\eta(x)$ again stands for the age distribution of the parent population at $t = 0$, the function $g(t)$ of (2.1) will obviously measure the rate of production of females at time t by members of the parent population. Thus we are again led to (1.1), with the difference, however, that this time either of the inequalities

$$(2.3) \qquad \int_0^\infty f(t)\,dt \gtreqless 1$$

may occur; the value of this integral shows the tendency of increase or decrease in the total population.

Theoretically speaking, $f(t)$ and $g(t)$ are two arbitrary non-negative functions. It is true that $g(t)$ is connected with $f(t)$ by (2.1); but, since the age distribution $\eta(x)$ is arbitrary, $g(t)$ can also be considered as an arbitrarily prescribed function.

It is hardly necessary to interpret the more general equation (1.3) in detail: it is the straightforward generalization of (1.1) to the case where the increase or decrease of the population is not necessarily a continuous process. This form

137

of the equation is frequently better adapted to practical needs. Indeed, the functions $f(t)$ and $g(t)$ are usually determined from observations, so that only their mean values over some time units (years) are known. In such cases it is sometimes simpler to treat $f(t)$ and $g(t)$ as discontinuous functions, using equation (1.3) instead of (1.1). For some advantages of such a procedure see section 7. It may also be mentioned that the most frequently (if not the only) special case of (1.1) studied is that where $g(t) = f(t)$. Now it is apparent from (2.1) that this means that all members of the parent population are of zero age: in this case, however, there is no continuous age-distribution $\eta(x)$. Instead we have to use a discontinuous function $\eta(x)$ and write (2.1) in the form of a Stieltjes integral. Thus discontinuous functions and Stieltjes integrals present themselves automatically, though in a somewhat disguised form, even in the simplest cases.

At this point a remark may be inserted which will prove useful for a better understanding later on (section 6). In the current literature we are frequently confronted not with (1.1) but with

$$(2.4) \qquad u(t) = \int_0^\infty u(t - x)f(x)\, dx,$$

together with the explanation that it is asked to find a solution of (2.4) which reduces, for $t < 0$, to a prescribed function $h(t)$. Now such a function, as is known, exists only under very exceptional conditions, and (2.4) is by no means equivalent to (1.1). The current argument can be boiled down to the following. Suppose first that the function $g(t)$ of (1.1) is given in the special form

$$(2.5) \qquad g(t) = \int_t^\infty h(t - x)f(x)\, dx,$$

where $h(x)$ is a non-negative function defined for $x < 0$. Since the solution $u(t)$ of (1.1) has a meaning only for $t > 0$, we are free to *define* that $u(-t) = h(-t)$ for $t > 0$. This arbitrary definition, then, formally reduces (1.1) to (2.4). It should be noted, however, that this function $u(t)$ does not, in general, satisfy (2.4) for $t < 0$, for $h(t)$ was prescribed arbitrarily. Thus we are not, after all, concerned with (2.4) but with (1.1), which form of the equation is, by the way, the more general one for our purposes. If there really existed a solution of (2.4) which reduced to $h(t)$ for $t < 0$, we could of course define $g(t)$ by (2.5) and transform (2.4) into (1.1) by splitting the interval $(0, \infty)$ into the subintervals $(0, t)$ and (t, ∞). However, as was already mentioned, a solution of the required kind does not exist in general. It will also be seen (section 6) that the true nature of the different methods and the limits of their applicability can be understood only when the considerations are based on the proper equation (1.1) and not on (2.4).

3. Existence of solutions.

THEOREM 1. *Let* $F(t)$ *and* $G(t)$ *be two finite non-decreasing functions which are continuous to the right*[5]. *Suppose that*

$$(3.1) \qquad\qquad F(0) = G(0) = 0,$$

and that the Laplace integrals[6]

$$(3.2) \qquad \varphi(s) = \int_0^\infty e^{-st} \, dF(t), \qquad \gamma(s) = \int_0^\infty e^{-st} \, dG(t)$$

converge at least for $s > \sigma \geq 0$[7]. *In case that* $\lim\limits_{s \to \sigma+0} \varphi(s) > 1$, *let* $\sigma' > \sigma$ *be the root*[8] *of the characteristic equation* $\varphi(s) = 1$; *in case* $\lim\limits_{s \to \sigma+0} \varphi(s) \leq 1$, *put* $\sigma' = \sigma$.

Under these conditions there exists for $t > 0$ *one and only one finite non-decreasing function* $U(t)$ *satisfying* (1.3). *With this function the Laplace integral*

$$(3.3) \qquad\qquad \omega(s) = \int_0^\infty e^{-st} \, dU(t)$$

converges for $s > \sigma'$, *and*

$$(3.4) \qquad\qquad \omega(s) = \frac{\gamma(s)}{1 - \varphi(s)}.$$

PROOF: A trivial computation shows that for any finite non-decreasing solution $U(t)$ of (1.3) and any $T > 0$ we have

$$\int_0^T e^{-st} \, dU(t) = \int_0^T e^{-st} \, dG(t) + \int_0^T e^{-sx} \, dF(x) \int_0^{T-x} e^{-st} \, dU(t);$$

---

[5] It is needless to emphasize that this restriction is imposed only to avoid trivial ambiguities.

[6] The integrals (3.2) should be interpreted as Lebesgue-Stieltjes integrals over *open* intervals; thus

$$\varphi(s) = \lim_{\epsilon \to +0} \int_\epsilon^\infty e^{-st} \, dF(t),$$

which implies that $\varphi(s) \to 0$ as $s \to \infty$. Alternatively it can be supposed that $F(t)$ and $G(t)$ have no discontinuities at $t = 0$. Continuity of $F(t)$ at $t = 0$ means that there is no reproduction at zero age. This assumption is most natural for our problem, but is by no means necessary. In order to investigate the case where $F(t)$ has a saltus $c > 0$ at $t = 0$, one should take the integrals (3.2) over the closed set $[0, \infty]$, so that

$$\varphi(s) = c + \lim_{\epsilon \to +0} \int_\epsilon^\infty e^{-st} \, dF(t).$$

It is readily seen that Theorem 1 and its proof remain valid if $0 < c < 1$. However, if $c > 1$, then (1.3) plainly has no solution $U(t)$. The continuity of $G(t)$ at $t = 0$ is of no importance and is not used in the sequel.

[7] The condition is formulated in this general way in view of later applications (cf., e.g., the lemma of section 4). In all cases of practical interest $\sigma = 0$.

[8] $\varphi(s)$ is, of course, monotonic for $s > \sigma$ and tends to zero as $s \to \infty$. In order to ensure the existence of a root of $\varphi(s) = 1$, it is sufficient to suppose that the saltus c of $F(t)$ at $t = 0$ is less than 1 (cf. footnote 6).

herein all terms are non-negative and hence by (3.2)

$$\int_0^T e^{-st}\,dU(t) \le \gamma(s) + \varphi(s) \int_0^T e^{-st}\,dU(t).$$

Now $\varphi(s) < 1$ for $s > \sigma'$, and hence it is seen that the integral (3.3) exists for $s > \sigma'$ and satisfies (3.4). On the other hand it is well-known that the values of $\omega(s)$ for $s > \sigma'$ determine the corresponding function $U(t)$ uniquely, except for an additive constant, at all points of continuity. However, from (1.3) and (3.1) it follows that $U(0) = 0$ and, since by (1.3) $U(t)$ is continuous to the right, the monotone solution $U(t)$ of (1.3), if it exists, is determined uniquely.

To prove the existence of $U(t)$ consider a function $\omega(s)$ defined for $s > \sigma'$ by (3.4). It is clear from (3.2) that $\varphi(s)$ and $\gamma(s)$ are completely monotone functions, that is to say that $\varphi(s)$ and $\gamma(s)$ have, for $s > \sigma$, derivatives of all orders and that $(-1)^n \varphi^{(n)}(s) \ge 0$ and $(-1)^n \gamma^{(n)}(s) \ge 0$. We can therefore differentiate (3.4) any number of times, and it is seen that $\omega^{(n)}(s)$ is continuous for $s > \sigma'$. Now a simple inductive argument shows that $(-1)^n \omega^{(n)}(s)$ is a product of $\{1 - \varphi(s)\}^{-(n+1)}$ by a finite number of completely monotone functions. It follows that $(-1)^n \omega^{(n)}(s) \ge 0$, so that $\omega(s)$ is a completely monotone function, at least for $s > \sigma'$. Hence it follows from a well-known theorem of S. Bernstein and D. V. Widder[9] that there exists a non-decreasing function $U(t)$ such that (3.3) holds for $s > \sigma'$. Moreover, this function can obviously be so defined that $U(0) = 0$ and that it is continuous to the right. Using $U(t)$ let us form a new function

(3.5)
$$V(t) = \int_0^t U(t-x)\,dF(x).$$

$V(t)$ is clearly non-negative and non-decreasing. It is readily verified (and, of course, well-known) that

$$\psi(s) \equiv \int_0^\infty e^{-st}\,dV(t) = \omega(s)\varphi(s).$$

It follows, therefore, from (3.4) that $\psi(s) = \omega(s) - \gamma(s)$, and this implies, by the uniqueness theorem for Laplace transforms, that $V(t) \doteq U(t) - G(t)$. Combining this result with (3.5) it is seen that $U(t)$ is a solution of (1.3).

THEOREM 2. *Suppose that $f(t)$ and $g(t)$ are measurable, non-negative and bounded in every finite interval $0 \le t \le T$. Let the integrals*

(3.6)
$$\varphi(s) = \int_0^\infty e^{-st} f(t)\,dt, \qquad \gamma(s) = \int_0^\infty e^{-st} g(t)\,dt$$

converge for $s > \sigma$. Then there exists one and only one non-negative solution $u(t)$

---

[9] This theorem has been repeatedly proved by several authors; for a recent proof cf. Feller [19].

of (1.1) which is bounded in every finite interval[10]. *With this function the integral*

$$(3.7) \qquad \omega(s) = \int_0^\infty e^{-st} u(t)\, dt$$

converges at least for $s > \sigma'$, where $\sigma' = \sigma$ if $\lim\limits_{s \to \sigma + 0} \varphi(s) \leq 1$, and otherwise $\sigma' > \sigma$ is defined as the root of the characteristic equation $\varphi(s) = 1$. For $s > \sigma'$ equation (3.4) holds.

If $f(t)$ is continuous except, perhaps, at a finite number of points then $u(t) - g(t)$ is continuous.

PROOF: Define $F(t)$ and $G(t)$ by (1.2). Under the present conditions these functions satisfy the conditions of Theorem 1, and hence (1.3) has a non-decreasing solution $U(t)$. Consider, then, an arbitrary interval $0 \leq t \leq T$ and suppose that in this interval $f(t) < M$ and $g(t) < M$. If $0 \leq t < t + h \leq T$ we have by (1.3)

$$0 \leq \frac{1}{h}\{U(t + h) - U(t)\}$$

$$\equiv \frac{1}{h}\{G(t + h) - G(t)\} + \frac{1}{h}\int_t^{t+h} U(t + h - x)f(x)\, dx$$

$$+ \frac{1}{h}\int_0^t \{U(t + h - x) - U(t - x)\}f(x)\, dx$$

$$\leq M + MU(T) + \frac{M}{h}\int_0^t \{U(t + h - x) - U(t - x)\}\, dx$$

$$= M + MU(T) + \frac{M}{h}\int_t^{t+h} U(y)\, dy - \frac{M}{h}\int_0^h U(y)\, dy$$

$$< M + 2MU(T).$$

Thus $U(t)$ has bounded difference ratios and is therefore an integral. The derivative $U'(t)$ exists for almost all t and $0 \leq U'(t) \leq M$. Accordingly we can differentiate (1.3) formally, and since $U(0) = 0$ it follows that $u(t) = U'(t)$ satisfies (1.1) for almost all t. However, changing $u(t)$ on a set of measure zero does not affect the integral in (1.1), and since $g(t)$ is defined for all t it is seen that $u(t)$ can be defined, in a unique way, so as to satisfy (1.1) and obtain (1.3). Since the solution of (1.3) was uniquely determined it follows that the solution $u(t)$ is also unique. Obviously equations (3.7) and (3.3) define the same function $\omega(s)$, so that (3.4) holds, and (3.7) converges for $s > \sigma'$.

Finally, if $f(t)$ has only a finite number of jumps, the continuity of $u(t) - g(t)$ becomes evident upon writing (1.1) in the form

$$u(t) - g(t) = \int_0^t u(x)f(t - x)\, dx.$$

---

[10] Without the assumptions of positiveness and boundedness this theorem reduces to a special case of a theorem by Paley and Wiener [21]; cf. section 1, p. 243.

4. **Asymptotic properties.** In this section we shall be concerned with the asymptotic behavior as $t \to \infty$ not of $u(t)$ itself but of the mean value $u^*(t) = \frac{1}{t} \int_0^t u(\tau)\, d\tau$. If $u(t)$ tends to the (not necessarily finite) limit C, then obviously also $u^*(t) \to C$, whereas the converse is not necessarily true. For the proof of the theorem we shall need the following obvious but useful

LEMMA: *If $u(t) \geq 0$ is a solution of (1.1) and if*

(4.1) $\qquad u_1(t) = e^{kt}u(t), \qquad f_1(t) = e^{kt}f(t), \qquad g_1(t) = e^{kt}g(t),$

then $u_1(t)$ is a solution of

$$u_1(t) = g_1(t) + \int_0^t u_1(t-x)f_1(x)\, dx.$$

THEOREM 3: *Suppose that using the functions defined in Theorem 2 the integrals*

(4.2) $\qquad \displaystyle\int_0^\infty f(t)\, dt = a, \qquad \int_0^\infty g(t)\, dt = b,$

are finite.

(i) *In order that*

(4.3) $\qquad u^*(t) = \dfrac{1}{t} \displaystyle\int_0^t u(\tau)\, d\tau \to C$

as $t \to \infty$, where C is a positive constant, it is necessary and sufficient that $a = 1$, and that the moment,

(4.4) $\qquad \displaystyle\int_0^\infty t\, f(t)\, d_t = m.$

be finite. In this case

(4.5) $\qquad\qquad\qquad\qquad C = \dfrac{b}{m}.$

(ii) *If $a < 1$ we have*

(4.6) $\qquad \displaystyle\int_0^\infty u(t)\, dt = \dfrac{b}{1-a}.$

(iii) *If $a > 1$ let σ' be the positive root of the characteristic equation $\varphi(s) = 1$ (cf. (3.2)) and put*[11]

(4.7) $\qquad \displaystyle\int_0^\infty e^{-\sigma' t} t\, f(t)\, dt = m_1.$

Then

(4.8) $\qquad \displaystyle\lim_{t\to\infty} \dfrac{1}{t} \int_0^t e^{-\sigma'\tau} u(\tau)\, d\tau = \dfrac{b}{m_1}.$

---

[11] (4.2) implies the finiteness of m_1.

REMARK: The case $a = 1$ corresponds in demography to a population of stationary size. In the theory of industrial replacement only the case $a = 1$ occurs; the moment m is the average lifetime of an individual. The case $a > 1$ corresponds in demography to a population in which the fertility is greater than the mortality. As is seen from (4.8), in this case the mean value of $u(t)$ increases exponentially. It is of special interest to note that in a population with $a < 1$ the integral (4.6) always converges.

PROOF: By (4.2) and (3.7)

$$(4.9) \qquad \lim_{s \to +0} \phi(s) = a, \qquad \lim_{s \to +0} \gamma(s) = b.$$

If $a < 1$, it follows from (3.4) that $\lim \omega(s) = b/(1 - a)$ is finite. Since $u(t) \geq 0$ this obviously implies that (4.6) holds. This proves (ii).

If $a = 1$ and m is finite, it is readily seen that

$$\lim_{s \to +0} \frac{1 - \varphi(s)}{s} = m,$$

and hence by (3.4)

$$\lim_{s \to +0} s\omega(s) = \lim_{s \to +0} \gamma(s) \lim_{s \to +0} \frac{s}{1 - \varphi(s)} = \frac{b}{m}.$$

By a well-known Tauberian theorem for Laplace integrals of non-negative functions[12] it follows that $u^*(t) \to \dfrac{b}{m}$. Conversely, if (4.3) holds it is readily seen that[13]

$$\lim_{s \to +0} s\omega(s) = C,$$

which in turn implies by (3.4) and (4.9) that

$$\lim_{s \to +0} \frac{1 - \varphi(s)}{s} = \frac{b}{C}.$$

This obviously means that the moment (4.4) exists and equals b/C. This proves (i).

Finally case (iii) reduces immediately to (ii) using the above lemma with $k = -\sigma'$. This finishes the proof.

---

[12] Cf. e.g. Doetsch [18], p. 208 or 210.

[13] Indeed, if (4.3) holds and if $U(t)$ is defined by (1.2), then there is a $M = M(\epsilon)$ such that $|U(t) - Ct| < M + \epsilon t$. Now

$$\varphi(s) = s \int_0^\infty e^{-st} U(t) \, dt,$$

and hence

$$s\varphi(s) - C = s^2 \int_0^\infty e^{-st}(U(t) - Ct) \, dt,$$

or

$$|s\varphi(s) - C| \leq s^2 \int_0^\infty e^{-st}(M + \epsilon t) \, dt = sM + \epsilon.$$

143

It may be remarked that the finiteness of the integrals (4.2) is by no means necessary for (4.3). ...

5. **Closer study of asymptotic properties.** In this section we shall deal almost exclusively with the most important special case, namely where

$$(5.1) \qquad\qquad \int_0^\infty f(t)\, dt = 1.$$

The question has been much discussed whether in this case necessarily $u(t) \to C$ as $t \to \infty$, which statement, if true, would be a refinement of (4.3). Hadwiger [2] has constructed a rather complicated example to show that $u(t)$ does not necessarily approach a limit. Now this can also be seen directly and without any computations. Indeed, if $u(t) \to C$ and if (5.1) holds, then obviously

$$\lim_{t \to \infty} \int_0^t u(t - x) f(x)\, dx = C,$$

and hence it follows from (1.1) that $g(t) \to 0$. In order that $u(t) \to C$ it is therefore necessary that $g(t) \to 0$, and this proves the assertion. In Hadwiger's example $\lim \sup g(t) = \infty$, which makes his computations unnecessary.

It can be shown in a similar manner that not even the condition $g(t) \to 0$ is sufficient to ensure that $u(t) \to C$. Some restriction as to the total variation of $f(t)$ seems both necessary and natural (conditions on the existence of derivatives are not sufficient). In the following theorem we shall prove the convergence of $u(t)$ under a condition which is, though not strictly necessary, sufficiently wide to cover all cases of any possible practical interest.

THEOREM 4: *Suppose that with the functions $f(t)$ and $g(t)$ of Theorem 2*

$$(5.2) \qquad \int_0^\infty f(t)\, dt = 1, \qquad \int_0^\infty g(t)\, dt = b < \infty.$$

Suppose moreover that there exists an integer $n \geq 2$ such that the moments

$$(5.3) \qquad\qquad m_k = \int_0^\infty t^k f(t)\, dt, \qquad\qquad k = 1, 2, \cdots, n,$$

are finite, and that the functions $f(t)$, $tf(t)$, $t^2 f(t)$, $\cdots$, $t^{n-2} f(t)$ are of bounded total variation over $(0, \infty)$. Suppose finally that

$$(5.4) \qquad \lim_{t \to \infty} t^{n-2} g(t) = 0 \quad and \quad \lim_{t \to \infty} t^{n-2} \int_t^\infty g(x)\, dx = 0.$$

Then

$$(5.5) \qquad\qquad \lim_{t \to \infty} u(t) = \frac{b}{m_1}$$

and

$$(5.6) \qquad\qquad \lim_{t \to \infty} t^{n-2} \left\{ u(t) - \frac{b}{m_1} \right\} = 0.$$

144

REMARK: As it was shown in section 4, the case where $\int_0^\infty f(t)\,dt > 1$ can readily be reduced to the above theorem by applying the lemma of section 4 with $k = \sigma'$, where σ' is the positive root of $\varphi(s) = 1$: it is only necessary to suppose that $e^{-\sigma' t}f(t)$ is of bounded total variation and that $e^{-\sigma' t}g(t) \to 0$. Obviously all moments of $e^{-\sigma' t}f(t)$ exist, so that the above theorem shows that $u_1(t) = e^{-\sigma' t}u(t)$ tends to the finite limit b'/m_1', where

$$b' = \int_0^\infty e^{-\sigma' t}g(t)\,dt, \qquad m_1' = \int_0^\infty e^{-\sigma' t}tf(t)\,dt.$$

Thus in this case and under the above assumptions $u(t) \sim \dfrac{b'}{m_1'}e^{\sigma' t}$, so that the renewal function increases exponentially as could be expected. If however

$$\int_0^\infty f(t)\,dt < 1,$$

$u(t)$ will in general *not* show an exponential character. If $f(t)$ is of bounded variation and has a finite moment of second order, and if $g(t) \to 0$, then it can be shown that $u(t) \to 0$. However, the lemma of section 4 can be applied only if the integral defining $\varphi(s)$ converges in some negative s-interval containing a value s' such that $\varphi(s') = 1$, and this is in general not the case.

PROOF: The proof of Theorem 4 will be based on a Tauberian theorem due to Haar[15]. With some specializations and obvious changes this theorem can be formulated as follows.

Suppose that $l(t)$ is, for $t \geq 0$, non-negative and continuous, and that the Laplace integral

$$(5.7) \qquad \lambda(s) = \int_0^\infty e^{-st}l(t)\,dt$$

converges for $s > 0$. Consider $\lambda(s)$ as a function of the complex variable $s = x + iy$ and suppose that the following conditions are fulfilled:

(i) For $y \neq 0$ the function $\lambda(s)$ (which is always regular for $x > 0$) has continuous boundary values $\lambda(iy)$ as $x \to +0$, for $x \geq 0$ and $y \neq 0$

$$(5.8) \qquad \lambda(s) = \frac{C}{s} + \psi(s),$$

where $\psi(iy)$ has finite derivatives $\psi'(iy), \cdots, \psi^{(r)}(iy)$ and $\psi^{(r)}(iy)$ is bounded in every finite interval;

(ii) $$\int_{-\infty}^{+\infty} e^{ity}\lambda(x + iy)\,dy$$

converges for some fixed $x > 0$ uniformly with respect to $t \geq T > 0$;
(iii) $\lambda(x + iy) \to 0$ as $y \to \pm\infty$, uniformly with respect to $x \geq 0$;
(iv) $\lambda'(iy), \lambda''(iy), \cdots, \lambda^{(r)}(iy)$ tend to zero as $y \to \pm\infty$;

145

(v) The integrals

$$\int_{-\infty}^{y_1} e^{ity} \lambda^{(r)}(iy)\, dy \quad \text{and} \quad \int_{y_2}^{\infty} e^{ity} \lambda^{(r)}(iy)\, dy$$

(where $y_1 < 0$ and $y_2 > 0$ are fixed) converge uniformly with respect to $t \geq T > 0$. Under these conditions

(5.9)
$$\lim t^r \{l(t) - C\} = 0.$$

Now the hypotheses of this theorem are too restrictive to be applied to the solution $u(t)$ of (1.1). We shall therefore replace (1.1) by the more special equation

(5.10)
$$v(t) = h(t) + \int_0^t v(t-x) f(x)\, dx,$$

where

(5.11)
$$h(t) = \int_0^t f(t-x) f(x)\, dx.$$

Plainly Theorem 2 can be applied to (5.10). It is also plain that $h(t)$ is bounded and non-negative and that (by (5.1))

(5.12)
$$\int_0^\infty h(t)\, dt = 1,$$

(5.13)
$$\chi(s) \equiv \int_0^\infty e^{-st} h(t)\, dt = \varphi^2(s).$$

Accordingly we have by Theorem 2

(5.14)
$$\zeta(s) \equiv \int_0^\infty e^{-st} v(t)\, dt = \frac{\varphi^2(s)}{1 - \varphi(s)}.$$

We shall first verify that $\zeta(s)$ satisfies the conditions of Haar's theorem with $r = n - 2$. For this purpose we write

(5.15)
$$f(t) = f_1(t) - f_2(t),$$

where $f_1(t)$ and $f_2(t)$ are non-decreasing and non-negative functions which are, by assumption, bounded:

(5.16)
$$0 \leq f_1(t) < M, \qquad 0 \leq f_2(t) < M.$$

(a) We show that $v(t)$ is continuous. Now by Theorem 2 the solution $v(t)$ of (5.10) is certainly continuous if $h(t)$ is continuous; however, that $h(t)$ is continuous follows directly from (5.11) and the fact that the functions

$$\int_0^t f_1(t-x) f(x)\, dx \quad \text{and} \quad \int_0^t f_2(t-x) f(x)\, dx$$

---

[15] Haar [20] or Doetsch [18], p. 269.

are continuous.

(b) In view of (5.1) the function $\varphi(s)$ exists for $x = \Re(s) \geq 0$. Obviously $|\varphi(x + iy)| < 1$ for $x > 0$. Now

$$1 - \varphi(iy) = \int_0^\infty (1 - e^{-iyt}) f(t)\, dt$$

$$= \int_0^\infty (1 - \cos yt)\, f(t)\, dt + i \int_0^\infty \sin yt \cdot f(t)\, dt,$$

and, since $1 - \cos yt \geq 0$ and $f(t) \geq 0$, the equality $\varphi(iy) = 1$ for $y \neq 0$ would imply that $f(t) = 0$ except on a set of measure zero. It is therefore seen that $\varphi(x + iy) \neq 1$ for all $x > 0$ and for $x = 0$, $y \neq 0$.

It follows furthermore from (5.3) that for $k = 1, \cdots, n$ and $x \geq 0$ the derivatives

$$\varphi^{(k)}(s) = \int_0^\infty (-t)^k e^{-st} f(t)\, dt$$

exist and that

$$\lim_{x \to +0} \varphi^{(k)}(x + iy) = \varphi^{(k)}(iy).$$

Finally, it is readily seen that in the neighborhood of $y = 0$ we have

$$\varphi(iy) = \int_0^\infty e^{-iyt} f(t)\, dt$$

(5.17)
$$= 1 - m_1 iy + \frac{m_2}{2} (iy)^2 - + \cdots$$

$$+ (-1)^{n-1} \frac{m_{n-1}}{(n-1)!} (iy)^{n-1} + O(|y|^n).$$

(c) From what was said under (b) it follows by (5.14) that $\mathfrak{s}(s)$ is regular for $x > 0$, and that $\mathfrak{s}(s), \mathfrak{s}'(s), \cdots, \mathfrak{s}^{(n)}(s)$ approach continuous boundary values as $s = x + iy$ approaches a point of the imaginary axis other than the origin. Now put

(5.18)
$$\psi(s) = \frac{\varphi^2(s)}{1 - \varphi(s)} - \frac{1}{m_1 s},$$

so that by (5.14)

(5.19)
$$\mathfrak{s}(s) = \frac{1}{m_1 s} + \psi(s).$$

For $x > 0$ and $x = 0$, $y \neq 0$ the function $\psi(x + iy)$ is obviously continuous; the derivatives $\psi'(iy), \cdots, \psi^{(n)}(iy)$ exist. To investigate the behavior of $\psi(iy)$ in the neighborhood of $y = 0$ put

(5.20)
$$P(y) = m_1 - \frac{m_2}{2}(iy) + - \cdots - (-1)^{n-1} \frac{m_{n-1}}{(n-1)!}(iy)^{n-2}.$$

By (5.17), (5.18) and (5.20)

(5.21)
$$\psi(iy) = \left[\frac{\{1 - iyP(y)\}^2}{P(y)} - \frac{1}{m_1}\right]\frac{1}{iy} + O(|y|^{n-2}).$$

147

Now the expression in brackets represents an analytic function of y which vanishes at $y = 0$. Hence $\psi(iy) = \mathfrak{P}(y) + O(|y|^{n-2})$, where $\mathfrak{P}(y)$ denotes a power series. It follows that the derivatives $\psi'(iy), \cdots, \psi^{(n-2)}(iy)$ exist for all real y (including $y = 0$) and are bounded for sufficiently small $|y|$: since they are continuous functions they are bounded in every finite interval.

(d). Next we show that there exists a constant $A > 0$ such that for sufficiently large $|y|$

(5.22)
$$|\varphi(x + iy)| < \frac{A}{|y|}$$

uniformly in $x \geq 0$. By (5.15)

(5.23)
$$\varphi(s) = \int_0^\infty \{\cos yt - i \sin yt\}e^{-xt}\{f_1(t) - f_2(t)\}\, dt.$$

Now $f_1(t)$ is non-decreasing and accordingly by the second mean-value theorem we have for any $T > 0$ and y

$$\int_0^T \cos yt \cdot f_1(t)\, dt = f_1(T) \int_\tau^T \cos yt\, dt = f_1(T) \frac{\sin Ty - \sin \tau y}{y},$$

where τ is some value between 0 and T (depending, of course, on y; at points of discontinuity, $f_1(T)$ should be replaced by $\lim_{t \to T-0} f_1(t)$). Hence by (5.16)

$$\left| \int_0^\infty \cos yt \cdot e^{-xt} \cdot f_1(t)\, dt \right| < \frac{2M}{|y|}.$$

Treating the other terms in (5.23) in a like manner, (5.22) follows.

Combining (5.22) with (5.14) it is seen that for sufficiently large $|y|$

$$|\zeta(s)| < \frac{2A^2}{y^2}$$

uniformly in $x \geq 0$. This shows that the assumptions (ii) and (iii) of Haar's theorem are satisfied for $\lambda(s) = \zeta(s)$. In order to prove that also conditions (iv) and (v) are satisfied it suffices to notice that the proof of (5.22) used only the fact that $f(t)$ is of bounded total variation. Now $\varphi^{(k)}(s)$ is the Laplace transform of $(-t)^k f(t)$, and, since $t^k f(t)$ is of bounded total variation for $k \leq n - 2$, it follows that

$$|\varphi^{(k)}(s)| = O(|y|^{-1}), \qquad k = 1, 2, \cdots, n - 2,$$

for sufficiently large $|y|$, uniformly in $x \geq 0$. Differentiating (5.14) k times it is also seen that

$$|\zeta^{(k)}(s)| = O(|y|^{-2}), \qquad k = 1, 2, \cdots, n - 2,$$

as $y \to +\infty$, uniformly with respect to $x \geq 0$.

This enumeration shows that $v(s) = l(t)$ and $\lambda(s) = \zeta(s)$ satisfy all hypotheses of Haar's theorem with $r = n - 2$ and $C = 1/m_1$. Hence

(5.24)
$$\lim_{t \to \infty} t^{k-2}\left\{v(t) - \frac{1}{m_1}\right\} = 0.$$

Returning now to (5.14) we get

$$\omega(s) = \gamma(s) + \gamma(s)\varphi(s) + \gamma(s)\varsigma(s),$$

or, by the uniqueness property of Laplace integrals,

(5.25)
$$u(t) = g(t) + \int_0^t g(x)f(t-x)\,dx + \int_0^t g(x)v(t-x)\,dx$$

$$= g(t) + u_1(t) + u_2(t)$$

(which relation can also be checked directly using (5.10)). Let us begin with the last term. We have by (5.2)

$$u_2(t) - \frac{b}{m_1} \equiv \int_0^t g(t-x)\left\{v(x) - \frac{1}{m_1}\right\}dx,$$

and hence

$$t^{n-2}\left|u_2(t) - \frac{b}{m_1}\right| \le 2^{n-2}\int_{t/2}^t g(t-x)x^{n-2}\left|v(x) - \frac{1}{m_1}\right|dx$$

$$+ t^{n-2}\int_{t/2}^t g(y)\left|v(t-y) - \frac{1}{m_1}\right|dy.$$

If t is sufficiently large we have by (5.24) in the first integral $x^{n-2}\left|v(x) - \frac{1}{m_1}\right| < \epsilon$.
In the second integral $v(t-y) - \frac{1}{m_1}$ is bounded, and hence by (5.4)

$$\lim_{t\to\infty} t^{n-2}\left|u_2(t) - \frac{b}{m_1}\right| = 0.$$

The same argument applies (even with some simplifications) also to the second term in (5.24); it follows that

$$\lim t^{n-2}u_1(t) = 0,$$

whilst $t^{n-2}g(t) \to 0$ by assumption (5.4). Now the assertion (5.6) of our theorem follows in view of (5.25) if the last three relationships are added. This finishes the proof of Theorem 4.

It seems that the solution $u(t)$ is generally supposed to oscillate around its limit b/m_1 as $t \to \infty$. It goes without saying that such a behavior is a priori more likely than a monotone character. It should, however, be noticed that there is no reason whatsoever to suppose that $u(t)$ always oscillates around its limit. Again no computation is necessary to see this, as shown by the following
EXAMPLE: Differentiating (1.1) formally we get

$$u'(t) = g'(t) + g(0)f(t) + \int_0^t u'(t-x)f(x)\,dx,$$

which shows that, if $g(t)$ and $f(t)$ are sufficiently regular, $u'(t)$ satisfies an integral equation of the same type as $u(t)$. Thus if

$$g'(t) + g(0)f(t) \ge 0$$

149

for all t, we shall have $u'(t) \geq 0$, and $u(t)$ is a monotone function. In particular, if $g'(t) + g(0)f(t) = 0$, then $u'(t) = 0$ and $u(t) = $ const. For example, let $f(t) = g(t) = e^{-t}$. Then $\varphi(s) = \gamma(s) = 1/(s + 1)$ and hence $\omega(s) = 1/s$, which is the Laplace transform of $u(t) = 1$. It is also seen directly that $u(t) \equiv 1$ is the solution. We have however the following

THEOREM 5[16]: *If the functions $f(t)$ and $g(t)$ of Theorem 4 vanish identically for $t \geq T > 0$, then the solution $u(t)$ of (1.1) oscillates around its limit b/m as $t \rightarrow \infty$.*

PROOF: For $t \geq T$ equation (1.1) reduces to

$$u(t) = \int_{t-T}^{t} u(t - x) f(x)\, dx,$$

and since $\int_{t-T}^{t} f(x)\, dx = 1$ it follows that the maxima of $u(t)$ in the intervals $nT < t < (n + 1)T$ form, for sufficiently large integers n, a non-increasing sequence. Similarly the corresponding minima do not decrease. Since $u(t) \rightarrow b/m_1$, by Theorem 4, it follows that the minima do not exceed b/m_1 and the maxima are not smaller than b/m_1.

6. **On Lotka's method.** Probably the most widely used method for treating equation (1.1) in connection with problems of the renewal theory is Lotka's method. As a matter of fact this method consists of two independent parts. The first step aims at obtaining the exact solution of (1.1) in the form of a series of exponential terms (this is achieved by an adaptation of a method which was used by P. Herz and Herglotz for other purposes. The second part of Lotka's theory consists of devices for a convenient approximative computation of the first few terms of the series. While restricting ourselves formally to Lotka's theory, it will be seen that some of the following remarks apply equally to other methods.

Lotka's method rests essentially on the fundamental assumption that the characteristic equation

(6.1) $$\varphi(s) = 1$$

has infinitely many distinct simple[17] roots s_0, s_1, $\cdots$, and that the solution $u(t)$ of (1.1) can be expanded into a series

(6.2) $$u(t) = \sum_k A_k e^{s_k t}$$

where the A_k are complex constants. The argument usually rests on an assumed completeness-property of the roots. Thus, starting from (2.4) it is required that

---

[16] Under some slight additional hypotheses and with quite different methods this theorem was proved by Richter [16].

[17] Hadwiger [3] objected to the assumption that all roots of (6.1) be simple. The modifications which are necessary to cover the case of multiple roots also will be indicated below.

(6.2) reduces to $h(t)$ for $t < 0$; in other words, that an arbitrarily prescribed function $h(x)$ be, for $x < 0$, respresentable in the form

$$(6.3) \qquad\qquad h(x) = \sum_k A_k e^{s_k x} \qquad\qquad (x < 0).$$

In practice we are, of course, usually not concerned with $h(t)$ but with $g(t)$ (cf. (2.5)), and according to Lotka's theory the coefficients A_k of the solution (6.2) of (1.1) can be computed directly from $g(t)$ in a way similar to the computation of the Fourier coefficients.

Lotka's method is known to lead to correct results in many cases and also to have distinct computational merits. On the other hand it seems to require a safer justification, since its fundamental assumptions are rarely realized. Thus clearly an arbitrary function $h(x)$ cannot be represented in the form (6.3): to see this it suffices to note that (6.1) frequently has only a finite number of roots (cf. also below). It should also be noted that, the series (6.3) having regularity properties as are assumed in Lotka's theory, any function representable in the form (6.3) is necessarily a solution of the integral equation (2.4), whereas the theory requires us to construct a solution $u(t)$ which reduces to an *arbitrarily* prescribed function $h(t)$ for $t < 0$, (which frequently is an empirical function, determined by observations). Nevertheless, it is possible to give sound foundations to Lotka's method so that it can be used (with some essential limitations and modifications) sometimes even in cases for which it originally was not intended. For this purpose it turns out to be necessary that all considerations be based on the more general equation (1.1), instead of (2.4) (cf. also section 2).

Before proceeding it is necessary to make clear *what is really meant by a root of* (6.1). The function $\varphi(s)$ is defined by (3.2), and the integral will in general converge only for s-values situated in the half-plane $\Re(s) > \sigma$. Usually only roots situated in this half-plane are considered . It is also argued that $\varphi(s)$ is, for real s, a monotone function, so that (6.1) has at most one real root: accordingly the terms of (6.2) are called "oscillatory components." However, the function $\varphi(s)$ can usually be defined by analytic continuation even outside the half-plane $\Re(s) > \sigma$, and, if this is done, (6.1) will in general also have roots in the half-plane $\Re(s) < \sigma$. It will be seen in the sequel that these roots play exactly the same role for the solution $u(t)$ as the other ones, and that the applicability of Lotka's method depends on the behavior of $\varphi(s)$ in the entire complex s-plane. ...

From now on we shall consistently denote by $\varphi(s)$ the function defined by the integral (3.4) and by the usual process of analytic continuation; accordingly we shall take into consideration *all* roots of (6.1). The main limitation of Lotka's theory can then be formulated in the following way: Lotka's method depends only on the function $g(t)$ and on the roots of (6.1). Now two different functions $f(t)$ can lead to characteristic equations having the same roots. Lotka's method would be applicable to both only if the corresponding two integral equations (1.1) had the same solution $u(t)$. This, however, is not necessarily the case. Thus, if Lotka's method is applied, and if all computations are correctly performed, and if the resulting series for $u(t)$ converges uniformly, there is no possibility of telling which equation is really satisfied by the resulting $u(t)$:

it can happen that one has unwittingly solved some unknown equation of type (1.1) which, by chance, leads to a characteristic equation having the same roots as the characteristic equation of the integral equation with which one was really concerned. ...

These preparatory remarks enable us to formulate rigorous conditions for the existence of an expansion of type (6.2). The following theorem shows the limits of Lotka's method, but at the same time it also represents an extension of it. In the formulation of the theorem we have considered only the case of absolute convergence of (6.2). This was done to avoid complications lacking any practical significance whatsoever. The conditions can, of course, be relaxed along customary lines.

THEOREM 6: *In order that the solution $u(t)$ of Theorem 2 be representable in form* (6.2), *where the series converges absolutely for $t \geq 0$ and where the s_k denote the roots of the characteristic equation[21]* (6.1), *it is necessary and sufficient that the Laplace transform $\omega(s)$ admit an expansion*

$$(6.6) \qquad \omega(s) \equiv \frac{\gamma(s)}{1 - \varphi(s)} = \sum \frac{A_k}{s - s_k}$$

and that $\Sigma |A_k|$ converges absolutely. The coefficients A_k are determined by

$$(6.7) \qquad A_k = -\frac{\gamma(s_k)}{\varphi'(s_k)}.$$

In particular, it is necessary that $\omega(s)$ be a one-valued function.

PROOF: All roots s_k of (6.1) satisfy the inequality $\Re(s_k) \leq \sigma'$, where σ' was defined in Theorem 2. It is therefore readily seen that in case $\Sigma |A_k|$ converges, the Laplace transform of (6.2) can be computed for sufficiently large positive s-values by termwise integration so that (6.6) certainly holds for sufficiently large positive s. Now with $\Sigma |A_k|$ converging, (6.6) defines $\omega(s)$ uniquely for all complex s (with singularities at the points s_k and the points of accumulation of s_k, if any). Since the analytic continuation is unique, it follows that (6.6) holds for all s. The series $\Sigma |A_k|$ must, of course, converge if (6.2) is to converge absolutely for $t = 0$, and this proves the necessity of our condition. Conversely, if $\omega(s) = \dfrac{\gamma(s)}{1 - \varphi(s)}$ is given by (6.6), and if $\Sigma |A_k|$ converges, then $\omega(s)$ is the Laplace transform of a function $u(t)$ defined by (6.2). Since the Laplace transform is unique, $u(t)$ is the solution of (1.1) by Theorem 2. The series (6.2) converges absolutely for $t \geq 0$ since $|A_k e^{s_k t}| \leq |A_k| e^{\sigma' t}$. Finally (6.7) follows directly from (6.6).

It is interesting to compare (6.7) with formulas (50) and (56) of Lotka's paper [8]. Lotka considers the special case $g(t) = f(t)$; in this case $\gamma(s_k) = \varphi(s_k) = 1$, and (6.7) reduces to $A_k = -\dfrac{1}{\varphi'(s_k)}$. If s_k lies in the domain of con-

---

[21] The number of roots may be finite or infinite. It should also be noted that it is not required that $s_k \to \infty$. If the s_k have a point of accumulation, $\omega(s)$ will have an essential singularity. That this actually can happen can be shown by examples.

vergence of the integral $\varphi(s) = \int_0^\infty e^{-st} f(t)\, dt$, that is, if $\Re(s_k) \geq \sigma$ then

(6.8)
$$\frac{1}{A_k} = \int_0^\infty e^{-st} t f(t)\, dt,$$

in accordance with Lotka's result. However, (6.8) becomes meaningless for the roots with $\Re(s_k) < \sigma$, whereas (6.7) is applicable in all cases.

Theorem 6 can easily be generalized to the case where the *characteristic equation has multiple roots*. The expansion (6.6) (which reduces to the customary expansion into partial fractions whenever $\omega(s)$ is meromorphic) is to be replaced by

(6.9)
$$\omega(s) = \sum_k \left\{ \frac{A_k^{(1)}}{s - s_k} + \frac{A_k^{(2)}}{(s - s_k)^2} + \cdots + \frac{A_k^{(m_k)}}{(s - s_k)^{m_k}} \right\},$$

where m_k is the multiplicity of the root s_k. This leads us formally to an expansion

(6.10)
$$u(t) = \sum_k e^{s_k t} \left\{ A_k^{(1)} + A_k^{(2)} \frac{t}{1!} + \cdots + A_k^{(m_k)} \frac{t^{m_k-1}}{(m_k - 1)!} \right\},$$

which now replaces (6.2). Generalizing Theorem 6 it is easy to formulate some simple conditions under which (6.1) will really represent a solution of (1.1). Other conditions which ensure that (6.9) is the transform of (6.10) are known from the general theory of Laplace transforms; such conditions usually use only function-theoretical properties of (6.9) and are applicable in particular when $\omega(s)$ is meromorphic. We mention in particular a theorem of Churchill [17] which can be used for our purposes.

7. **On the practical computation of the solution.** There are at hand two main methods for the practical computation of the solution of (1.1). One of them has been developed by Lotka and consists of an approximate computation of a few coefficients in the series (6.2). The other method uses an expansion

(7.1)
$$u(t) = \sum_{n=0}^\infty u_n(t),$$

where $u_n(t)$ represents the contribution of the nth "generation" and is defined by x

(7.2)
$$u_0(t) = g(t), \qquad u_{n+1}(t) = \int_0^t u_n(t - x) f(x)\, dx.$$

Now the Laplace transform of $u_{n+1}(t)$ is $\gamma(s)\varphi^n(s)$, and hence (7.2) corresponds to the expansion

(7.3)
$$\omega(s) = \frac{\gamma(s)}{1 - \varphi(s)} = \gamma(s) \sum_{n=0}^\infty \varphi^n(s).$$

In practice the functions $g(t)$ and $f(t)$ are usually not known exactly. Frequently their values are obtained from some statistical material, so that only their integrals over some time units, e.g. years, are actually known or, in other

153

words, only the values

$$(7.4) \qquad f_n = \frac{1}{\delta} \int_{n\delta}^{(n+1)\delta} f(t)\, dt, \qquad g_n = \frac{1}{\delta} \int_{n\delta}^{(n+1)\delta} g(t)\, dt,$$

are given, where $\delta > 0$ is a given constant. Ordinarily in such cases some theoretical forms (e.g. Pearson curves) are fitted to the empirical data and equation (1.1) is solved with these theoretical functions. Now such a procedure is sometimes not only very troublesome, but also somewhat arbitrary. Consider for example the limit of $u(t)$ as $t \to \infty$; this asymptotic value is the main point of interest of the theory and all practical computations. However, as has been shown above, this limit depends only on the moments of the first two orders of $f(t)$ and $g(t)$, and, unless the fitting is done by the method of moments, the resulting value will depend on the special procedure of fitting. Accordingly it will sometimes happen that it is of advantage to use the empirical material as it is, and this can, at least in principle, always be done.

If only the values (7.4) are used it is natural to consider $f(t)$ and $g(t)$ as step-functions defined by

$$(7.5) \qquad \left. \begin{aligned} f(t) &= f_n, \\ g(t) &= g_n, \end{aligned} \right\} \qquad \text{for} \quad n\delta \le t < (n+1)\delta.$$

In practice only a finite number among the f_n and g_n will be different from zero: accordingly the Laplace transforms $\gamma(s)$ and $\varphi(s)$ reduce to trigonometrical polynomials, so that the analytic study of $\omega(s) = \dfrac{\gamma(s)}{1 - \varphi(s)}$ becomes particularly simple. Lotka's method can be applied directly in this case.

For a convenient computation of (7.1) it is better to return to the more general equation (1.3), instead of (1.1). The summatory functions $F(t)$ and $G(t)$ should not be defined by (1.2) in this case, but simply by

$$(7.6) \qquad F(t) = \sum_{n=0}^{\left[\frac{t}{\delta}\right]} f_n, \qquad G(t) = \sum_{n=0}^{\left[\frac{t}{\delta}\right]} g_n.$$

It is readily seen that the solution $U(t)$ of (1.3) can be written in the form $U(t) = \sum_{n=0}^{\infty} U_n(t)$, where

$$U_0(t) = G(t), \qquad U_{n+1}(t) = \int_0^t U_n(t - x)\, dF(x);$$

in our case $U_n(t)$ will again be a step-function with jumps at the points $k\delta$, the corresponding saltus being

$$u_0^{(k)} = g_k, \qquad u_{n+1}^{(k)} = \sum_{r=0}^{k} u_n^{(k-r)} f_r.$$

Thus we arrive at exactly the same result as would have been obtained if the integrals (7.2) had been computed, starting from (7.4), by the ordinary methods

154

for numerical integration of tabulated functions. It is of interest to note that this method of approximate evaluation of the integrals (7.2) leads to the *exact values of the renewal function* of a population where all changes occur in a discontinuous way at the end of time intervals of length δ in such a way that each change equals the mean value of the changes of the given population over the corresponding time interval.

REFERENCES

I. Papers on the integral equation of renewal theory

Note: Lotka's paper [8] contains a list of 74 papers on the subject published before 1939. The following list is to bring Lotka's list up to June 1941; however no claims to completeness are made.

[1] A. W. BROWN, A note on the use of a Pearson type III function in renewal theory," *Annals of Math. Stat.* Vol. 11 (1940), pp. 448–453.

[2] H. HADWIGER, "Zur Frage des Beharrungszustandes bei kontinuierlich sich erneuernden Gesamtheiten," *Archiv f. mathem. Wirtschafts- und Sozialforschung*, Vol. 5 (1939), pp. 32–34.

[3] H. HADWIGER, "Über die Integralgleichung der Bevölkerungstheorie," *Mitteilungen Verein. schweizer Versicherungsmathematiker (Bull. Assoc. Actuaires suisses)*, Vol. 38 (1939), pp. 1–14.

[4] H. HADWIGER, "Eine analytische Reproduktionsfunktion für biologische Gesamtheiten," *Skand. Aktuarietidskrift* (1940), pp. 101–113.

[5] H. HADWIGER, "Natürliche Ausscheidefunktionen für Gesamtheiten und die Lösung der Erneuerungsgleichung," *Mitteilungen Verein. Schweiz. Versich.-Math.*, Vol. 40 (1940), pp. 31–39.

[6] H. HADWIGER and W. RUCHTI, "Über eine spezielle Klasse analytischer Geburtenfunktionen," *Metron*, Vol. 13 (1939), No. 4, pp. 17–26.

[7] A. LINDER, "Die Vermehrungsrate der stabilen Bevölkerung," *Archiv f. mathem. Wirtschafts- und Sozialforschung*, Vol. 4 (1938), pp. 136–156.

[8] A. LOTKA, "A contribution to the theory of self-renewing aggregates, with special reference to industrial replacement," *Annals of Math. Stat.*, Vol. 10 (1939), pp. 1–25.

[9] A. LOTKA, "On an integral equation in population analysis," *Annals of Math. Stat.*, Vol. 10 (1939), pp. 144–161.

[10] A. LOTKA, "Théorie analytique des associations biologiques II," *Actualités Scientifiques No. 780*, Paris, 1939.

[11] A. LOTKA, "The theory of industrial replacement," *Skand. Aktuarietidskrift* (1940), pp. 1–14.

[12] A. LOTKA, "Sur une équation intégrale de l'analyse démographique et industrielle," *Mitt. Verein. Schweiz. Versich.-Math.*, Vol. 40 (1940), pp. 1–16.

[13] H. MÜNZNER, "Die Erneuerung von Gesamtheiten," *Archiv f. math. Wirtschafts- u. Sozialforschung*, Vol. 4 (1938).

[14] G. A. D. PREINREICH, "The theory of industrial replacement," *Skand. Aktuarietidskrift* (1939), pp. 1-19.

[15] E. C. RHODES, "Population mathematics, I, II, III," *Roy. Stat. Soc. Jour.*, Vol. 103 (1940), pp. 61–89, 218–245, 362–387.

[16] H. RICHTER, "Die Konvergenz der Erneuerungsfunktion," *Blätter f. Versicherungsmathematik*, Vol. 5 (1940), pp. 21–35.

[16a] H. HADWIGER, "Über eine Funktionalgleichung der Bevölkerungstheorie und eine spezielle Klasse analytischer Lösungen," *Bl. f. Versicherungsmathematik*, Vol. 5 (1941), pp. 181-188.

[16b] G. A. D. PRIENREICH, "The present status of renewal theory," Waverly Press, Baltimore (1940).

155

II. Other papers quoted

[17] R. V. CHURCHILL, "The inversion of the Laplace transformation by a direct expansion in series and its application to boundary-value problems," *Math. Zeits.*, Vol. 42 (1937), pp. 567-579.

[18] G. DOETSCH, *Theorie und Anwendung der Laplace Transformation.* J. Springer, Berlin, 1937.

[19] W. FELLER, "Completely monotone functions and sequences," *Duke Math. Jour.*, Vol. 5 (1939), pp. 661-674.

[20] A. HAAR, "Über asymptotische Entwicklungen von Funktionen," *Math. Ann.*, Vol. 96 (1927), pp. 69-107.

[21] R. E. A. C. PALEY and N. WIENER, "Notes on the theory and application of Fourier transforms, VII. On the Volterra equation," *Amer. Math. Soc. Trans.*, Vol. 35 (1933), pp. 785-791.

17. A New Method for Calculating Lotka's r — The Intrinsic Rate of Growth in a Stable Population

ANSLEY J. COALE (1957)

Population Studies 11: 92—94.

The determination of the intrinsic rate of increase in a stable population involves finding the real root r_o of the following integral equation :

$$\int_0^\omega e^{-ra}p(a)m(a)da=1 \tag{1}$$

where r is the intrinsic rate of growth, $p(a)$ is the probability of surviving to age a, and $m(a)$ is the probability of giving birth to a female child at age a.

Lotka uses the first three terms of a Taylor expansion to obtain a quadratic equation in r yielding a very close approximation to the exact root.

The equation is :[2]

$$\frac{1}{2}\beta\, r^2+\alpha\, r-\log_e R_0=0 \tag{2}$$

$$\text{where } \alpha=\frac{R_1}{R_0}$$

$$\beta=\alpha^2-\frac{R_2}{R_0}$$

$$\text{and } R_n=\int_0^\omega a^n\, p(a)m(a)da$$

The calculation required is fairly laborious. Typically one calculates R_0, R_1, and R_2 by dividing the childbearing age interval into five-year groups and approximating the integral expression for R_n by a finite sum. Three such sums must be calculated, and then the quadratic equation (2) solved.

Lotka points out that the correctness of his solution can be verified by substituting the value of r in Equation (1), and approximating the integral by forming a summation by five-year age groups. The summation should, of course, have a value of unity. This method of verification suggests a method of finding a root for (1) that is somewhat less laborious and at least as precise.

In brief, the simpler computation involves substituting a rough approximation for r in Equation (1), ascertaining how much the resulting integral *diverges* from unity, and using the amount of divergence to determine how the approximate value of r must be adjusted.

The rough approximation can be obtained from the relation

$$R_0=e^{rT} \tag{3}$$

(where T is the mean length of generation), and the fact that T normally takes on only a limited range of values, with an average of about 29 years. An assumption that the value of T is 29 years gives a first approximation of r-as $\left(\dfrac{\log_e R_0}{29}\right)$

---

[1] The author proposed in an earlier note a still simpler method of calculating approximate intrinsic rates. This earlier method produced an approximation that was adequate for most purposes but not nearly as close as that produced by the procedure here outlined. *Cf.* Coale, A. J., " The Calculation of Approximate Intrinsic Rates ", *Population Index*, vol. xxi, April 1955, pp. 94–97.

[2] Dublin, Louis I., and Lotka, Alfred J., " On the True Rate of Natural Increase ", *Journal of the American Statistical Association*, vol. xx, no. 151, September 1925.

that may be designated r_1. Substituting r_1 in (1) yields an integral slightly larger or smaller than unity—smaller if r_1 is greater than the correct value of r.

$$\int_0^\omega e^{-r_1 a}\, p(a)m(a)da = 1 + \delta \tag{4}$$

The problem is to infer from δ—the divergence of the integral from unity—the difference between r_1 and the root r_0 of (1). Again following Lotka, let

$$y = \int_0^\omega e^{-ra}\, p(a)m(a)da, \text{ then}$$

$$\frac{dy}{dr} = -\int_0^\omega a\, e^{-ra}\, p(a)m(a)da \tag{5}$$

$$= -Ay, \text{ where } A \text{ is the average age of childbearing}$$
in the stable population.

But since $y = 1$ when $r = r_0$ (the root of Equation (1)),

$$\left(\frac{dy}{dr}\right)_{r=r_0} = -A \tag{6}$$

Hence for values of r differing only slightly from r_0, the integral in (1) will differ from unity by an amount $-A \triangle r$, where $\triangle r$ is the error in r. Conversely, the indicated adjustment in r in order to yield a value of unity for the integral is

$$\triangle r = -\frac{\delta}{A} \tag{7}$$

The mean age of childbearing (A) differs from the mean length of generation (T) by an amount that depends on r,[1] but that in no case exceeds 0·6 of a year, or about 2% of T. Hence 29 years is also a good approximation for A.

Thus r can be estimated by adding $\dfrac{\delta}{29}$ (with due regard for the sign of δ) to the first approximation, r_1. Or, by taking note of the fact that the error in assuming that A equals 29 years is nearly the same as in assuming T is 29 years, and that the relative error in the latter is $\dfrac{\triangle T}{T} \cong -\dfrac{\triangle r}{r}$, a more exact determination is obtained by the following expression :

$$r = r_1 + \frac{\delta}{29 - \dfrac{\delta}{r_1}} \tag{8}$$

---

[1] The difference between A and T is (neglecting higher powers of r) $\dfrac{1}{2}\beta r$, with β defined as $\left(\dfrac{R_1}{R_0}\right)^2 - \dfrac{R_2}{R_0}$ (cf. page 92). β lies between about 35 and 50 years, with a typical value of 42. If r is as great as 3%, T and A will differ by about 0·6 years. (See Dublin and Lotka, *op. cit.*)

As a check of the validity of the technique here described, it was used to recompute a number of values of r already computed by Lotka's procedure. The agreement is excellent. In fact, if the mean length of generation is within about a year of 29 years (as it almost always is), (8) should usually give a closer approximation than the algebraic solution of (2). However, when the intrinsic rate of increase for mid-nineteenth century Sweden is computed, Lotka's procedure yields a value of $9 \cdot 65$ per thousand and (8) gives a value of $9 \cdot 68$. In this instance, the mean length of generation is about $32 \cdot 3$ years, which is unusually long. The value of r that brings the integral in (1) within one part in ten thousand of unity is $9 \cdot 655$ per thousand. But even in this instance (where r_1 is a relatively poor first approximation) the technique outlined here gives an error of only $2 \cdot 5$ per 100,000 in r, or of less than three-tenths of 1% of the correct value.

This method of computation has two virtues to recommend it—it is less laborious than the solution of the quadratic equation while achieving about the same precision, and it has the pedagogical advantage of staying close to fundamental ideas of stable population theory.

Equation (1) follows from the fact that the stable age distribution is

$$c\,(a)da = be^{-ra}\,p(a)da \tag{9}$$

where $c(a)da$ is the fraction of the population in the interval between a and $a + d\,a$, and b is the intrinsic birth rate; and the further fact that the number of births per unit of time must equal the product of the number of women at each childbearing age and the age-specific fertility rate for each age; or

$$b = \frac{B}{N} = \int_0^\omega c\,(a)m(a)da \tag{10}$$

Equation (3) relates the rate of increase *per generation* (the net reproduction rate) in the stable population to the annual rate of increase. No further concepts (such as Thielean semi-invariants) need be introduced.

18. The Fundamental Theorem of Natural Selection

R. A. Fisher (1958 (1930))

From *The Genetical Theory of Natural Selection*, pp. 25—30. New York: Dover.

The total reproductive value, due to Fisher, is of great importance in stable population theory. It is defined as

$$V = \int_0^\infty P(x)\,v(x)\,dx$$

$$= \int_{t=0}^{\beta} \int_{x=0}^{\beta-t} e^{-rt} P(x) \frac{l(t+x)}{l(x)} m(t+x)\,dx\,dt,$$

where $P(x)$ is the observed population between ages x and $x+dx$ at time 0, $v(x)$ is the reproductive value, and standard notation r, $m(x)$ is used in place of Fisher's m and b_x for the intrinsic growth rate and probability of giving birth in the age interval x to $x+dx$, respectively. The fraction $l(t+x)/l(x)$ is the probability that an individual age x at time 0 survives t years to his $(t+x)$th birthday. In words, V is the backward projection of fertility accruing to an observed population to find the size of a birth cohort that would be reproductively equivalent to it. Its application is shown in Feller (1941, paper 16 above).

161

The Malthusian parameter of population increase

If we combine the two tables giving the rates of death and reproduction, we may, still speaking in terms of human populations, at once calculate the expectation of offspring of the newly-born child. For the expectation of offspring in each element of age dx is $l_x b_x dx$, and the sum of these elements over the whole of life will be the total expectation of offspring. In mathematical terms this is

$$\int_0^\infty l_x b_x dx,$$

where the integral is extended from zero, at birth, to infinity, to cover every possible age at which reproduction might conceivably take place. If at any age reproduction ceases absolutely, b_x will thereafter be zero and so give automatically the effect of a terminating integral.

The expectation of offspring determines whether in the population concerned the reproductive rates are more or less than sufficient to balance the existing death rates. If its value is less than unity the reproductive rates are insufficient to maintain a stationary population, in the sense that any population which constantly maintained the death and reproduction rates in question would, apart from temporary fluctuations, certainly ultimately decline in numbers at a calculable rate. Equally, if it is greater than unity, the population biologically speaking is more than holding its own, although the actual number of heads to be counted may be temporarily decreasing.

This consequence will appear most clearly in its quantitative aspect if we note that corresponding to any system of rates of death and reproduction, there is only one possible constitution of the population in respect of age, which will remain unchanged under the action of this system. For if the age distribution remains unchanged the relative rate of increase or decrease of numbers at all ages must be the same; let us represent the relative rate of increase by m; which will also represent a decrease if m is negative. Then, owing to the constant rates of reproduction, the rate at which births are occurring at any epoch will increase proportionately to e^{mt}. At any particular epoch, for which we may take $t = 0$, the rate at which births were occurring x years ago will be proportional to e^{-mx}, and this is the rate at which births were occurring at the time persons now of age x were being born. The number of persons in the infinitesimal age interval

162

dx will therefore be $e^{-mx}l_x dx$, for of those born only the fraction l_x survive to this age. The age distribution is therefore determinate if the number m is uniquely determined. But knowing the numbers living at each age, and the reproductive rates at each age, the rate at which births are now occurring can be calculated, and this can be equated to the known rate of births appropriate to $t=0$. In fact, the contribution to the total rate, of persons in the age interval dx, must be $e^{-mx}l_x b_x dx$, and the aggregate for all ages must be

$$\int_0^\infty e^{-mx}l_x b_x dx,$$

which, when equated to unity, supplies an equation for m, of which one and only one real solution exists. Since e^{-mx} is less than unity for all values of x, if m is positive, and is greater than unity for all values of x, if m is negative, it is evident that the value of m, which reduces the integral above expressed to unity, must be positive if the expectation of offspring exceeds unity, and must be negative if it falls short of unity.

The number m which satisfies this equation is thus implicit in any given system of rates of death and reproduction, and measures the relative rate of increase or decrease of a population when in the steady state appropriate to any such system. In view of the emphasis laid by Malthus upon the 'law of geometric increase' m may appropriately be termed the Malthusian parameter of population increase. It evidently supplies in its negative values an equally good measure of population decrease, and so covers cases to which, in respect of mankind, Malthus paid too little attention.

In view of the close analogy between the growth of a population supposed to follow the law of geometric increase, and the growth of capital invested at compound interest, it is worth noting that if we regard the birth of a child as the loaning to him of a life, and the birth of his offspring as a subsequent repayment of the debt, the method by which m is calculated shows that it is equivalent to answering the question—At what rate of interest are the repayments the just equivalent of the loan? For the unit investment has an expectation of a return $l_x b_x dx$ in the time interval dx, and the present value of this repayment, if m is the rate of interest, is $e^{-mx}l_x b_x dx$; consequently the Malthusian parameter of population increase is the rate of interest at which the present value of the births of offspring to be expected is equal to unity at the date of birth of their parent. The actual values of the parameter of population increase, even in sparsely populated dominions, do not, however, seem to approach in magnitude the rates of interest earned by money, and negative rates of interest are, I suppose, unknown to commerce.

Reproductive value

The analogy with money does, however, make clear the argument for another simple application of the combined death and reproduction rates. We may ask, not only about the newly born, but about persons of any chosen age, what is the present value of their future offspring; and if present value is calculated at the rate determined as before, the question has the definite meaning—To what extent will persons of this age, on the average, contribute to the ancestry of future generations? The question is one of some interest, since the direct action of Natural Selection must be proportional to this contribution. There will also, no doubt, be indirect effects in cases in which an animal favours or impedes the survival or reproduction of its relatives; as a suckling mother assists the survival of her child, as in mankind a mother past bearing may greatly promote the reproduction of her children, as a foetus and in less measure a sucking child inhibits conception, and most strikingly of all as in the services of neuter insects to their queen. Nevertheless such indirect effects will in very many cases be unimportant compared to the effects of personal reproduction, and by the analogy of compound interest the present value of the future offspring of persons aged x is easily seen to be

$$v_x = \frac{e^{mx}}{l_x} \int\limits_x^\infty e^{-mt} l_t b_t dt.$$

Each age group may in this way be assigned its appropriate reproductive value. Fig. 2 shows the reproductive value of women according to age as calculated from the rates of death and reproduction current in the Commonwealth of Australia about 1911. The Malthusian parameter was at that time positive, and as judged from female rates was nearly equivalent to $1\frac{1}{4}$ per cent. compound interest; the rate would be lower for the men, and for both sexes taken together, owing to the excess of men in immigration. The reproductive value,

164

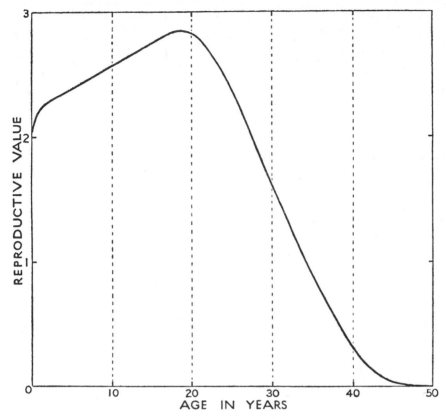

Fig. 2. Reproductive value of Australian women.

The reproductive value for female persons calculated from the birth- and death-rates current in the Commonwealth of Australia about 1911. The Malthusian parameter is $+0\cdot01231$ per annum.

which of course is not to be confused with the reproductive rate, reaches its maximum at about $18\frac{1}{2}$, in spite of the delay in reproduction caused by civilized marriage customs; indeed it would have been as early as 16, were it not that a positive rate of interest gives higher value to the immediate prospect of progeny of an older woman, compared to the more remote children of a young girl....

A property that well illustrates the significance of the method of valuation, by which, instead of counting all individuals as of equal value in respect of future population, persons of each age are assigned an appropriate value v_x, is that, whatever may be the age constitution of a population, its total reproductive value will increase or decrease according to the correct Malthusian rate m, whereas counting all heads as equal this is only true in the theoretical case in which the population is in its steady state. For suppose the number of persons in the age interval dx is $n_x dx$; the value of each element of the population will be $n_x v_x dx$; in respect of each such group there will be a gain in value by reproduction at the rate of $n_x b_x v_o dx$, a loss by death of $n_x \mu_x v_x dx$, and a loss by depreciation of $-n_x dv_x$, or in all

$$n_x \{ (b_x v_o - \mu_x v_x)\, dx + dv_x \},$$

but by differentiating the equation by which v_x is defined, it appears that

$$\frac{1}{v_x}\frac{dv_x}{dx} + \frac{1}{l_x}\frac{dl_x}{dx} - m = \frac{-l_x b_x e^{-mx}}{\dfrac{v_x}{v_o} l_x e^{-mx}} = -\frac{b_x v_o}{v_x},$$

or that

$$dv_x - \mu_x v_x dx + b_x v_o dx = m v_x dx.$$

Consequently the rate of increase in the total value of the population is m times its actual total value, irrespective of its constitution in respect of age. A comparison of the total values of the population at two census epochs thus shows, after allowance for migration, the genuine biological increase or decrease of the population, which may be entirely obscured or reversed by the crude comparison of the number of heads. The population of Great Britain, for example, must have commenced to decrease biologically at some date obscured by the war, between 1911 and 1921, but the census of 1921 showed a nominal increase of some millions, and that of 1931 will, doubtless in less degree, certainly indicate a further spurious period of increase, due to the accumulation of persons at ages at which their reproductive value is negligible.

19. How the Age Distribution of a Human Population is Determined

Ansley J. Coale (1957)

Cold Spring Harbor Symposia on Quantitative Biology 22: 83—88.

This analysis of age distribution will be restricted to closed populations of human females; restricted to closed populations because to consider the effects of migration would be unduly complicated; restricted to human populations because the author is a demographer rather than a biologist; and restricted to females because differences in age composition and in age at parenthood between the sexes make a combined treatment awkward. If data were equally available, a completely similar analysis could be made for males, with only one major reservation: The special effects of war mortality on males of military age would require special analysis.

I. STABLE POPULATIONS

The forces affecting the shape of the age distribution are most readily visualized in the special case of an unchanging distribution. Lotka (Dublin and Lotka, 1925; Lotka, 1939) has shown that constant mortality and fertility schedules will ultimately produce a constant age distribution, and a constant rate of growth. This constant age distribution can be calculated as the product of a factor representing relative number of survivors to each age from birth (the life table) and a factor reflecting the continuously growing (or shrinking) number of births. Specifically:

$$(1) \qquad c(a) = be^{-ra}p(a)$$

where $c(a)$ is the fraction at any given age a, b is the birth rate (which Lotka shows is constant under the specified conditions) r is the growth rate, and $p(a)$ is the fraction who survive from birth to age a. No matter what the initial age distribution (provided the childbearing ages are amply represented) a constant age schedule of fertility $(m(a))$ and probability of surviving $(p(a))$ eventually establish the distribution given in equation (1). Actually, a period of 50 to 100 years is adequate to produce a very close approximation to the stable population.

The basis for equation (1) can be simply explained. The birth rate (b) is the proportion at age zero (set $a = 0$, and the other factors are each unity). The term e^{-ra} relates the size at birth of the cohort now aged a to the current birth cohort, and $p(a)$ allows for the attrition of mortality.

Since $b = \dfrac{1}{\displaystyle\int_0^\omega e^{-ra}p(a)\,da}$ (where ω is the

highest age attained) the stable age distribution is wholly determined by the growth rate r, and the survivorship function $p(a)$. The value of r in turn can be calculated from the second fundamental equation of stable population theory.

$$(2) \qquad \int_0^\omega e^{-ra}p(a)m(a)\,da = 1$$

The real root of this integral equation is the stable growth rate, while the complex roots determine how the stable population is approached from arbitrary initial conditions.

We now turn to the role of fertility and mortality in determining the shape of the stable age distribution. The role of mortality will be described first.

Mortality and the stable age distribution

Two rather surprising conclusions emerge when the effect of mortality schedules on the stable age distribution is examined:

(a) The effect of alternative mortality schedules is relatively minor. Roughly similar age distributions result from a given fertility schedule in conjunction with a very high mortality life table on the one hand, or a very low mortality life table on the other. Figure 1 illustrates this point. The life table of Sweden for 1860, with a life expectancy of 45.4 years, and that for 1946 to 50, with a life expectancy of 71.6 years produce about the same stable age distributions when combined with the same fertility schedule.

(b) Among life tables reflecting experience so far recorded, it is nearly universally true that a more favorable mortality schedule—with a higher life expectancy—will yield a stable population with a higher proportion under 15. In a vast majority of contrasting life tables, the lower mortality life table will produce a distribution with a lower average age; and at least half the time, lower mortality will produce a smaller fraction over 65.

These conclusions are partly analytical and partly empirical. A full analysis of the contrasting stable age distributions associated with different mortality schedules is somewhat laborious (Coale, 1956). A simple qualitative argument is enough to dispel the common belief that lower mortality inevitably means an older population. If in one life table the probability of surviving for a year at each age exceeded the corresponding probability in another life table by a fixed proportion (for example, 1%) the two life tables would produce precisely the same age distributions. The tendency toward more survivors with advancing age would be exactly offset by a faster rate of growth that tends to make each cohort *smaller* than the next younger.

This point can be proved by assuming a sudden shift to a life table with a one per cent higher probability of surviving at each age. The year following this change, there would be one per cent more one-year olds surviving from birth, one per cent more two-year olds, etc. Improved survivorship would produce one per cent more persons at every age above 0. But since the increase would yield one per cent more persons at every childbearing age, with constant fertility there would be one per cent more births—persons at age 0—as well. The whole population would be one per cent larger; and the age distribution would be unaffected.

The common view that lower mortality means an older population takes account of only part of the effect of lower death rates. It is immediately clear that lower death rates produce more old people. However, lower death rates also produce more parents, more births, and more children. Whether the dominant effect on the stable age distribution of a lower level of mortality is to enlarge the upper end of the age distribution through higher survivorship, or to tilt the age distribution more steeply through more rapid growth depends on the relative age pattern of mortality in the two life tables (Coale, 1956).

168

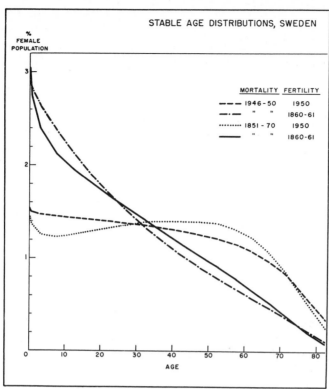

FIGURE 1.

A particularly illuminating way of comparing the age pattern of mortality is to compute the proportionate difference in the probability of surviving for one year at each age. Let the probability of surviving from age a to $a + 1$ be $\pi'(a)$ in the better life table, and $\pi(a)$ in the higher mortality table, and consider the ratio $\frac{\pi' - \pi}{\pi}(a)$. We have already shown that a constant value of $\frac{\pi' - \pi}{\pi}(a)$ implies the same age distributions. We can subtract the minimum value of $\frac{\pi' - \pi}{\pi}$ from the value at other ages—the subtracted portion is equivalent to no difference in mortality—and consider only the residual. A large excess of $\frac{\pi' - \pi}{\pi}$ above the minimum at ages below childbearing means there is a large difference in the *growth rates* of the two stable distributions. Moreover, since a large excess $\frac{\pi' - \pi}{\pi}$ at the youngest ages implies more survivors at *all* subsequent ages, the *proportion* of survivors above—say age 50 in the better life table would not be much larger than in the poorer one. In short, an excess of $\frac{\pi' - \pi}{\pi}$ above the minimum at young ages means that the better mortality schedule tends to have a *younger* age distribution.

On the other hand, an above-the-minimum $\frac{\pi' - \pi}{\pi}(a)$ at ages over 50 implies no difference in the long term growth rate (since the reproductive ages are not involved) but does mean more old age survivors in the better life table. The net result is a higher fraction at ages above 50.

The effects of three special instances of percent difference in the probability of surviving have been described in a non-rigorous fashion. These particular patterns have been emphasized because differences among the great majority of recorded life tables can be approximated as the sum of three components—a minimum per cent difference in the probability of surviving from age 5 to 50, above minimum differences below age 5, and above minimum differences above age 50 (see Fig. 2). The general pattern of $\frac{\pi' - \pi}{\pi}(a)$ is roughly U-shaped, declining from a high at age zero to near its minimum by age 5. It is relatively constant until age 50 or 60 where it frequently but not universally rises. The central portion (from 5 to 50) contributes nothing to the differences in stable age distribution, serving only to diminish the effect of the differences represented by the two legs of the U, and to raise the growth rate of the population.

The two legs of the U-shaped pattern work more or less in opposition in causing differences in

169

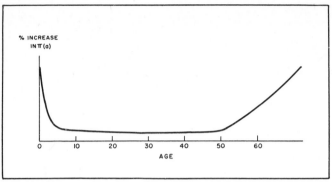

% INCREASE
IN π (a)

AGE

FIGURE 2.

age distribution. If there were no right leg, the age distribution with the better life table could be approximated at all ages except those under 5 by a sort of pivoting of the higher mortality distribution on its average age, with increased fractions at less than the average age, and reduced fractions above. If there were no left leg, rising per cent differences in survivorship over age 50 would mean higher fractions above 50, and lower fractions below. When the two legs operate together, the net effect is to reduce the differences each would cause alone. If the left leg is large, and the right leg is small, the effect of growth rates will dominate; and the low mortality population may have smaller proportions at all ages above the average. This domination is typical when a life table of high mortality is compared with a moderate or low mortality table. However, when mortality of 20 years ago in the most advanced areas is compared with current mortality, the right leg—improved survivorship at the older ages—takes on more importance. There is not much room for a left leg in a comparison of future life tables with the best current ones. A two or three per cent further improvement in the probability of surviving to age 5 in Sweden would raise this probability to one.

There is no logical necessity that mortality differences should produce only minor differences in stable age distributions. This result arises from the age structure of differences among life tables observed to date. However, a contributing factor in the typical U-shaped age structure of differences in the probability of surviving is that large differences can last only when there is room for them. Differences of the order of those observed for age 0 could not occur at those ages where even with high general mortality levels the risk of death is moderate.

Fertility levels and the stable age distribution

The role of fertility in shaping the stable age distribution is at once simpler than that of mortality, and quantitatively much more important, when account is taken of the range of mortality and fertility observed in the world. The simplicity and the quantitative importance have a common source: the fact that differences in fertility operate in a single direction in affecting the relative shape

of the age distribution. While higher probabilities of surviving simultaneously flatten out $p(a)$ in equation (1) and make e^{-ra} steeper, higher probabilities of giving birth affect only the growth rate—making it larger, of course, and making the exponential factor in equation (1) contribute to a more rapid taper in the distribution.

If a high fertility stable age distribution is compared with a low fertility distribution with the same mortality, the two will be found to intersect at the mean of the average ages. The higher fertility population will have higher proportions at ages below the average, of course.

Figure 1 makes it clear that fertility differences can produce profoundly different stable age distributions. This fact together with the relatively slight influence of mortality on the stable age distribution means that a schedule of fertility by age is sufficient to give at least a fair approximation to the stable age distribution even if mortality rates by age are not known. One would simply use whatever female life table was lying around in calculating the stable population. It must be admitted that there is a much better chance of a close fit if some hint about infant mortality is visible.

One final comment on a common sense basis for understanding the powerful influence of fertility on the stable age distribution. The general fertility rate establishes the ratio of the area in the age distribution from ages 17 to 44 to the zero ordinate of the distribution. If fertility is twice as high, this ratio must be cut in half irrespective of mortality. The exact inverse relation of this ratio to fertility is a consequence of the fact that the zero ordinate is proportional to births, and the area from 17 to 44 to the number of women of childbearing age. Thus the fertility level clamps a vice on the relation of the beginning of the distribution to an area near the middle.

Several straightforward conclusions emerge from this consideration of stable age distributions:

(1) Sustained high fertility (average completed size of family, 6 or more children) produces a young population with a median age well below 25 years, more than 40 per cent of the population under 15, and no more than three or four per cent over 65.

(2) Sustained low fertility (average completed size of family below 2.5 children) pro-

170

duces an old population with a median age above 35, no more than 20 per cent of the population under 14, and at least 15 per cent over 65.

(3) In general, the approximate form of the age distribution is determined by the level of fertility. The level of mortality has more or less second order effects on the distribution. The general quality of these effects has been to date that low mortality yields a slightly larger fraction at ages up to at least 15 but as high as the average age, and somewhat smaller fractions either at all higher ages or until the age 50 or higher. More often than not, sustained low mortality yields a slightly lower average age.

II. Varying Fertility and Mortality and the Age Distribution

We turn now to a brief consideration of how an age distribution is determined under a regime of continuously changing mortality and fertility rates.

The general equations corresponding to equation (1) are:

(3) $n(a, t) = B(t - a)p(a, t)$ where $B(t - a)$ is the number of births a years before time t, $n(a, t)$ is the number of persons at age a at time t, and $p(a, t)$ is the proportion of those born at $(t - a)$ who survive to achieve age a at time t, and

(4) $$c(a, t) = \frac{n(a, t)}{\int_0^\omega n(a, t)\, da}.$$

The equation corresponding to (2) is:

(5) $$B(t) = \int_0^\omega n(a, t)m(a, t)\, da$$ where $m(a, t)$ is the probability of bearing a female child at age a and time t.

These equations of course do not lead to Lotka's tidy solutions since we here permit mortality and fertility to vary with time. In fact, they do not even give much clue to the role of fertility and mortality in determining age distributions, since equation (3) requires us to know $B(t - a)$, and equation (5) tells us that $B(t - a)$ depends on the age distribution at time $(t - a)$ as well as on fertility rates at time $(t - a)$. It would appear that to account for the present age distribution, one needs to know:

(a) an age distribution at some past date, and
(b) schedules of mortality and fertility since that date.

However, as the date of the past age distribution is made more remote, its form makes less and less difference to the shape of the current age distribution. It seems intuitively plausible, in fact, that if the course of fertility and mortality were known since $t - \infty$, the proportionate age distribution would be wholly determined at time t no matter what the distribution at $t - \infty$, assuming of course that the initial distribution was not one—for example with no one under 50—headed for extinction. Lotka (1939) shows this statement to hold in the special case of endlessly unchanging fertility and mortality. But the same factors that cause the transient effects of a particular initial age distribution to disappear from the stable population would also operate for *any* observed time path of fertility and mortality rate. After a a suitably long period the effect of an initial age distribution is swamped by the cumulative effect

of the time pattern of vital rates. To put the point more concretely, any age distribution with persons at every age could be assumed for 100 years ago in place of the actual distribution. If such an assumed population were projected to the present with observed fertility and mortality rates, the proportionate age distribution would differ negligibly from the actual. In short, the age distribution of a closed population is determined by the mortality and fertility rates of recent history.

Another result from stable population analysis is suggestive for our general case—the conclusion that mortality differences have only second order effects on the age distribution. An increase or decrease in mortality tends to decrease or increase all cohorts, implying a small effect of mortality on the immediate age distribution as well as the stable. The short run transient age distribution effect of a mortality change may differ somewhat in form from the long run effects as expressed by stable age distributions, but the magnitude should be small in both instances.

We shall proceed with the provisional hypothesis that changes over time in mortality schedules do not have major effects on the age distribution. If this hypothesis is valid, a current age distribution could be closely approximated by calculating what the distribution would have been had observed fertility rates and *unchanging* mortality rates prevailed for the last 80 or 90 years. I have tried such a calculation using Swedish data. The results are shown in Figure 3. The proportionate distribution based on mortality unchanged for 90 years indeed does come close to the actual distribution. Moreover, the differences between the census population and the hypothetical are nearly identical to the differences between two stable populations with the same fertility, one based on an 1860 Swedish life table, and the other on a life table for 1946 to 50 (compare Fig. 1). The principal effect of projecting with unchanged mortality is to produce smaller fractions at ages under 25 and over 70. The reason is that since 1860 there have been disproportionate improvements in the probability of surviving in infancy and in the older ages. If the projection from 1860 is calculated with mortality unchanging at 1946 to 50 levels, the result is a distribution virtually indistinguishable from the census distribution. If in the 90 years before 1950 mortality had remained as high as in 1860 the result would have been a much smaller female population with an age distribution similar to the actual; if in these 90 years mortality had always been as low as in 1950, the result would have been a much *larger* female population with an age distribution nearly identical to the actual.

Figure 3 also shows the distribution that would have resulted with observed mortality risks during the 90 years before 1950, and with fertility assumed constant at 1860 levels. This figure makes clear what is the major determinant of an age distribution—the course of fertility. I wish I were now able to present a short, simple explanation of the effects of fertility on the age distribution. It is clear that fertility determines the age distribution, but an attempt to explain the relationship precisely soon runs afoul of major complications.

The basic difficulty is that when high fertility produces a brief series of unusually large cohorts, for example, these cohorts not only exceed their

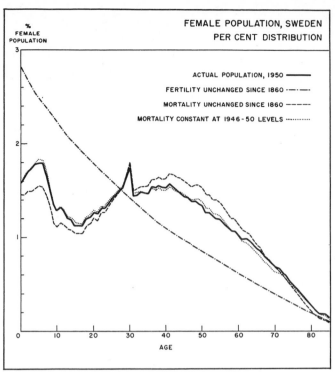

FIGURE 3.

neighbors through their lifetime (this is a simple effect), but also when they pass through the child-bearing ages they produce a diffused "echo" of larger birth cohorts. It is easy to give instructions about how to compute these consequences, but impossible to describe them simply in terms of the resultant age distribution.

We can make these observations about the effects of fertility on age distribution:

(1) Current and recent high fertility produce a younger population than would low fertility, and *vice versa*.

(2) Transitory waves of unusually high or low fertility create humps and hollows that move out through the age distribution as the cohorts move through life. In fact all of the notches and knobs in the Swedish female age distribution of 1950 can be traced back to unusual birth crops.

(3) A long period of high fertility, or a period of rising fertility creates a section of the age distribution that tapers rapidly with age. Conversely a long period of low fertility, or a period of falling fertility creates a relatively flat (or even rising) section of the age distribution.

(4) When a cohort of unusual size reaches the childbearing ages, it sets up an attenuated and flattened out "echo" in the number of births. It is this second generation effect that makes the analysis of the relation of the age distribution to the course of fertility so complex.

Finally I turn to one or two practical or at least worldly observations arising from this anslysis.

First, the rising fraction of the aged in western countries has not resulted from lowered death rates but almost wholly from a long history of declining fertility. However, future improvements in mortality may make this statement obsolete. Second, improved mortality will not reduce the "burden of dependency" imposed on low income areas because of their age distributions. To the contrary, mortality reduction can be expected to lead to a somewhat higher fraction in dependent ages. If the burden is to be reduced, fertility must be lowered. Third, the remarkable variation in fertility in the U. S. during the past 20 years has produced a very irregular age distribution. The succession of variously sized cohorts will doubtless have interesting implications in the next half century. Thus in the next 20 years the number of persons 20 to 24 is due to increase by at least 75 per cent, while the number 40 to 44 can be expected to decline by about three per cent.

REFERENCES

Coale, Ansley, J., 1956, The effects of changes in mortality and fertility on age composition. Milbank Mem. Fund Quart. *34:* 79–114.
Dublin, Louis I., and Lotka, Alfred J., 1925, On the true rate of natural increase. J. Amer. Statis. Assoc. *20:* 305–327.
Lorimer, Frank, 1951, Dynamics of age structure in a population with initially high fertility and mortality. Population Bull. U. N. *1:* 31–41.
Lotka, Alfred J., 1939, Théorie analytique des associations biologiques. Deuxième partie: Analyse démographique avec application particulière à l'espèce humaine. Paris, Hermann et Cie.

20. On the Reproduction of Organisms with Overlapping Generations

W. R. Thompson (1931)

From *Bulletin of Entomological Research* 22. Excerpts are from pages 147—152, 154—160, 163, 168—170.

We omit the final section of Thompson's paper, in which he shows the effects on population size of (a) lengthening the pre-reproductive interval, (b) lengthening the period of reproduction, (c) increasing the number of offspring produced per day, and (d) compacting the reproductive period while leaving the number of offspring constant. The effects can be summarized in the approximate equality, due to Lotka (1939, pp. 70—71)

$$r = \frac{\mu - \sqrt{\mu^2 - 2\sigma^2 \ln R_0}}{\sigma^2} = \frac{\ln R_0}{\mu - \dfrac{\sigma^2 r}{2}}$$

where r is the intrinsic rate of growth, R_0 the net reproduction rate, and μ and σ^2 the mean and variance of the net maternity function. Where the net maternity function takes the form of a normal distribution the equalities are exact.

Much of Thompson's work can be translated to matrix form, where his generation law G becomes the first row of a projection matrix whose sub-diagonal terms (survival) are unity. Thus, the generation law: $G = (aT^1 + bT^2 + cT^3 + dT^4)$, becomes the projection matrix M:

$$
\begin{array}{cccc}
T^1 & T^2 & T^3 & T^4
\end{array}
$$

$$
\begin{bmatrix} a & b & c & d \\ 1 & 0 & 0 & 0 \\ 0 & 1 & 0 & 0 \\ 0 & 0 & 1 & 0 \end{bmatrix}
\begin{bmatrix} B(T^0) \\ 0 \\ 0 \\ 0 \end{bmatrix}
=
\begin{bmatrix} B(T^1) \\ B(T^0) \\ 0 \\ 0 \end{bmatrix}
$$

$$\quad M \qquad\qquad B^{(0)} \qquad\quad B^{(1)}$$

which we show post multiplied by the initial forebears $B(T^0)$ to yields births $B(T^1)$ and total population $B^{(1)}$ at time 1. (Mortality can be subsumed in the generation law or in the subdiagonals, hence the form shown here is quite general.) Matrix formulation facilitates the calculation of characteristic roots and vectors but was not yet adopted when Thompson wrote. It does not lend itself as conveniently to the separation of births by generation or to an appreciation of generation overlap, which constitute important contributions of Thompson's work.

INTRODUCTION.

In the great majority of organisms, the reproducing individual does not engender all of its progeny simultaneously. Between the beginning and the end of the reproductive period there is usually a certain interval of time, during which the production of offspring continues in a manner depending, on the one hand, upon the specific characteristics of the organism, and on the other, upon the conditions under which it lives. From this simple fact follow some interesting consequences.

If the successive generations are separated by a long interval of time and conditions during this interval are of such a nature that the organism is inactive and in the condition of obligatory dormancy, known as diapause, then, in spite of the increase of the population, the generations will remain separate and distinct, and a comparison of the total populations in the successive reproductive periods will enable us to determine the rate of increase.

On the other hand, if the interval between generations is short and favourable to activity and the organism does not pass through a period of obligatory dormancy at this time, then, as the species increases, the successive generations will come to overlap, and the amount of overlapping will increase progressively as time goes on.

This is primarily due to the fact that when there is reproduction over a period of time-intervals, in a typical individual of the species, and conditions remain constant, then the number of time-intervals during which births occur will increase from generation to generation. In this way there will be produced a lengthening of the reproductive period for the whole mass of individuals ; and this will cause a diminution of the intervals between stages in the same generations, and between generations, so that individuals belonging to different stages of development, and to different generations, will be found simultaneously in the field.

On account of these facts the calculation of the number of progeny and of the number of individuals in successive stages, existing at any given moment, presents serious difficulties, which make both the practical and the theoretical treatment of the subject laborious. Some years ago I began to study the matter and succeeded in getting the process of calculation into a form in which the required data could be obtained, but the method of calculation was extremely tedious. The results obtained were accordingly submitted, through the kindness of Professor M. Greenwood, of the Department of Epidemiology and Vital Statistics of the London School of Hygiene and Tropical Medicine, to Mr. H. E. Soper. Within a few days I received from Mr. Soper a communication, in which he explained and developed in detail an extremely elegant method of dealing with the problem. It is this method which I have utilised in this paper. I have added such explanations as seem to me necessary in order to make the process of reasoning clear to those who, like myself, have little knowledge of mathematics, and have worked out a series of examples to show how the method may be employed in entomological problems. The credit for the solution utilised is entirely due to Mr. Soper, and I am glad to have this occasion of thanking him for the kind and patient interest he has shown in the problems submitted to him. ...

The first part of the paper is devoted to an exposition of the problem ; the second part to the methods which can be used in dealing with it ; and the third part to an application of these methods. The mathematical theory of the method is given by Mr. Soper in the Appendix.

The object of the paper is, in the first place, to enable the entomologist to investigate the natural control of insects with reproduction of this type ; and, in the second place, to show how variation in the length of the pre-reproductive and reproductive periods, and in the production of offspring, affect the increase in population. When the length of life and of the pre-reproductive and reproductive periods

174

have been determined, and the average daily production of progeny is known, we can calculate from the formulae given the daily births or the daily population of individuals of all ages, when no causes of mortality intervene. By comparing the figures with those obtained in an actual experiment, we can then determine when causes of mortality begin to operate and get some idea as to the nature and effect of the controlling factors which come into play at any given date. Organisms with overlapping generations are particularly suitable for work of this kind, because of their rapidity of reproduction, but the complex way in which their populations increase has, hitherto, constituted a very serious obstacle to their study. The formulae developed by Mr. Soper enormously simplify the problem and constitute, I believe, a very important contribution to our methods of study of the process of natural control.

I.

Let us suppose, to begin with, that we have an *initial organism* or stem-mother, producing d offspring per day during a certain period, which we may call k days for the moment ; and let us suppose that the reproducing individual dies on the day it has produced its last offspring. Then, in the first generation, if we put $k=4$, for the sake of simplicity, we shall have births occurring as follows : —

Day ...	1	2	3	4	5	6	7	8	9	10
G.I ...	d	d	d	d						

Now suppose that both the organisms produced in the first generation and the environmental conditions under which they live remain identical. The group of organisms produced on the second day will reproduce a day later than those produced on Day 1 ; the group produced on the third day, two days later than those produced on Day 1, and so on. The second generation, produced by the reproduction of the groups occurring on the successive days, will thus be as follows, since each individual of the group of d produces d offspring :—

	Day	1	2	3	4	5	6	7	8	9	10
G.II	... 1	d^2	d^2	d^2	d^2						
G.II	... 2		d^2	d^2	d^2	d^2					
G.II	... 3			d^2	d^2	d^2	d^2				
G.II	... 4				d^2	d^2	d^2	d^2			
Total		$1d^2$ ·	$2d^2$ ·	$3d^2$ ·	$4d^2$ ·	$3d^2$ ·	$2d^2$ ·	$1d^2$			

In the next generation the process will obviously continue in the same way ; during the first reproductive period $1d^2$ individuals will produce d offspring per day ; during the second period $2d^2$ will produce d each per day ; during the third period $3d^2$ will produce d per day, and so on—the result being as follows :—

	Day	1	2	3	4	5	6	7	8	9	10
G.III	... 1	d^3	d^3	d^3	d^3						
	2		$2d^3$	$2d^3$	$2d^3$	$2d^3$					
	3			$3d^3$	$3d^3$	$3d^3$	$3d^3$				
	4				$4d^3$	$4d^3$	$4d^3$	$4d^3$			
	5					$3d^3$	$3d^3$	$3d^3$	$3d^3$		
	6						$2d^3$	$2d^3$	$2d^3$	$2d^3$	
	7							d^3	d^3	d^3	d^3
Total		$1d^3$ ·	$3d^3$ ·	$6d^3$ ·	$10d^3$ ·	$12d^3$ ·	$12d^3$ ·	$10d^3$ ·	$6d^3$ ·	$3d^3$ ·	$1d^3$

175

When we compare the series of births in the successive generations the first thing that we notice is that the number of days (or time-intervals) on which births occur becomes greater in each generation. Thus, as an organism increases in numbers, the length of time during which births occur in the field will also increase, from generation to generation, and so will the length of time during which the successive developmental stages are found, without there being, necessarily, any deviation whatever from the specific reproductive habits, or any change in the environmental conditions ; or, in other words, any real irregularity in the course of events. ...

Furthermore, since the number of time-intervals during which births occur increase from generation to generation, it is also evident that, as time goes on, the interval between the end of one generation and the beginning of another will become smaller and smaller, and then disappear, after which the successive generations will overlap. The overlap will become greater and greater as the generations succeed one another. Finally, the number of generations overlapping in a given time-interval will also become greater, though the rate of increase is not a smooth and regular one.

Thus, suppose the pre-reproductive period of a species is of two days (including day of birth) and the reproductive period also two days, each individual producing d births per day, the succession of events will be as follows :—

The first and second generations do not overlap, but the second and third overlap on one day, the third and fourth on two days, the fourth and fifth on three days, the fifth and sixth on four days, the nth and $(n+1)$th on $(n-1)$ days. On the first four days individuals of only one of the two generations produced are found on any one day ; on the fifth day individuals of G.II and G.III co-exist ; on the seventh and eighth days individuals of G.III and G.IV, on the ninth and tenth days individuals of G.IV and G.V ; on the eleventh day individuals of G.IV, V, and VI, and on the seventeenth day individuals of G.VI, VII, VIII, and IX.

II.

It will be evident from what has preceded that the direct calculation of the progeny produced day by day in the successive generations, to say nothing of the total number of progeny produced by all generations taken together, or, to take the most difficult case of all, of the total population, including both the newly born and the older individuals, is a very complex matter. In order to deal with the problem it is, therefore, necessary to simplify the operation.

It is, of course, possible to represent the increase in population by using the compound interest formula, according to which the number of individuals existing at any given moment is given by the equation

$$N_1 = N_0 \epsilon^{Kt}$$

Where N_1 is the population at the moment considered, N_0 the initial population, t the number of time-intervals (hours, days, years, etc.), ϵ the incommensurable number $2 \cdot 71828$ . . ., and K a constant depending upon the particular case considered. This formula is, however, not satisfactory for experimental work. In the first place, the constant K cannot be determined until the growth of the population under certain definite conditions has been studied during a considerable period ; in the second place, no intelligible significance can be attached to the constant after its value has been determined ; in the third place, the growth of the population is considered in this formula to be at every moment proportional to the size of the population, which is not true except with large numbers and over long periods, and cannot be safely taken as a basis for the examination of experimental data. Thus, using the figures for births in the last example given in this paper, we obtain values from K ranging from $0 \cdot 652$ to $0 \cdot 693$. The former value is valid when the population reaches a high level, but under such conditions accurate analytical experimental work is not possible. The number of individuals at the end of five time-intervals might be anything from 208 to 255. It is therefore necessary to deal with the reproductive process as a discontinuous phenomenon. ...

Suppose the ancestor or stem-mother is born on T^0, and that we represent the successive days or "time-intervals" on which reproduction takes place by T^1, T^2, T^3, T^4, . . . meaning the first, second, third, fourth, etc., days, and suppose one offspring is produced each day. Suppose the reproductive period is one of four days, and the pre-reproductive period is of one day, including the day of birth, then the first generation, $= G_1$ may be written—

$$G_1 = 1\,T^1 + 1\,T^2 + 1\,T^3 + 1\,T^4.$$

Now the one individual born on T^1 will begin to reproduce on T^2 and from thenceforward during four days; the one individual born on T^2 will begin to reproduce on T^3 and from thenceforward during four days; and similarly, for the individuals born on T^3 and T^4, which begin reproducing on T^4 and T^5 respectively. Thus, writing the series of births for G^2 we have—

$$\begin{aligned}
G_2 = &\ 1\,T^2 + 1\,T^3 + 1\,T^4 + 1\,T^5 \\
&\ 1\,T^3 + 1\,T^4 + 1\,T^5 + 1\,T^6 \\
&\ 1\,T^4 + 1\,T^5 + 1\,T^6 + 1\,T^7 \\
&\ 1\,T^5 + 1\,T^6 + 1\,T^7 + 1\,T^8 \\
\hline
&\ 1\,T^2 + 2\,T^3 + 3\,T^4 + 4\,T^5 + 3\,T^6 + 2\,T^7 + 1\,T^8
\end{aligned}$$

which is—

$$= (T^1 + T^2 + T^3 + T^4)^2.$$

In G^3, the one individual produced on T^2 will reproduce on four succeeding days, and similarly, for the two produced on T^3, the three produced on T^4 and so on; and if these series of births are written out and added as before, we obtain—

$$\begin{aligned}
G_3 = &\ 1\,T^3 + 3\,T^4 + 6\,T^5 + 10\,T^6 + 12\,T^7 + 12\,T^8 + 10\,T^9 + 6\,T^{10} + 3\,T^{11} + 1\,T^{12} \\
= &\ (T^1 + T^2 + T^3 + T^4)^3.
\end{aligned}$$

It is clear that, in general, we have—

$$G_n = (T^1 + T^2 + T^3 + T^4)^n,$$

and that by multiplying this out, or, as the mathematicians say, "expanding it," we shall obtain a series of T's multiplied by numbers, or coefficients in which the index of T designates the *day* and the coefficient gives the number of births on that day in the generation considered.

From this simple case all others may readily be derived without any alteration in the reasoning.

Thus, suppose the pre-reproductive period is of three days, including the day of birth, and the reproductive life is of four days, as before. If the original ancestor or stem-mother was born on T^0 (an assumption made in the first case), then its first offspring would be produced, not on T^1 but on T^3, and the first generation would be written—

$$G_1 = (1\,T^3 + 1\,T^4 + 1\,T^5 + 1\,T^6) ;$$

then the individual produced on T^3 would begin reproducing on T^6, and we should obtain, by proceeding as before, a series for the second generation equivalent to—

$$(1\,T^3 + 1\,T^4 + 1\,T^5 + 1\,T^6)^2.$$

Finally, if, instead of one birth on each day, we assume d on each day, we should obtain—

$$G_1 = d(T^1 + T^2 + T^3 + T^4) ;$$
$$G_2 = d^2(T^1 + T^2 + T^3 + T^4)^2 ;$$
$$\dotsb$$
$$G_n = d^n(T^1 + T^2 + T^3 + T^4)^n ;$$

while if the number of births varied from day to day, so that we had—

$$G^1 = (a\,T^1 + b\,T^2 + c\,T^3 + d\,T^4) ; \text{ we shall have}$$
$$\dotsb$$
$$G_n = (a\,T^1 + b\,T^2 + c\,T^3 + d\,T^4)^n.$$

Suppose that on one or more days in the reproductive period no progeny are produced. This is allowed for simply by omitting these days. Thus, in the last series mentioned, let $b=0$. The formula for the nth generation will then be :—

$$G_n=(aT^1+cT^3+dT^4)^n. \ ...$$

What has preceded covers fully the series of births (or production of offspring) in each generation considered separately; but it does not give us any information as to the overlapping of the generations, nor does it enable us to calculate directly the *total* number of births produced by the individuals of all the different generations which happen to be reproducing on any given day; and that individuals belonging to different generations will be producing simultaneously we have already seen....

If, taking the successive generations in the example already studied, we write them down and add, we have—

$$G_1+G_2+G_3=(T^1+T^2+T^3+T^4)+(T^1+T^2+T^3+T^4)^2$$
$$+(T^1+T^2+T^3+T^4)^3.$$

It is obvious, from what has preceded, that by expanding these expressions and adding up all the T^1, T^2, T^3 .... T^{12} we shall get the number of births produced by all the individuals of all the generations existing simultaneously on any given day, up to and including the third generation.

If we designate—

$$T^1+T^2+T^3+T^4 \text{ by } G$$

so that we have

$$G_1=G$$
$$G_2=G^2$$
$$..........$$
$$G_n=G^n,$$

we can write

$$G_1+G_2+G_3=G+G^2+G^3$$
$$=\frac{G(1-G^3)}{1-G}$$

by the usual rule for the summation of a geometrical series.

But this formula does not represent the *total* number of individuals produced up to the day on which G_3 ends (T^{12}), because by that time generations G_4 to G_{12} inclusive will already have begun.

Let us, therefore, take the sum of $G_1+G_2+G_3+...........$to n generations. We have, including the original ancestor or stem-mother—

$$1+G_1+G_2+G_3+.........+G_n=1+G+G^2+G^3+.........+G^n$$
$$=\frac{(1-G^n)}{(1-G)}.$$

We have also $G=T^1+T^2+T^3+T^4$, which may be written—

$$\frac{T-T^5}{1-T}$$

as division will prove.

Substituting, we have—

$$1+G_1+.........+G_n=\frac{1-\left\{\dfrac{T^1-T^5}{1-T}\right\}^n}{1-\left\{\dfrac{T^1-T^5}{1-T}\right\}}$$

$$=\frac{(1-T)\ (1-\{\ T^1+T^2+T^3+T^4\}^n)}{1-2T+T^5}$$

$$=\frac{(1-T)\ (1-T^n-\text{higher powers of } T)}{1-2T+T^5}$$

$$=\frac{(1-T-T^n-\text{higher powers of } T)}{1-2T+T^5}.$$

Now here, exactly as in the case of—

$$\left\{\frac{T-T^5}{1-T}\right\}.$$

we have only to divide the numerator by the denomination to get a series in which the index of T will designate the day and the coefficient of T the total number of births for that day.

If we suppose n is infinitely large, then T^n and the powers of T following it may be neglected, and the expression becomes—

$$\frac{(1-T)}{(1-2T+T^5)}$$

which is the same thing as

$$1+G+G^2+G^3+\ldots\ldots+G^n$$

summed to infinity as a diminishing series

$$=\frac{1}{1-G}.$$

When we substitute—

$$\left\{\frac{T-T^5}{1-T}\right\}$$

for G.

The same method of procedure, up to this point, may obviously be used whatever the initial value of T, whatever the number of terms in the series for G_1, and whatever the daily birth rate, provided it is constant and regular.

Thus we have—

$$G_1=(T^3+T^4+T^5+T^6+T^7)\;;\quad\text{this can be written}$$
$$G_1=T^3\frac{(1-T^5)}{1-T}=\frac{T^3-T^8}{1-T};$$

so that $G^1+G^2+G^3+\ldots\ldots\ldots$

$$=\frac{1-T}{1-T-T^3+T^8}$$

If we have—

$$G_1=(dT^3+dT^4+dT^5+dT^6+dT^7)=\frac{d(T^3-T^8)}{(1-T)}$$

the formula will be—

$$1+G_1+G_2+G_3+\ldots\ldots\quad\text{etc.}=\frac{1-T}{(1-T-dT^3+dT^8)}$$

If the number of births on the successive days of the period of reproduction is not the same, suppose these births to be a, b, c, d, as before. We then have—

$$G=aT^1+bT^2+cT^3+dT^4$$
$$G_2=(aT^1+bT^2+cT^3+dT^4)^2=G^2\text{ etc., and}$$
$$1+G_1+G_2+\ldots\ldots=\frac{1}{1-G}$$

$$=\frac{1}{1-(aT^1+bT^2+cT^3+dT^4)}$$

which will be valid even when there are no births on one or more days of the reproductive period, or, in other words, when one or more of the coefficients of T are made equal to 0. ...

It is evident that if the organism continues to live after it has ceased to reproduce, this will still be valid, so that if the life is of l days, the series of births must be multiplied by—

$$\frac{(T^1-T^l)}{(1-T)}$$

to get the population series. Thus, if the formula of the births series is—

$$\frac{1-T}{1-2T+T^5}$$

the formula for the population series will be—

$$\frac{1-T^l}{1-2T+T^5}$$

The rule for finding the successive members of the series is precisely the same as before, after the term T^l has been obtained.

The series given by the formulae includes, of course, only the individuals which are capable of reproducing themselves, *i.e.*, only the females ; to obtain the total number of individuals, each member of the series must be multiplied by a figure representing the ratio of total population to females, *e.g.*, by 2, if the numbers of males and females in the progeny are equal, on the average ; by 4, if there are three times as many males as females, etc.

Provided we know the length of the reproductive period we can readily obtain from observations giving either the series of daily births, or the series giving the total number of individuals produced up to and including each day, the formula for individual reproduction.

It has already been shown that if the generation formula is represented by G, the series of daily births is given by the formula—

$$\frac{1}{1-G}.$$

Thus, if the series be represented by S, we have—

$$S = \frac{1}{1-G}, \text{ whence}$$

$$G = 1 - \frac{1}{S}.$$

Thus, suppose the series of daily births, due to an individual reproductive period of four days, is—

$$1+1T^1+3T^2+8T^3+21\,T^4+\ldots\ldots\ldots\ldots\ldots\ldots\ldots\ldots ,$$

we obtain for the generation formula, by dividing the above series into 1 and subtracting the result from 1 :—

$$G=1T^1+2T^2+3T^3+4T^4.$$

APPENDIX.

PROBLEMS OF SIMPLE PROPAGATION WHOSE SOLUTION MAY BE FACILITATED BY THE USE OF OBJECTIVE SYMBOLS.

By H. E. SOPER,

Division of Epidemiology and Vital Statistics, London School of Hygiene and Tropical Medicine.

The problem of finding the resultant at any time, from given independent incoming births, or outgoing, *i.e.*, destroyed, births of the operation of a given law of reproduction and the converse problems of how to find the independent birth happenings, knowing the resultant births and the law, and how to find the law knowing the independent birth happenings and resultant births, and many such problems, are susceptible of analysis by a process, akin to that known as " generating functions " (akin also to chemical formulation), but which is more aptly described as one that employs objective symbols to indicate and array the differing characters of the things enumerated.

In the present instance we enumerate " events " ; and the " time of occurrence " is the characteristic that we desire to set against any event. Let then T be the symbol, whose power or index has the duty of indicating the number of units of time (which unit we shall call a day for convenience although any other period may be chosen), from any zero time, at which the event happens. Hence T^0 (or 1), T^1 (or T), T^2, ... T^t, ... indicate zero time and times 1, 2, ... t, ... days after zero and an array or formula as,

$$F_0 = 2 + 5T + 11T^2 + 25T^3 + \dots , \qquad \dots \qquad \dots \qquad \dots \quad (i)$$

will be read as " 2 events at zero time, 5 events one day after, 11 events two days after, 25 events three days after zero ... ."

Suppose the event is a birth and suppose each event gives rise to other like events, that is births, at known intervals, then the law of reproduction can also be expressed in symbols and, for instance, the law,

$$G = 5T^3 + 10T^5 + 3T^8, \qquad \dots \qquad \dots \qquad \dots \qquad \dots \quad (ii)$$

would indicate 5 births 3 days after, 10 births 5 days after, and 3 births 8 days after zero, which can be the birth of the reproducing organism.

In this case it will be clear that if F_0 stand for a births formula as (i) and G for a generation formula as (ii), then

$$F^1 = F_0 \times G \qquad \dots \qquad \dots \qquad \dots \qquad \dots \qquad \dots \quad (iii)$$

will formulate, in the time symbol T, the births F_1 in the first generation arising from the original births F_0.

Here it may be pointed out that T is an objective symbol only, without the quantitative connotation of such a symbol as t in algebra, and that other large letters as F and G stand for arrays of the symbol T such as (i) or (ii). We interpret $11T^2$ in statistics in much the same way as we should interpret $11H^2$ in chemistry, but in statistics the convention leads to a useful calculus, based upon such algebraic renderings as (iii).

The following elementary propositions may now be enunciated.

Proposition I.—*The independent or original incoming births being* F_0 *and the generation law being* G *to find the births in the first and each succeeding generation and the total births.*

The original births are	F_0.
The 1st generations births are	$F_0 G$.
The 2nd ,, ,, ,, ...	$F_0 G^2$.
....	
The r-1th ,, ,, ,, ...	$F_0 G^{r-1}$.
The total births to the r-1th generation are ...	$F_0 \dfrac{1-G^r}{1-G}.$
The total births to unlimited generations are ...	$\dfrac{F_0}{1-G}.$ *

Proposition II.—*To find the population under these conditions when each birth survives* s *days or unit intervals.*

A birth at zero time gives rise to a population 1 in the succeeding interval and 1 in each succeeding interval to s intervals. This succession may be represented by

$$1 + T + T^2 + \dots + T^{s-1} = \frac{1-T^s}{1-T}.$$

* We say that G^r is negligible when $r = \infty$, not because G $<$ 1, but because we do not concern ourselves with the exact enumeration of events at infinite time.

Hence from births F formulated as occurring at successive *instants of time* 0 1 2 ...... there results a population in the succeeding *intervals of time*,† which we may call the 0 1 2 ... intervals, formulated as,

$$F\frac{1-T^s}{1-T}$$

Instants at which births occur $\qquad T^0 \qquad\qquad T^1 \qquad\qquad T^2$

Intervals in which population is enumerated $\quad T^0 \qquad\qquad T^1 \qquad\qquad T^2$

Proposition III.—*To find the independent incoming births* F_0 *that must be postulated to give rise to the observed total births* F *under the known generation law* G.

Since, by Proposition I, $\quad F = \dfrac{F_0}{1-G}$,

it follows that, $\qquad\qquad\qquad F_0 = F(1-G)$.

Proposition IV.—*To find the generation law, given the independent incoming or originating births* F_0 *and the observed total births* F.

Since, by Proposition I, $\quad F = \dfrac{F_0}{1-G}$,

it follows that, $\qquad\qquad\qquad G = 1 - \dfrac{F_0}{F}$.

Proposition V.—*A series of events being expressed as an array* F *in the symbol* T, *to find the slope of the curve and its inflexion and to find the successive sums or totals, these also being expressed in array form.*

The slope or backward first difference is $(1-T)F$.

The inflexion or backward 2nd difference is $(1-T)^2 F$.

The sum to (and including) the instant is $\dfrac{F}{1-T}$.

The second, third, etc., sums are $\quad \dfrac{F}{(1-T)^2}$, $\dfrac{F}{(1-T)^3}$, etc.

The succeeding items follow from the first, and the first is obvious since if

$$F = a + bT + cT^2 + \ \ldots\ldots$$

then

$$(1-T)F = a + (b\text{-}a)T + (c\text{-}b)T^2 + \ \ldots\ldots \qquad .$$

Many other propositions could be stated, but the process is so elementary that the student will be able, with the help of the above examples and using similar algebraical renditions and transformations, to evolve many desired results.

† It would perhaps have accorded better with our common conceptions had populations been put at instants and births in intervening intervals of time.

21. The Population Consequences of Life History Phenomena

Lamont C. Cole (1954)

From *Quarterly Review of Biology* 29. Excerpts are from pages 107—115.

The part of Cole's article included here outlines the relationship of Thompson's work (1931, paper 20 above) to stable population theory. Omitted sections discuss the relationships between the rate of increase, numbers and spacing of offspring, and age structure, with their implications for species survival.

Quite recently a number of ecologists have recognized the importance of a knowledge of the value of r for non-human populations and have computed its value for various species by employing empirical values of age-specific birth rates and survivorship (Leslie and Ranson, 1940; Birch, 1948; Leslie and Park, 1949; Mendes, 1949; Evans and Smith, 1952). While Chapman's term "biotic potential" would seem to have ecological merit as the name for this parameter r it has been variously called by Lotka the "true," the "incipient," the "inherent," and the "intrinsic" rate of increase, and by Fisher (1930) the "Malthusian parameter" of population increase. Probably for the sake of stabilizing nomenclature it is advisable to follow the majority of recent writers and refer to r as "the intrinsic rate of natural increase."

In the works of Dublin and Lotka (1925), Kuczynski (1932), and Rhodes (1940) on human populations and in the papers mentioned above dealing with other species, the value of r has typically been determined by some application of three fundamental equations developed by Lotka (1907a, b; Sharpe and Lotka, 1911). He showed that if the age-specific fecundity (b_x) and survivorship (l_x) remained constant, the population would in time assume a fixed or "stable" age distribution such that in any interval of age from x to $x + dx$ there would be a fixed proportion (c_x) of the population. Once this stable age distribution is established the population would grow exponentially according to our formula (1') and with a birth rate per head, β. Then the following equations relate these quantities:

$$\int_0^\infty e^{-rx} l_x b_x \, dx = 1 \tag{3}$$

and

$$\int_0^\infty e^{-rx} l_x \, dx = 1/\beta \tag{4}$$

$$\beta e^{-rx} l_x = c_x. \tag{5}$$

While the use of formulas (3), (4), and (5) to compute the value of r often presents practical difficulties owing to the difficulty of approximating the functions l_x and b_x by a mathematical function, and also because the equations usually must be solved by iterative methods, it may fairly be stated that Lotka's pioneer work establishing these relationships provided the methods for interpreting the relationships between life history features and their population consequences.

However, the exceedingly important ecological questions of what potential advantages might be realized if a species were to alter its life history features have remained largely unexplored. Doubtless, as already noted, this is largely to be explained by a certain suspicion felt by biologists toward analyses such as those of Lotka, which seem to involve assumptions very remote from the realities of life histories as observed in the field and laboratories. A particularly pertinent statement of this point of view is that of Thompson (1931), who recognized the great practical need for methods of computing the rate of increase of natural populations of insects adhering to particular life history patterns but who insisted that the reproductive process must be dealt with as a discontinuous phenomenon rather than as a compound interest phenomenon such as that of formula (1'). His methods of computation were designed to give the exact number of individuals living in any particular time period and, while he recognized that the population growth can be expressed in an exponential form such as (1'), he rejected its use on these grounds:

"In the first place, the constant (r) cannot be determined until the growth of the population under certain definite conditions has been studied during a considerable period; in the second place, no intelligible significance can be attached to the constant after its value has been determined; in the third place, the growth of the population is considered in this formula to be at every moment proportional to the size of the population, which is not true except with large numbers and over long periods and cannot be safely taken as a basis for the examination of experimental data."

In the following sections of the present paper an effort will be made to reconcile these two divergent points of view and to show under what conditions Thompson's "discontinuous" approach and the continuous methods lead to identical results. Practical methods of computation can be founded on either scheme, and there are circumstances where one or the other offers distinct advantages. It is hoped that a theoretical approach to population phenomena proceeding from exact computational methods will clarify the meaning of some of the approximations made in deriving equations such as (3), (4), and (5) by continuous methods, and will stimulate students of ecology to a greater interest in the population consequences of life history phenomena.

Before proceeding to a discussion of potential population growth, one point which has sometimes

184

caused confusion should be mentioned. This concerns the sex ratio and the relative proportions of different age classes in the growing population. Once stated, it is obvious that if a population is always growing, as are the populations in the models used for determining potential population growth, then each age and sex class must ultimately come to grow at exactly the same rate as every other class. If this were not the case the disproportion between any two classes would come to exceed all bounds; the fastest growing class would continue indefinitely to make up a larger and larger proportion of the total population. It is thus intuitively recognizable that with fixed life history features there must ultimately be a fixed sex ratio and a stable age distribution. In discussing potential population growth it is often convenient to confine our attention to females or even to a restricted age class, such as the annual births, while recognizing that the ultimate growth rate for such a restricted population segment must be identical to the rate for the entire population.

SIMPLEST CASES OF POPULATION GROWTH

Non-overlapping generations

The simplest possible cases of population growth from the mathematical point of view are those in which reproduction takes place once in a lifetime and the parent organisms disappear by the time the new generation comes on the scene, so that there is no overlapping of generations. This situation occurs in the many plants and animals which are annuals, in those bacteria, unicellular algae, and protozoa where reproduction takes place by fission of one individual to form two or more daughter individuals, and in certain other forms. Thus in the century plants (*Agave*) the plant dies upon producing seeds at an age of four years or more, the Pacific salmon (*Oncorhynchus*) dies after spawning, which occurs at an age of two to eight years (two years in the pink salmon *O. gorbuscha*), and cicadas breed at the end of a long developmental period which lasts from two years (*Tibicen*) to 17 years in *Magicicada*. For many other insects with prolonged developmental stages such as neuropterans and mayflies potential population growth may be considered on the assumption that generations do not overlap.

In these cases, perhaps most typically illustrated in the case of annuals, the population living in any year or other time interval is simply the number of births which occurred at the beginning of that interval. Starting with one individual which is replaced by b offspring each of which repeats the life history pattern of the parent, the population will grow in successive time intervals according to the series: $1, b, b^2, b^3, b^4, \cdots b^x$. Hence the number of "births," say B_x, at the beginning of any time interval, T_x, is simply b^x which is identical with the population, P_x, in that interval of time. If the population starts from an initial number P_0 we have:

$$P_x = P_0 b^x \qquad (6)$$

which is obviously identical with the exponential formula (1'), $P_x = Ae^{rx}$, where the constant A is precisely P_0, the initial population size, and $r = \ln b$; the intrinsic rate of increase is equal to the natural logarithm of the litter size.

If litter size varies among the reproductive individuals, with each litter size being characteristic of a fixed proportion of each generation, it is precisely correct to use the average litter size, say $\bar{b}$, in the computations, so that we have $r = \ln \bar{b}$. Furthermore, if not all of the offspring are viable, but only some proportion, say l_1, survive to reproduce, we shall have exactly $r = \ln \bar{b}^{l_1}$. Thus, mortality and variations in litter size do not complicate the interpretation of population growth in cases where the generations do not overlap. On the other hand, even in species which reproduce only once, if the generation length is not the same for all individuals, this will lead to overlapping generations, and the simple considerations which led to formula (6) will no longer apply. In other words, we can use an average figure for the litter size b but not for the generation length x. It will be shown in the next section, however, that the more general formula (1') is still applicable.

In these simplest cases the assumption of a geometric progression as the potential form of population growth is obviously correct, and numerous authors have computed the fantastic numbers of offspring which could potentially result from such reproduction. For example, according to Thompson (1942), Linnaeus (1740?) pointed out that if only two seeds of an annual plant grew to maturity per year, a single individual could give rise to a million offspring in 20 years. (In all editions available to the present writer this interesting essay of Linnaeus' is dated 1743, and the number of offspring at the end of twenty years is stated by the curious and erroneous figure 91,296.)

That is, $P_{20} = 2^{20} = e^{20 \ln 2} = 1,048,576$. Additional examples are given by Chapman (1935, p. 148).

Formulas (1') or (6) may, of course, also be used in an inverse manner to obtain the rate of multiplication when the rate of population growth is known. For the example given by Molisch (1938, p. 25), referring to diatoms reproducing by binary fission where the average population was observed to increase by a factor of 1.2 per day, we have $1.2 = e^{x \ln 2}$, where x is the number of generations per day. Solving for $1/x$, the length of a generation, we obtain $1/x = \dfrac{\ln 2}{\ln 1.2} = \dfrac{.69315}{.18232} = 3.8$ days.

Overlapping generations

Interest in computing the number of offspring which would be produced by a species adhering to a constant reproductive schedule dates back at least to Leonardo Pisano (= Fibonacci) who, in the year 1202, attempted to reintroduce into Europe the study of algebra, which had been neglected since the fall of Rome. One of the problems in his *Liber Abbaci* (pp. 283–84 of the 1857 edition) concerns a man who placed a pair of rabbits in an enclosure in order to discover how many pairs of rabbits would be produced in a year. Assuming that each pair of rabbits produces another pair in the first month and then reproduces once more, giving rise to a second pair of offspring in the second month, and assuming no mortality, Fibonacci showed that the number of pairs in each month would correspond to the series

$$1, 2, 3, 5, 8, 13, 21, 34, 55, \text{ etc.},$$

where each number is the sum of the two preceding numbers. These "Fibonacci numbers" have a rather celebrated history in mathematics, biology, and art (Archibold, 1918; Thompson, 1942; Pierce, 1951) but our present concern with them is merely as a very early attempt to compute potential population growth.

Fibonacci derived his series simply by following through in words all of the population changes occurring from month to month. One with sufficient patience could, of course, apply the same procedure to more complicated cases and could introduce additional variables such as deductions for mortality. In fact, Sadler (1830, Book III) did make such computations for human populations. He was interested in discovering at what ages persons would have to marry and how often they would have to reproduce to give some of the rates of population doubling which had been postulated by Malthus (1798). To accomplish this, Sadler apparently employed the amazingly tedious procedure of constructing numerous tables corresponding to different assumptions until he found one which approximated the desired rate of doubling.

Although we must admire Sadler's diligence, anyone who undertakes such computations will find that it is not difficult to devise various ways of systematizing the procedure which will greatly reduce the labor of computation. By far the best of these methods known to the present writer is that of Thompson (1931), which was originally suggested to him by H. E. Soper.

In the Soper-Thompson approach a "generation law" (G) is written embodying the fixed life history features which it is desired to consider. The symbol T^x stands for the xth interval of time, and a generation law such as $G = 2T^1 + 2T^2$ would be read as "two offspring produced in the first time interval and two offspring produced in the second time interval." This particular generation law might, for example, be roughly applicable to some bird such as a cliff swallow, where a female produces about four eggs per year. Concentrating our attention on the female part of the population, we might wish to compute the rate of population growth which would result if each female had two female offspring upon attaining the age of one year and had two more female offspring at the age of two years. The fundamental feature of the Thompson method is the fact that the expression:

$$\frac{1}{1 - G} \qquad (7)$$

is a generating function which gives the series of births occurring in successive time intervals. In the algebraic division the indices of the terms T^1, T^2, etc., are treated as ordinary exponents and the number of births occurring in any time interval T^x is simply the coefficient of T^x in the expansion of expression (7). Thus, for our example where $G = 2T^1 + 2T^2$ we obtain:

$$\frac{1}{1 - 2T^1 - 2T^2} = 1 + 2T^1 + 6T^2 +$$
$$16T^3 + 44T^4 + 120T^5 + 328T^6 + \cdots,$$

showing that one original female birth gives rise to 328 female offspring in the sixth year. The

186

series could be continued indefinitely to obtain the number of births any number of years hence. However, in practice it is not necessary to continue the division. In the above series the coefficient of each term is simply twice the sum of the coefficients of the two preceding terms; hence the generation law gives us the rule for extending the series. $G = 2T^1 + 2T^2$ instructs us to obtain each new term of the series by taking twice the preceding term plus twice the second term back. In the case of the Fibonacci numbers we would have $G = T^1 + T^2$, telling us at once that each new term is the sum of the two preceding it.

From the birth series we can easily obtain the series enumerating the total population. If each individual lives for λ years, the total population in T^x will be the sum of λ consecutive terms in the expansion of the generating function. Multiplying formula (7) by the length of life expressed in the form $1 + T^1 + T^2 + T^3 + \cdots + T^{\lambda-1}$ will give the population series. In our above example if we assume that each individual lives for three years, although, as before, it only reproduces in the first two, we obtain for the population

$$\frac{1 + T^1 + T^2}{1 - 2T^1 - 2T^2} = 1 + 3T^1 + 9T^2 +$$

$$24T^3 + 66T^4 + 180T^5 + 492T^6 + \cdots,$$

a series which still obeys the rule $G = 2T^1 + 2T^2$.

Thompson's method for obtaining the exact number of births and members of the population in successive time intervals is very general. As in the case of non-overlapping generations, the coefficients in the generation law may refer to average values for the age-specific fecundity. Also the length of the time intervals upon which the computations are based can be made arbitrarily short, so that it is easy to take into account variations in the age at which reproduction occurs. For the above example, time could have been measured in six-month periods rather than years so that the generation law would become $G = 2T^2 + 2T^4$, with the same results already obtained.

Furthermore, the factor of mortality can easily be included in the computations. For example, suppose that we wish to determine the rate of population growth for a species where the females have two female offspring when they reach the age of one, two more when they reach the age of two, and two more when they reach the age of three. Neglecting mortality, this would give us the

generation law $G = 2T^1 + 2T^2 + 2T^3$. If we were further interested in the case where not all of the offspring survive for three years, the coefficients in the generation law need only be multiplied by the corresponding survivorship values. For example, if one-half of the individuals die between the ages of one and two, and one half of the remainder die before reaching the age of three we would have $l_1 = 1$, $l_2 = \frac{1}{2}$, $l_3 = \frac{1}{4}$, and the above generation law would be revised to $G = 2T^1 + T^2 + \frac{1}{2}T^3$. The future births per original individual would then be

$$\frac{1}{1 - G} = 1 + 2T^1 + 5T^2 + 25/2T^3 + 31T^4 +$$

$$151/2T^5 + \cdots.$$

Very generally, if the first reproduction for a species occurs at some age α and the last reproduction occurs at some age ω, and letting b_x and l_x represent respectively the age-specific fecundity and survivorship, we may write the generation law as:

$$G = l_\alpha b_\alpha T^\alpha + l_{\alpha+1} b_{\alpha+1} T^{\alpha+1} + \cdots$$

$$+ l_\omega b_\omega T^\omega = \sum_{x=\alpha}^{\omega} l_x b_x T^x. \quad (8)$$

Therefore, in the Thompson method we have a compact system of computation for obtaining the exact number of births and the exact population size at any future time, assuming that the significant life history features (α, ω, l_x, and b_x) do not change.

Not all of the possible applications of Thompson's method have been indicated above. For example, formula (7) may be used in an inverse manner so that it is theoretically possible to work back from a tabulation of births or population counts made in successive time intervals and discover the underlying generation law. Formulas (7) and (8), together with the procedure of multiplying the birth series by the length of life expressed as a sum of T^x values, provide the nucleus of the system and offer the possibility of analyzing the potential population consequences of essentially any life-history phenomena. The system has the merit of treating the biological units and events as discontinuous variates, which, in fact, they almost always are. The members of populations are typically discrete units, and an event such as reproduction typically occurs at a point in time with no spreading out or overlapping between successive litters. While survivorship, l_x, as a population quantity, is most realistically regarded as continuously changing in time, the product $l_x b_x$ which enters our computations by way of formula (8) is typically discontinuous because of the discontinuous nature of b_x.

It is quite obvious that equations of continuous variation such as (1') are often much more convenient for purposes of computation than the series of values obtained by expanding (7). This is especially true in dealing with the life histories of species which have long reproductive lives. In writing a generation law for man by (8) we should have to take α at least as small as 15 years and ω at least as great as 40 years, since for the population as a whole reproduction occurs well outside of these extremes and it would certainly be unrealistic to regard b_x as negligibly small anywhere between these limits. Thus there would be at least 25 terms in our generation law, and the computations would be extremely tedious. By selecting special cases for study it is sometimes possible greatly to simplify the procedures. For example, if one is interested in the case where there is no mortality during the reproductive span of life and where the litter size is a constant, say b, the expression for the generation law (8) can be simplified to:

$$G = bT^{\alpha} + bT^{\alpha+1} + \cdots bT^{\omega} = \frac{bT^{\alpha} - bT^{\omega+1}}{1 - T}.$$

Since one can also write the length of life as

$$1 + T^1 + T^2 + \cdots + T^{\lambda-1} = \frac{1 - T^{\lambda}}{1 - T},$$

the generating function for the total population simplifies to

$$\frac{1 - T^{\lambda}}{1 - T - bT^{\alpha} + bT^{\omega+1}}.$$

This last formula is much more convenient for computations than one containing 25 terms or so in the denominator, but it applies only to a very special case and is much less convenient than formula (1'). Consequently, great interest attaches to these questions: can (1') be used as a substitute for (7)? (i.e., does Thompson's method lead to a geometric progression?) and, if it is so used, can the constants, particularly r, be interpreted in terms of life-history features?

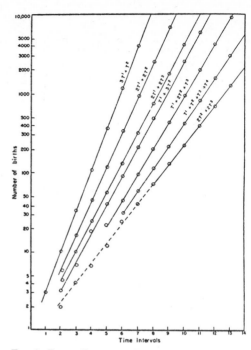

FIG. 1. EXACT VALUES OF POPULATION GROWTH IN TERMS OF BIRTHS PER UNIT TIME UNDER SEVERAL GENERATION LAWS, WHEN EACH FEMALE HAS A TOTAL OF FOUR FEMALE OFFSPRING

In each case it is assumed that a single female exists at time zero and produces her four progeny on or before her fourth birthday. The plotted points represent exact values as determined by Thompson's method. To the extent that the points for any generation law fall on a straight line in this logarithmic plot, they can be represented by the exponential growth formula (1'), and the slope of each line is a measure of the intrinsic rate of natural increase (r).

THE GENERALIZATION OF THOMPSON'S METHOD

Fig. 1 shows the exact values, as determined by Thompson's method, of the birth series arising from several generation laws (life-history patterns) which have in common the feature that in each case every female produces a total of four female offspring in her lifetime and completes her reproductive life by the age of four "years." The number of births is plotted on a logarithmic scale, hence if it can be represented by formula (1'), $P = Ae^{rx}$ or, logarithmically, $\ln P = \ln A + rx$, the points should fall on a straight line with slope proportional to r. It is apparent from Fig. 1 that after the first few time intervals the points in each case are well represented by a straight line. Therefore, except in the very early stages, formula (1') does

188

give a good representation of potential population growth. The question remains, however, as to whether we can meet Thompson's objection to (1') and attach any intelligible significance to the constants of the formula. From Fig. 1 it is obvious that the lines do not, if projected back to time 0, indicate exactly the single individual with which we started. Thus, in these cases the constant A cannot be precisely P_0 as was the case with non-overlapping generations.

Before proceeding to interpret the constants of formula (1') for the case of overlapping generations it will be well to notice one feature of Fig. 1 which is of biological interest. In the literature of natural history one frequently encounters references to the number of offspring which a female can produce per lifetime, with the implication that this is a significant life-history feature. The same implication is common in the literature dealing with various aspects of human biology, where great emphasis is placed on the analysis of total family size. From Fig. 1 it will be seen that this datum may be less significant from the standpoint of contributions to future population than is the age schedule upon which these offspring are produced. Each life history shown in Fig. 1 represents a total production of four offspring within four years of birth, but the resulting rates of potential population growth are very different for the different schedules. It is clear that the cases of most rapid population growth are associated with a greater concentration of reproduction into the early life of the mother. This is intuitively reasonable because we are here dealing with a compound interest phenomenon and should expect greater yield in cases where "interest" begins to accumulate early. However, the writer feels that this phenomenon is too frequently overlooked in biological studies, possibly because of the difficulty of interpreting the phenomenon quantitatively.

In seeking to reconcile the continuous and discontinuous approaches to potential population growth, let us first note that Thompson's discontinuous method corresponds to an equation of finite differences. We have seen above that the generation law gives us a rule for indefinitely extending the series representing the population size or the number of births in successive time intervals by adding together some of the preceding terms multiplied by appropriate constants. If we

let $f_{(x)}$ represent the coefficient of T^x in the expansion of the generating function (7) and, for brevity, write in (8) $V_x = l_x b_x$, then our population series obeys the rule:

$$f_{(x)} = V_\alpha f_{(x-\alpha)} + V_{\alpha+1} f_{(x-\alpha-1)} + \cdots + V_\omega f_{(x-\omega)}, \qquad (9)$$

which may be written in the alternative form,

$$f_{(x+\omega)} - V_\alpha f_{(x+\omega-\alpha)} - V_{\alpha+1} f_{(x+\omega-\alpha-1)} - \cdots - V_\omega f_{(x)}, = 0. \qquad (10)$$

Thus for our "cliff swallow" example, where we had $G = 2T^1 + 2T^2$ we have

$$f_{(x)} = 2f_{(x-1)} + 2f_{(x-2)} \text{ or,}$$
$$f_{(x+2)} - 2f_{(x+1)} - 2f_{(x)} = 0.$$

Formula (10) represents the simplest and best understood type of difference equation, a homogeneous linear difference equation with constant coefficients. It is outside the scope of the present paper to discuss the theory of such equations, which has been given, for example, by Jordan (1950). By the nature of our problem as summarized in formula (9), all of our V_x values are either equal to zero or are positive real numbers and all of the signs of the coefficients in (9) are positive: features which considerably simplify generalizations. By virtue of these facts it can be shown that there is always a "characteristic" algebraic equation corresponding to (10). This is obtained by writing ρ^x for $f_{(x)}$ and dividing through by the ρ value of smallest index. This gives

$$\rho^\omega - V_\alpha \rho^{\omega-\alpha} - V_{\alpha+1} \rho^{\omega-\alpha-1} \cdots - V_\omega = 0 \qquad (11)$$

an algebraic equation which has the roots ρ_1, ρ_2, etc.

The general solution of the corresponding difference equation (10) is

$$f_{(x)} = C_1 \rho_1^x + C_2 \rho_2^x + \cdots + C_n \rho_n^x \qquad (12)$$

where the C's are constants to be determined by the initial conditions of the problem. Formula (12) is precisely equivalent to Thompson's method and is a general expression for the number of births or the population size in any future time interval.

As an example we may consider the case where $G = 2T^1 + 2T^2$. The difference equation, as already noted, is $f_{(x+2)} - 2f_{(x+1)} - 2f_{(x)} = 0$ and the characteristic algebraic equation is $\rho^2 - 2\rho - 2 = 0$ which is a quadratic equation with the roots $\rho_1 = 1 + \sqrt{3}$, and $\rho_2 = 1 - \sqrt{3}$. Hence the general

189

solution is $f_{(x)} = C_1(1 + \sqrt{3})^x + C_2(1 - \sqrt{3})^x$. To determine the constants C_1 and C_2 we look at the beginning a of the seriesnd note that we have $f_{(0)} = 1$ and $f_{(1)} = 2$. Substituting these values in the general solution we obtain $C_1 = \dfrac{\sqrt{3} + 1}{2\sqrt{3}}$ and $C_2 = \dfrac{\sqrt{3} - 1}{2\sqrt{3}}$. Therefore, the general expression for the number of births in time interval T^x is

$$f_{(x)} = \frac{\sqrt{3} + 1}{2\sqrt{3}}(1 + \sqrt{3})^x + \frac{\sqrt{3} - 1}{2\sqrt{3}}(1 - \sqrt{3})^x$$

which can be simplified to $f_{(x)} = \dfrac{\rho_1^{x+1} - \rho_2^{x+1}}{\sqrt{3}} = \rho_1^x + \rho_1^{x-1}\rho_2 + \cdots + \rho_2^x$.

In order to have the difference equation (12) correspond to the equation of exponential growth (1′), the ratio between populations in successive time intervals must assume a constant value giving

$$\frac{f_{(x+1)}}{f_{(x)}} = e^r. \tag{13}$$

By the nature of our problem, as already noted, the potential population growth is always positive, so that any limit approached by the ratio $\dfrac{f_{(x+1)}}{f_{(x)}}$ must be a positive real number.

It is beyond the scope of the present paper to discuss the conditions, for difference equations in general, under which this ratio does approach as a limit the largest real root of the characteristic algebraic equation. (See, for example, Milne-Thompson, 1933, chap. 17). Dunkel (1925) refers to the homogeneous equation with real constant coefficients corresponding to our formulas (10) and (11). The algebraic equation (11) has a single positive root which cannot be exceeded in absolute value by any other root, real or complex. Using (12) to express the ratio between successive terms, we have

$$\frac{f_{(x+1)}}{f_{(x)}} = \frac{C_1\rho_1^{x+1} + C_2\rho_2^{x+1} + \cdots + C_n\rho_n^{x+1}}{C_1\rho_1^x + C_2\rho_2^x + \cdots + C_n\rho_n^x}. \tag{14}$$

If we let ρ_1 represent the root of (11) of greatest absolute value and divide both numerator and denominator of (14) by $C_1\rho_1^x$ we obtain

The expressions in parentheses are all less than unity, on the assumption that ρ_1 is the largest root, and the entire expression in brackets approaches unity as x increases. Consequently we have, for x large

$$\frac{f_{(x+1)}}{f_{(x)}} \sim \rho_1 \sim e^r. \tag{16}$$

This then explains the shape of the potential birth and population series as illustrated in Fig. 1. In the very early stages population growth is irregular, because the expressions in (12) and (15) involving the negative and complex roots of (11) are still large enough to exert an appreciable influence. As x increases, the influence of these other roots becomes negligible and the population grows exponentially, conforming to (16). In considering potential population growth we are concerned with the ultimate influence of life-history features, and the equation of geometric progression or compound interest does actually represent the form of potential population growth. We are interested only in the single positive root of (11) for the purpose of determining the constant r, and this can readily be computed with any desired degree of precision by elementary algebraic methods.

Having established the relationship of formula (13) or (16), it is easy to reconcile Thompson's discontinuous approach to population growth with Lotka's continuous approach, as exemplified by formulas (3), (4), and (5).

Employing formula (9) we may write the ratio between populations in successive time intervals as

$$\frac{f_{(x+1)}}{f_{(x)}} = V_\alpha \frac{f_{(x-\alpha+1)}}{f_{(x)}} + V_{\alpha+1}\frac{f_{(x-\alpha)}}{f_{(x)}} + \cdots + V_\omega\frac{f_{(x-\omega+1)}}{f_{(x)}}.$$

Substituting the relationship given by (13), this becomes

$$e^r = V_\alpha e^{-r(\alpha-1)} + V_{\alpha+1}e^{-r\alpha} + \cdots + V_\omega e^{-r(\omega-1)}, \text{ or}$$

$$1 = V_\alpha e^{-r\alpha} + V_{\alpha+1}e^{-r(\alpha+1)} + \cdots + V_\omega e^{-r\omega}.$$

$$\frac{f_{(x+1)}}{f_{(x)}} = \rho_1 \left[\frac{1 + \dfrac{C_2}{C_1}\left(\dfrac{\rho_2}{\rho_1}\right)^{x+1} + \dfrac{C_3}{C_1}\left(\dfrac{\rho_3}{\rho_1}\right)^{x+1} + \cdots + \dfrac{C_n}{C_1}\left(\dfrac{\rho_n}{\rho_1}\right)^{x+1}}{1 + \dfrac{C_2}{C_1}\left(\dfrac{\rho_2}{\rho_1}\right)^{x} + \dfrac{C_3}{C_1}\left(\dfrac{\rho_3}{\rho_1}\right)^{x} + \cdots + \dfrac{C_n}{C_1}\left(\dfrac{\rho_n}{\rho_1}\right)^{x}} \right]. \tag{15}$$

Replacing V_z by its equivalent, $l_z b_z$, this is

$$1 = \sum_{z=\alpha}^{\omega} e^{-rz} l_x b_x. \qquad (17)$$

Formula (17) is the precise equivalent in terms of finite time intervals of Lotka's equation (3) for infinitesimal time intervals. In Lotka's equation, as in (17), the limits of integration in practice are α and ω since b_x is zero outside of these limits. Formula (17) was in fact employed by Birch (1948) as an approximation to (3) in his method of determining r for an insect population. The only approximation involved in our derivation of (17) is the excellent one expressed by formula (13); otherwise the formula corresponds to Thompson's exact computational methods. It is hoped that recognition of this fact will make some of the approaches of population mathematics appear more realistic from the biological point of view.

Formulas (4) and (5), originally due to Lotka, are also immediately derivable from the relationship (13). In any time interval, T_z, we may say that the population members aged 0 to 1 are simply the births in that interval, say B_z. The population members aged 1 to 2 are the survivors of the births in the previous interval, that is $l_1 B_{z-1}$, or employing (13), $l_1 B_z e^{-r}$. Quite generally, the population members aged between z and $z + 1$ are the survivors from the birth z intervals previous, or $l_z B_z e^{-rz}$. If λ is the extreme length of life for any population members ($l_{\lambda+1} = 0$) we have for the total population

$$P_z = B_z(1 + l_1 e^{-r} + l_2 e^{-2r} + \cdots$$

$$+ l_\lambda e^{-r\lambda} = B_z \sum_{z=0}^{\lambda} e^{-rz} l_x.$$

The birth rate per individual, β, is B_z/P_z, therefore,

$$1/\beta = \sum_0^\lambda e^{-rz} l_x \qquad (18)$$

which is the equivalent in finite time intervals of Lotka's equation (4). Also the proportion, c_z, of the population in the age range z to $z + 1$ is $\dfrac{l_z B_z e^{-rz}}{P_z}$ which is simply,

$$c_z = \beta e^{-rz} l_z. \qquad (5)$$

LIST OF LITERATURE

ARCHIBOLD, R. C. 1918. A Fibonacci series. *Amer. math. Monthly*, 25: 235–238.

BIRCH, L. C. 1948. The intrinsic rate of natural increase of an insect population. *J. Anim. Ecol.*, 17: 15–26.

CHAPMAN, R. N. 1938. *Animal Ecology.* McGraw-Hill, New York.

DUBLIN, L. I., and A. J. LOTKA. 1925. On the true rate of natural increase as exemplified by the population of the United States, 1920. *J. Amer. statist. Ass.*, 20: 305–339.

DUNKEL, O. 1925. Solutions of a probability difference equation. *Amer. math. Monthly*, 32: 354–370.

EVANS, F. C., and F. E. SMITH. 1952. The intrinsic rate of natural increase for the human louse, *Pediculus humanus* L. *Amer. Nat.*, 86: 299–310.

FIBONACCI, L. (1857). *Liber Abbaci di Leonardo Pisano publicati da Baldasarre Boncompagni.* Tipografia delle Scienze mathematiche e fisiche, Roma.

FISHER, R. A. 1930. *The Genetical Theory of Natural Selection.* Oxford Univ. Press, London.

JORDAN, C. 1950. *Calculus of Finite Differences*, 2nd ed. Chelsea, New York.

KUCZYNSKI, R. R. 1932. *Fertility and Reproduction.* Falcon Press, New York.

LESLIE, P. H., and R. M. RANSON. 1940. The mortality, fertility, and rate of natural increase of the vole (*Microtus agrestis*) as observed in the laboratory. *J. Anim. Ecol.*, 9: 27–52.

——, and T. PARK. 1949. The intrinsic rate of natural increase of *Tribolium castaneum* Herbst. *Ecology*, 30: 469–477.

LINNAEUS, C. 1743. Oratio de telluris habitabilis. In *Amoenitates Academicae seu Dissertationes Variae....* Editio tertia curante J. C. D. Schrebero, 1787, Vol. 2: 430–457. J. J. Palm, Erlangae.

LOTKA, A. J. 1907a. Relation between birth rates and death rates. *Science*, 26: 21–22.

——. 1907b. Studies on the mode of growth of material aggregates. *Amer. J. Sci.*, 24: 119–216.

MALTHUS, T. R. 1798. *An Essay on the Principle of Population as it Affects the Future Improvement of Society, with Remarks on the Speculations of Mr. Godwin, M. Condorcet, and Other Writers.* Printed for J. Johnson in St. Paul's Churchyard, London.

MENDES, L. O. 1949. Determinação do potencial biotico da "broca do café"—*Hypothenemus hampei* (Ferr.)—e consideracões sôbre o crescimento de sua população. *Ann. Acad. bras. Sci.*, 21: 275–290.

MILNE-THOMPSON, L. M. 1933. *The Calculus of Finite Differences.* Macmillan, London.

MOLISCH, H. 1938. *The Longevity of Plants.* E. H. Fulling, Lancaster Press, Lancaster, Pa.

PIERCE, J. C. 1951. The Fibonacci series. *Sci. Monthly*, 73: 224–228.

RHODES, E. C. 1940. Population mathematics. *J. R. statist. Soc.*, 103: 61–89, 218–245, 362–387.

SADLER, M. T. 1830. *The Law of Population: A Treatise, in Six Books; in Disproof of the Super-fecundity of Human Beings, and Developing the Real Principle of their Increase.* Murray, London.

SHARPE, F. R., and A. J. LOTKA. 1911. A problem in age distribution. *Phil. Mag.*, 21: 435–438.

THOMPSON, D'ARCY W. 1942. *On Growth and Form,* new ed. Cambridge Univ. Press, Cambridge.

THOMPSON, W. R. 1931. On the reproduction of organisms with overlapping generations. *Bull. ent. Res.*, 22: 147–172.

Attempts at Prediction and the Theory They Stimulated

The tools needed for population projections, age-specific fertility and mortality rates, were in existence by the early part of the 19th century. They remained unused almost to the 20th. Edwin Cannan's (1895) article, which begins this chapter, both established life table rates as the necessary base for competent projections and demonstrated that if care were taken the rates could reasonably be adjusted in anticipation of future changes.

Independently of Cannan's work and later contributions by A. L. Bowley (1924, 1926), P. K. Whelpton (1936) also developed component population projections. Whelpton in his projections for the United States divided total population into ethnic groupings recognized by the census, handling each separately, and considered immigration as an independent component; apart from fertility he followed procedures he and W. S. Thompson had previously worked out (Whelpton 1928; Thompson and Whelpton 1933). The estimates were almost immediately overtaken by events, a war that was not widely expected in 1936 and a postwar baby boom not expected even when it hit. Yet for the few years that the future was orderly it accorded with Whelpton's calculations, and the clear logic of separating age groups has never since been at issue. We include Whelpton's 1936 article as paper 23. Methods it has replaced are discussed by Cannan and also in chapter 4.

The use of matrices in population projections was first suggested by Harro Bernardelli (1941), paper 24, who noted from the sizes of Burmese birth cohorts that birth peaks occurred at intervals of a generation and that these might carry important economic and social implications, a question examined more recently by Richard Easterlin (1966) for its possible relationship to Kuznets cycles. From a projection matrix formed by holding fertility and mortality constant Bernardelli isolated the characteristic roots of the net maternity function for the discrete case, identifying the dominant root as the intrinsic growth rate of the population and relating the other roots to oscillations in the birth rate. By mischance, Bernardelli utilized a matrix concentrating fertility into a single age category. This gives rise to the exceptional result, which he took to be a general case and drew several spurious conclusions from, that oscillations in the birth rate and hence in the age structure might be maintained or amplified over time rather than becoming damped.

In a paper that followed Bernardelli's address by three months, E. G. Lewis (1942) presented a better but still essentially preliminary exploration of the projection matrix, which we include here as paper 25. This is followed by P. H. Leslie's (1945) definitive article: Leslie develops the use of matrices fully, with attention to the several types of information they make accessible and their relationship to the earlier theory developed by Lotka. The present widespread use of projection matrices is an extension of his work. A generalized inverse for the Leslie matrix to permit limited backward projection is given by Greville and Keyfitz (1974). We follow Leslie's article with an extension due to Leon Tabah (1968), in which he shows how matrices may be applied to migration and labor force projections as well as to overall population change. The method is also explained in Rogers (1968, pp. 6—15).

From the development of the projection matrix we turn to the concepts of strong and weak ergodicity, whose initial proofs were developed through matrix formulation but which otherwise are a continuation of the discussion in chapter 2. Paper 28 extracts from Alvaro Lopez (1961) a statement of ergodicity. The concluding article by Beresford Parlett (1970) compactly summarizes matrix development of ergodicity and general stable population theory.

The reader might return from Parlett to the opening article of the chapter to appreciate the impressive achievements in discrete stable theory and determination of characteristic roots and vectors that have followed from Cannan's introduction of simple but correct principles for population projection.

22. The Probability of a Cessation of the Growth of Population in England and Wales during the Next Century

EDWIN CANNAN (1895)

From *The Economic Journal* 5. Excerpts are from pages 505—514.

Many things about which we are habitually obliged to form estimates are of a much more speculative character than the growth of population, and the estimates which we do form about them in many cases actually depend on our estimate of the probable growth of population. There is not a builder nor a town council in the country that is not obliged to prophesy every month what the growth of population in a particular district is likely to be, and it was the speculations of the Metropolitan Water Commission as to the population of London in 1931 that inspired me to make this contribution to the literature of the subject. The real question is not whether we shall abstain altogether from estimating the future growth of population, but whether we shall be content with estimates which have been formed without adequate consideration of all the data available, and can be shown to be founded on a wrong principle.

The generally accepted principle is that of 'as the increase has been in the past so it will be in the future.' This is susceptible of more than one interpretation. All we really know of the increase in the past is contained in the totals obtained by the censuses. A schoolboy whose arithmetic is described as 'v. g.' would probably boldly disregard all the intermediate censuses and divide the total increase of 20,109,989 between 1801 and 1891 by ninety years. The quotient of 223,444 he would call 'the average annual increase,' and say that in order to get the probable increase of population in a certain number of years after 1891 you must simply multiply 223,444 by that number of years. ...

A person with a slight smattering of statistics will probably say that the schoolboy's method is quite wrong, and that in arguing from the past to the future, you ought to consider not the number which has been added to the population, but the factor by which the population has been multiplied. You ought to consider not the absolute amount, but the rate of the increase. But which rate? For there is a different rate in every one of the nine decades. Are we to take the rate from 1801 to 1891, disregarding the intermediate censuses, or are we to somehow deduce some-

---

[1] This article contains the substance of a paper read before Section F of the British Association at Ipswich.

thing which may be called an average rate from the nine different rates? The Metropolitan Water Commission groped for an answer to this question and found none. A very able mathematician to whom I gave the figures of the ten censuses tells me that they are so irregular that no law of increase of the smallest value can be deduced from them. The Registrar-General's method in forming what are called the 'official estimates' cuts the knot by disregarding all the decades except the last. The only conceivable argument in favour of this course is the allegation that the immediate past being nearer the future than the long past, affords a better basis for estimating future probability. By itself this would justify the *reductio ad absurdum*, 'if the last ten years are better than the last ninety, the last year must be better than the last ten,' but 'temporary fluctuations' have of course to be considered, and it may perhaps be asserted that ten years is sufficient, while twenty years is more than sufficient, to give an average unaffected by such fluctuations.

Which rate of multiplication be taken, however, does not very much matter, since any rate that can possibly be deduced from the census figures gives ridiculous results. Carried back into the past the assumption of a uniform rate of increase equal to that which prevailed in the decade 1881 to 1891 necessitates the supposition that in A.D. 525 there were scarcely two persons in England, and carried forward into the future it gives a population of a thousand millions in less than 300 years. After 350 years more there would be only just room for the people to lie down on the ground. Yet the rate of increase in the decade 1881—90 is the lowest of the nine.

These periods, it may be urged, are long. But applied to short periods the assumption of a uniform rate of increase equal to that shown by the last decade leads to just as absurd results as when it is applied to long periods. As is well known, the Registrar-General, whose annual reports reveal no acquaintance with the very accurate statistics of immigration and emigration now collected by the Board of Trade, ignores his own figures of births and deaths, on the ground that in estimating a population immigration and emigration cannot be disregarded. The official estimates of the population in the middle of each of the following ten years which he forms after each census on the assumption that the rate of increase shown by it will continue, are allowed to stand in each of his annual reports till the next census, and would, we must suppose, be allowed to stand if a plague swept away half the people. When the next census is taken they are, of course, found to be wrong. [At the last census the error amounted to 701,843 at the end of the decade, while Dr. Longstaff had in the pages of this Journal estimated the population within 10,251.[1]] They are then corrected on the assumption that throughout the decade population annually increased at the rate observed during the whole decade. Now whenever the rate of increase in a decade is less than in the preceding decade, these 'corrected' figures, on which the 'corrected' birth, marriage, and death-rates are calculated, present a very odd and most unnatural series of increases. Thus they represent the increase for 1871-2 to have been 307,901, and make the annual increase grow gradually till in 1879-80 it amounted to 342,799. Then comes the census year, which stands by itself, as three-quarters of it lie in one decade and the other quarter in another ; it is allotted an increase of 331,854. After this the new period begins in earnest with an increase of only 288,800 in 1881-2 ; and the rising process starts again and continues till the increase of 1890-1 amounts to 318,912. In short, the increase grows steadily till it reaches the census of 1881, then it comes down with a crash only to begin immediately to rise again. It must always happen under this system that the addition to the population in the last half of a decade is greater than the addition in the first half, even when the addition in each decade is less than the addition in the preceding decade. The truth is that every estimate of population, past, present, and to come, ought to be founded on a consideration of the factors on which the growth or decline o. population is dependent—births, deaths, immigration, and emigration. The number of births and deaths, and of immigrants and emigrants, is now so well known, that if two government departments, the Board of Trade and the General Registry Office, would only recognise each other's existence, the population at the present time, or at any point in recent years, could be given within ten thousand of the actual number.

[1] ECONOMIC JOURNAL, vol. i., p. 382.

Population of England and Wales, 1851–1951.

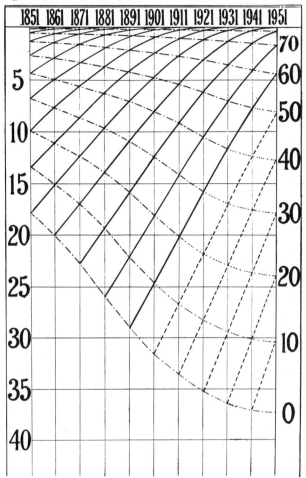

The population in each census year is measured down from the top, the figures on the left indicating the number of millions. The lines sloping downwards from left to right divide the population at each census into persons over 80 ; persons between 70 and 80 ; between 60 and 70 ; and so on down to between 0 and 10. The ages are shown by the figures on the right. The lines sloping upwards from left to right divide the persons born in each decade from those born before and after it. They thus show graphically how each generation becomes smaller as it passes from under ten years of age at one census to over ten but under twenty at the next, and then to over twenty but under thirty, and so on.

In estimating future population the most important data we have to rely upon are the ages of the people as taken at recent censuses. These are shown measured down from the top on the first five perpendicular lines of the diagram opposite. The lines sloping downwards from left to right divide the total population according to each ten years of age from eighty downwards, so that the top layer of persons includes all over eighty years of age, the next all between seventy and eighty, the next all between sixty and seventy, and so on down to the children under ten. Now as the number of immigrants at any particular decade of age is probably always exceeded by the number of emigrants of that age, the people who are between ten and twenty at one census are (with the exception that some immigrants are substituted for some emigrants) the survivors in England of those who were under ten at the preceding census, and similarly with each ten years of age. So if we call the persons born between one census and the next a generation, the lines sloping upwards from left to right divide each generation from the next, and show how each generation becomes smaller and smaller, till it finally becomes extinct on the death of some centenarian.

Now, provided that the rate of mortality and loss by migration remain the same at each age period, it is possible to calculate with absolute accuracy from the observed decrease of a generation between any two censuses, how much the next generation will decrease when it arrives at the same age. For example, if it be known that everyone over 100 in 1881 died before 1891, it may be confidently assumed that everyone over 100 in 1891 will die before 1901. If it be known that the people between fifty and sixty in 1881 lost about 27 per cent. of their numbers before 1891, it may be taken for granted that the people between fifty and sixty in 1891 will lose the same proportion between 1891 and 1901. Nothing is requisite to get the total population over ten at the next census except to work out by proportion sums the population in each ten years of age and add up the results.

Before, however, we can apply this method, it is necessary to inquire whether we are justified in assuming that the rate of mortality and loss by migration will not change. As to the future of mortality and loss by migration every man has a right to his own opinion, but no one can expect to prove anything. We may therefore fairly treat it as an even chance whether the rate of loss is likely to increase or diminish. In the past a decrease of the rate of loss by mortality has been counteracted by an increase in the rate of loss by migration, so that estimates in which this method has been used have been surprisingly accurate. ...

The question is, therefore, what is the probable number of births? Here we border on considerations of a more speculative character, but we have still something statistical to rely upon.

The fact that it is the custom to calculate birth-rates as a rate per thousand on the whole population makes it natural to say that the number of births depends in the first place upon the population, so that if the population increases rapidly the births will increase rapidly. The continuous lines of the diagram show that the population over certain ages will increase rapidly for many decades, so that we might consequently expect a considerable increase in the number of births. But of course, as a matter of fact the increase of population over and under certain ages has obviously no tendency to increase the number of births. The number of old women and children may be doubled without making it the least more likely that the births will increase until the children have grown to marriageable age. The number of births is more likely indeed to be somewhat repressed, since the old people and children form a burden upon the shoulders of those in the prime of life. It is consequently much more true to say that the number of births depends in the first place upon the number of men and women between certain ages. For practical purposes the ages of twenty and forty are sufficiently near the mark, and they are much more convenient than the ages for each sex which would have to be taken if perfect accuracy were required. Now from 1853 to 1876 the number of births, after allowing for some deficiency in registration, increased rapidly and was almost uniformly just about 12 per cent. on the number of persons between twenty and forty. From 1876 onwards the number of births has been almost stationary, and the rate per cent. on the persons between twenty and forty has consequently been rapidly declining. In ten years it had fallen to 11 per cent.; by 1890 it had further fallen to 10 per cent.; in 1891 it went up to about 10·4 per cent.; in the next two years it was 10 per cent.; in 1894 it descended to 9·8 per cent.

Now if the future births were estimated at this rate on the number of persons between twenty and forty we should have no grounds for expecting a cessation of the growth of population, though the growth would be at a much less rate than heretofore. But the statistics make it probable that the birth-rate on persons between twenty and forty will continue to fall. It must be remembered that the effect of the births having been nearly stationary for twenty years will be to change considerably the age distribution of the group of people between twenty and forty. A larger proportion of them will be at the higher ages. It seems at first sight paradoxical to say that the persons between twenty and forty can be older at one time than another, but it is really quite simple. The persons between 0 and 100 years old obviously do not average fifty years, and in just the same way the persons between twenty and forty do not average thirty. What the exact average age will be depends chiefly on the variation in the number of births between twenty and forty years before. As the number of persons born rose rapidly year by year before 1876, the number of persons becoming twenty years of age in each year has been rising rapidly and will continue to rise rapidly till 1896. After that it will still continue to rise because there has been a reduction of juvenile mortality, but the rise will be far less rapid. The consequence will be that while in 1891 30 per cent. of the twenty to forty group were under twenty-five, in 1911 not more than $27\frac{1}{3}$ per cent. will be so.

This increase in the average age of persons in the prime of life, or, to put the same thing in other words, this diminution in the increase of the number of persons reaching marriageable age in each year, must tend somewhat to reduce the birth-rate. How much effect should be attributed to it would not be very difficult to calculate if the necessary statistics of the ages of parents were forthcoming, but unfortunately none such exist.

If, however, we take into account not only the diminution to be expected from this cause, but also the diminution to be expected from the working of the enormously strong economic and social forces which have brought about the diminution of the last twenty years, it seems a very moderate hypothesis to suppose that the rate of births on the number of persons between twenty and forty may fall to a little below 9 per cent. by 1901 and to the neighbourhood of 8 per cent. by 1911. This, with a further slight diminution to a little below 8 per cent. by 1921 is all that is necessary in order to keep the number of births stationary at the level of 1881–90. Adopting then this hypothesis, I have continued the lines of the diagram so as to show what the future of the population will be if the rate of mortality and loss by migration at each age, and also the absolute number of births, remain the same as in 1881–90. It will be seen that the increase of population, large at first, becomes less and less, till it is trifling in 1941–51. It would continue, but always growing less and less, till about 1995, when the last survivor of the period before 1891 would disappear, and the population would then stand at its maximum of 37,376,000.

I have no desire to stake my reputation as a prophet on the growth of population following exactly the line shown in the diagram, and ceasing to increase in 1991. I am only prepared to assert confidently that the line shown is a much more probable one than that which might be laid down by the ' official ' method, and which would shoot through the bottom of the diagram between 1921–31 and encircle the globe before the diagram was widened very many yards.. Whether the cessation of the growth of population is reached, as I personally should expect, before 1991, or afterwards, it must be reached at last, and if it is reached without any violent changes in mortality, migration, or natality, it will necessarily be reached by a curve of increase closely approximating to that laid down in the diagram. The value of the diagram lies not in its prediction of a maximum population of thirty-seven millions, but in the fact that it shows how a cessation of growth may be reached within no very long period without any violent or unnatural changes.

23. An Empirical Method for Calculating Future Population

P. K. WHELPTON (1936)

From *Journal of the American Statistical Association* 31. Excerpts are from pages 457—459, 461—463, 465—471, 473.

We omit Whelpton's adjustments for underenumeration of children in the census and his discussion of immigration levels, and limit the analysis to his discussion of the non-immigrant white population.

IN VIEW OF the large number of estimates of the future population of the United States that had been made prior to 1926 and the variety of results that had been obtained, it may be wondered why the Scripps Foundation began to work in this field at that time. It should be emphasized that we had no feeling of possessing superior prophetic power. We did not expect to prepare more accurate estimates than some of those already available, but did believe we could present much valuable material in addition to the customary figures of total population. For example, a growing demand was being manifested for information regarding probable changes in the makeup of the population in years to come. Furthermore, it seemed worth while to obtain data showing the effect on population growth and composition of differences in future trends of birth rates, death rates, and immigration—differences that seemed reasonable in view of past trends and present conditions in the United States and other nations.

The method we developed was based on a consideration of births, deaths, immigration, and emigration by age periods.[1] It was necessary to consider native-born whites, foreign-born whites, Negroes, and other colored by sex in order to secure the detailed data desired regarding the future composition of the population. Using various trends of birth rates, death rates, and migration by nativity, age and sex showed the effect on future growth and composition of differences in these factors and permitted the reader to choose the results of the combination that seemed most reasonable to him rather than confining him to the current choice of the writers. This paper will be confined to the procedure used in the calculations recently completed for the National Resources Committee, which differs in minor respects from that used previously.
...

DEATH RATES AND TRENDS

An examination of birth and death data by age indicated that little if any increased accuracy would be obtained by dealing with one-year age periods and time intervals instead of five-year and that the additional labor involved would not be justified. The procedure adopted was to compute death rates by five-year age periods and time intervals for native whites, foreign-born whites and Negroes by sex, and subtract these from unity to obtain survival rates by five-year age periods

---

* Revision of a paper read before a joint meeting of the Population Association of America and the American Statistical Association, New York City, December 31, 1935.
[1] Although the method followed was worked out independently, much of it was found afterward to have been used by Bowley, A. L.: "Births and Population of Great Britain," *The Journal of the Royal Economic Society*, vol. 34 (1924), pp. 188–192.

and time intervals, e.g., the number of native white women living to be 15–19 on April 1, 1935 out of 1,000 who were 10–14 on April 1, 1930. The steps in detail for the population in the 1930 death registration states are as follows: (1) Estimate the deaths for each race and nativity group from April 1 to December 31, 1930 and January 1 to March 31, 1935 on the assumption that the seasonal variation in deaths at all ages applies to deaths at each age. (2) For each calendar year or fraction thereof from April 1, 1930 to March 31, 1935, draw summation curves based on deaths by age at death, and read deaths by age on April 1, 1930, e.g., persons dying between the ages of $43\frac{1}{4}$ and $48\frac{1}{4}$ in 1933 were between 40 and 45 on April 1, 1930. (3) Add deaths from April 1, 1930 to March 31, 1935 by age of persons on the former date. (4) Divide these sums by the number of persons (in thousands) enumerated by the census, except that deaths of children born from April 1, 1930 to March 31, 1935 are divided by the number of births (in thousands) during that time interval. (5) Subtract these five-year-age-period-five-year-time-interval death rates from 1,000 to obtain five-year-age-period-five-year-time-interval survival rates. The death rates for native white males are shown in Table I. Because the death registration states of 1930 contained such a large proportion of the nation's population it was assumed that the age specific rates of this area in 1930–34 represented those of the United States.

With five-year-age-period-five-year-time-interval survival rates obtained according to mortality conditions of 1930–34 the next problem was to determine the changes that would be likely to occur in them during future years. Past trends in death rates and life expectancy were wanted as one guide, and were obtained from data in various life tables, extending back to 1900–02 for the original death registration states and to 1789 for Massachusetts. The outstanding facts are the large decreases in death rates occurring at younger ages and the small decreases or occasional increases at older ages. ...

As a second guide in determining the future death rates to be used in these calculations it seemed appropriate to study the country with the world's most favorable mortality conditions, namely, New Zealand. Past trends there have been similar to those in the United States, but more progress has been made in lowering death rates at ages below 65 and especially in the 40's and below 10. ...

A third type of information, not statistical but helpful nevertheless, comes from the fields of medicine and public health. There is no question but that all the knowledge now available about controlling various causes of death is not being put into practice to the extent that it will be in the future. Furthermore, it is almost inconceivable that additions to this fund of knowledge are suddenly going to cease. ...

After considering the facts that could be brought together three trends were laid out for native white death rates up to 1980. According to the least favorable assumption—with highest death rates—the expectation of life at birth will be raised from 62.4 in 1930–34 to 67 in 1980, through death rates declining somewhat below 1931 New Zealand rates at ages below 40, and as low as those rates at older ages. The most favorable assumption—with lowest death rates—envisaged somewhat greater improvement in mortality conditions at the younger ages and much more at middle life, with the expectation of life rising to 73 in 1980. The medium trend, which now seems most probable to us, lengthened life expectancy at birth to 70 years, keeping death rates at younger ages nearer the low than the high assumptions, and at older ages about midway between the two. Death rate trends for foreign-born whites, Negroes, and other colored were assumed to be similar to those for native whites, but with slightly larger decreases. The unfavorable differential of these groups compared with native whites was diminished by one-fourth in the high death rate trends, one-half in the medium, and three-fourths in the low.

BIRTH RATE TRENDS

For births as for deaths it was desired to have rates for five-year age periods and time intervals in making the actual computations. Because of the necessity of using ratios of children to women in determining long-time trends of United States birth rates in the past it was decided to use the methodology of the former and divide births during a five-year interval by women living at the end of the interval.

To obtain rates for the base period, births registered from April 1, 1930 to March 31, 1935 in the 1930 registration states were increased to allow for nonregistration, white births being divided by .96 and colored births by .86. Using summation curves, the grouping of births by age, color and nativity of mother was changed from age at birth of child to age on April 1, 1935. The number of women in the 1930 birth registration states on this later date was calculated by multiplying the number enumerated in the 1930 census by the five-year-age-period-five-year-time-interval survival rates described above. Dividing births to native and foreign-born white women by the number of women in the proper year and age periods gave preliminary rates for the 1930 birth registration states. Since these states contained 95 per cent of the nation's native white population in 1930 these native white rates were used for the United States.

To obtain rates for the other groups several minor adjustments were necessary. ...

Since it is not customary to use five-year-age-period-five-year-time-interval birth rates or their sum a brief interpretation may be helpful. The sum of these rates is approximately equal numerically to the sum of annual rates by one-year age periods or five times the sum of annual rates by five-year age periods, the total fertility of Kuczynski.[17] For example, the sum of the five-year-age-period-five-year-time-interval rates and the total fertility rates in the United States during 1930–34 were as follows: native whites, 2,177 and 2,158; foreign-born whites, 2,434 and 2,404; and Negroes, 2,750 and 2,728. The sum of the five-year-age-period-five-year-time-interval rates also indicates approximately the number of births per 1,000 women living to age 50. ...

Choosing trends which these five-year-age-period-five-year-time-interval birth rates might be expected to follow in the United States during future years presented more difficulties than the similar choice for death rate trends. Although the experience of foreign countries probably is less helpful in this case than the other, it was believed to merit attention. For this purpose total fertility rates calculated for many foreign countries by Kuczynski[20] were compared with those for the United States. Low as the recent rate of 2,158 is for native whites in the United States, rates for Australia and New Zealand are almost identical (2,195 and 2,151), while those for all the Western European countries and two of the Eastern (Austria and Latvia) are even lower. Wherever the immediate post-war rates and the recent rates can be compared a striking drop is shown, in most cases continuing a slower decline that was going on before the war. Judging from what has happened in other nations, therefore, there is little basis for expecting the downward trend in the United States to end suddenly and a rise to set in. ...

In addition to this international comparison the long-time trend in the United States during past years was desired before making assumptions regarding the future. Because the birth registration area was not organized until 1915 it was necessary to estimate this from census data. ...

---

[17] Kuczynski, R. R.: *The Measurement of Population Growth* (London: Sidgwick & Jackson, 1935), p. 117.

[20] Kuczynski, R. R.: *op. cit.*, pp. 122–124.

As would be expected from earlier studies of ratios of children to women,[23] these estimates show there has been a large and almost uninterrupted decline of age specific birth rates since the beginning of the nineteenth century. White rates during 1875–79 were more than twice as large as those during 1930–34, while during 1795–99 they were over $3\frac{1}{2}$ times as large. ... In only one ten-year interval does there appear to have been an increase in these rates (from 1845–49 to 1855–59) every other ten-year interval showing a decided decrease. Although the numerical declines have become smaller as the rates have become lower, the percentage declines in ten years have shown little tendency to diminish, the 18 per cent drop from 1915–19 to 1925–29 being the largest in the series, and the 17 per cent drop in the *five*-year interval from 1925–29 to 1930–34 being nearly as large.

Similar rates for native and foreign-born white women cannot be estimated accurately prior to 1905–09, nor for Negro women prior to 1845–49. Since these dates the decline of native white rates has been somewhat less than that of all whites, that of Negro rates about the same (although concentrated in different decades) and that of foreign-born white rates much more rapid.

Because birth rates are more subject to human control than death rates there is less basis for predicting their future course by extrapolating the local trend during past years, or by accepting the present situation in some foreign countries as the goal that will be reached here at some future time. Recent studies indicate that the practice of contraception is an exceedingly important immediate factor governing the size of the birth rate.[24] Judging from the improvements in the technique of contraception and the wider diffusion through the population of information regarding birth control that have taken place in the last decade or two further declines in birth rates are to be expected. However, a wide number of conditions govern the extent to which birth control is practiced by those acquainted with the methods, and little is known about how or to what degree they may be influenced. As a consequence of the legitimate differences in opinions regarding what is ahead it seems desirable to calculate the future population

[23] Willcox, Walter F.: *Proportion of Children in the United States*, Washington, G.P.O. (1905)· Bull. 22.
———: "The Change in the Proportion of Children in the United States and in the Birth Rate in France during the Nineteenth Century," *American Statistical Association Publications*, vol. 12, n.s. no. 93 (March 1911), pp. 490–499.
Whelpton, P. K.: "Industrial Development and Population Growth," *Social Forces*, vol. 6, nos. 3–4 (March–June 1928), pp. 458–467; 629–638.
[24] Pearl, Raymond: "Contraception and Fertility in 4945 Married Women. A Study of Family Limitation," *Human Biology*, vol. 6, no. 2 (May 1934), pp. 355–401.
Stix, Regine K., and Frank W. Notestein: "Effectiveness of Birth Control, "*Quarterly Bulletin of the Milbank Memorial Fund*, vol. 13, no. 2 (April 1935), pp. 162–178.

according to more divergent assumptions of trends of birth rates than of death rates.

Careful consideration of the information available regarding birth rates and factors affecting them lead us to adopt as the high birth rate assumption an abrupt stopping of the past decline, and a continuation of the 1930–34 values during subsequent years. For the low assumption, the past downward trend was continued at a decreasing pace, rates for native whites reaching a level in 1980 at 1,500 births per 1,000 women living through the childbearing period compared with 2,177 during 1930–34. ...   This represents a decline of 31 per cent from 1930–34 to 1980 as compared with 34 per cent from 1905–09 to 1930–34. The 1980 rate is approximately that of England in 1933, and only a little under that of California and Washington, D.C., in 1929–31. It would be attained if nine-tenths of the women living to age 50 marry and one-sixth of those marrying bear no children as at present, and if the remainder are equally divided into those bearing 1, 2, and 3 children.

For the medium assumption the past downward trend was continued at a slower pace, reaching a level in 1980 at 1,900 births per 1,000 native white women living through the childbearing period. This rate would prevail if present rates for marriage and infertility of married women continued, and if fertile women were about equally divided between those bearing 1, 2, 3, and 4 children. It would represent a decline in rate of 12.7 per cent in the next fifty years compared with a 34 per cent drop in the last 25 years. The United States rate in 1980 would be approximately that of Sweden, and also of Massachusetts, Connecticut, Washington, and Oregon in 1929–31, and significantly higher than the rate of New York and New Jersey. In the low assumption the 1930–34 differentials between native whites and other groups were reduced by 50 per cent, in the medium by 25 per cent. ...

THE FUTURE POPULATION

If the three assumptions for the future course of birth rates and three for death rates that have been described are combined in all possible ways with and without immigration, 18 series showing the future population of the United States will be obtained. Only 6 will be shown, the extremes and the middle ground with and without immigration[26] (Table IV). *A* will be referred to as the low series, *D*

---

[26] In all cases immigration is assumed to begin in 1940.

TABLE IV
FUTURE POPULATION OF THE UNITED STATES IF FERTILITY, MORTALITY, AND
IMMIGRATION FOLLOW STATED TRENDS 1935 TO 1980*
(Thousands)

Year	Low Fertility High Mortality		Medium Fertility Medium Mortality		High Fertility Low Mortality	
	No Immigration A	100,000† Immigrants B	No Immigration C	100,000† Immigrants D	No Immigration E	200,000† Immigrants F
1930‡	122,775.1	122,775.1	122,775.1	122,775.1	122,775.1	122,775.1
1935	127,354.3	127,354.3	127,354.3	127,354.3	127,354.3	127,354.3
1940	131,157.0	131,157.0	131,993.1	131,993.1	132,613.2	132,613.2
1945	134,138.7	134,647.9	136,447.4	136,956.6	138,209.7	139,228.4
1950	136,176.6	137,245.7	140,560.9	141,644.8	143,898.4	146,086.2
1955	137,171.9	138,840.3	144,093.1	145,808.1	149,353.0	152,852.1
1960	137,088.7	139,375.3	146,986.5	149,371.6	154,563.5	159,481.2
1965	136,026.3	138,938.0	149,340.8	152,421.2	159,719.5	166,135.2
1970	134,048.9	137,588.3	151,170.0	154,969.3	164,835.8	172,837.0
1975	131,221.1	135,387.4	152,432.9	156,977.4	169,779.0	179,471.1
1980	127,570.9	132,356.7	153,022.3	158,335.2	174,330.3	185,823.5

* See text for description of trends assumed. The calculated numbers of children 0–4 and hence of total persons are reduced to correspond with underenumeration of children in the census.
† Net annually, beginning with 1940.
‡ Census.

as the medium, and *F* as the high. According to the low series the population will cease to grow between 1955 and 1960, and from then to 1980 will decline at an accelerated pace. Its maximum size will be about 137,300,000 and by 1980 it will be back to the 1935 figure (Table IV). If the high series is followed growth will continue until after 1980, although at a decreasing rate, and the population on that date will amount to about 185,800,000. At the present time it seems to us that the medium series will be closer to what happens than the high or low, though probably somewhat too high. According to this series, the population will reach its maximum of about 160,000,000 soon after 1980, and then begin to dwindle numerically. ...

Making comparisons between 1935 and 1980 on a percentage basis, the proportion of children and youths (under 20) will decline almost one-third from 36.6 per cent to 25.8 per cent, the proportion in the best working ages (20–39) will decrease slightly from 31.8 per cent to 27.6 per cent, the middle age group (40–64) will have a relative rise of over one-fourth from 25.7 per cent to 32.6 per cent, and elders (65 and over) will more than double relatively, jumping from 5.9 per cent to 14 per cent. If persons in the first and last group are considered as dependents and those in the second and third as producers, the number of dependents per 100 producers will decline significantly, from 73.9 to 66.1. Even if the middle aged are counted only half as valuable from a production standpoint as those 20–39 the ratio of dependents to producers will still decline from 95.2 to 90.6.

If population growth follows the low series these changes in composition will be much more striking. But even if the high series is followed the makeup of the 1980 population will differ sufficiently from that of the present so that making adjustments to the new conditions may present serious difficulties.

Although much time and thought have been devoted to these empirical calculations of future population it is fully realized that this does not insure obtaining results which show accurately the growth and composition that will occur. It is hoped, however, that the undertaking is worth while because of the information it may contribute and the thinking it may stimulate regarding the relationship of birth rates, death rates, and immigration to the size and composition of the population in years to come, and also regarding the changes in growth and makeup which now appear probable and which will raise problems of great significance.

24. Population Waves

Harro Bernardelli (1941)

From *Journal of the Burma Research Society* 31, Part 1. Excerpts are from pages 1—4, 6—7.

We omit Bernardelli's discussions of economic cycles and his further comments on cycles in human and animal populations, which are of uneven quality. The appendix referred to in the text is also omitted: it introduces a projection matrix and indicates the derivation of its characteristic roots.

1. In thinking of population growth usually the notion is applied—consciously or unconsciously—that such growth takes place along a smooth curve: either in the form of Malthus' geometric progression, the interest law, or according to Verhulst-Pearl's logistic equation, or along such curves as have been worked out recently by Dr. Kuczynski and his collaborators in order to trace the decline of the European race. Practical statisticians, no doubt, realise that populations actually show considerable deviations from such models, but they have tended on the whole to consider such "irregularities" as "random variations" which are best smoothed out.

Having worked on a variety of problems which all involve assumptions about the manner in which the numbers of a population change, it appears to me now rather doubtful that the preference given to smooth curves is justified even as a first approximation. Smoothing always means introducing a definite hypothesis about the nature of the causes that determine the value of the variable under consideration; in eliminating the irregular oscillations typical for a population graph, we assume, in fact, that there are two sets of causes operating: a fundamental, basic set of causes which, left alone, would produce the picture of our smooth curve, and a second set—more irregular and haphazard in occurrence and intensity—which accounts for the superimposed aberrations.

The justification of the procedure of smoothing thus depends essentially on the validity of the hypothesis that the basic causes will manifest themselves smoothly. If it can be shown that this is not the case, that the basic causes themselves are bound, in fact, to generate oscillations, then the operation of smoothing will lose its rationale. Its application will merely blur the picture, and possibly even block the way for a correct understanding of the causal effects which we wish to elucidate.

2. The oscillations of which I am talking can easily be demonstrated in any Census Report. The following table, taken from the Burma Census 1931, gives the age-distribution of 10,000 persons of each sex for the indigenous population

I

Age Group	Males				Females			
	1931	1921	1911	1901	1931	1921	1911	1901
0— 5	1397	1265	1354	1433	1414	1286	1372	1442
5—10	1282	1271	1355	1301	1236	1260	1344	1277
10—15	1186	1223	1236	1143	1134	1146	1140	1046
15—20	916	1002	899	877	1002	1082	966	961
20—25	886	862	775	828	997	941	864	917
25—30	838	777	760	828	848	801	792	850
30—35	776	722	767	799	737	687	727	756
35—40	604	600	656	626	560	539	584	552
40—45	532	571	566	539	504	558	552	521
45—50	412	446	409	407	397	406	374	369
50—55	361	410	373	383	358	418	384	395
55—60	283	252	241	244	278	251	242	241
60—65	239	274	278	—	229	264	278	—
65—70	130	133	133	—	130	126	128	—

of this country[1]. The figures have been compiled for the indigenous inhabitants only, because they reveal best the pure effects of population growth, and are not, in any way, affected by migration.

The table shows a remarkable instability of the age-structure. Its characteristic features can be studied most effectively by averaging for the four censuses the frequencies in each age-group and recording for each census the deviations of the actual frequencies from their mean. The result is shown in Fig. 1.

It will be observed that there was in 1901 an abnormally high frequency of children in the age-group 0—5 which subsequently produced a wave of survivors in the later census records. Similarly the low frequency for the age-groups 10—20 in 1901 can be traced as a depression in the waves for the following three decades. The waves for females (shown in the Census Report 1931 page 73ff.) run exactly parallel. It is obvious that the new wave crest of children which is seen to develop in 1931 originates from the survivors of the wave of female children in 1901.

These waves have a sufficiently large amplitude to exert a significant influence on the economic and social conditions of the country. This became painfully evident during the Great Depression, in 1929 and the succeeding years. There was during those years an unusually large proportion of indigenous males in the age-group 15—30 which forms, of course, the most active element of a population from which the new entries into the labour market are drawn (See Section ABC of the 1921 and DEF of the 1931 wave). With agriculture disorganised by the impact of the unprecedented depression, a large number of these persons—who normally would have found accommodation in agricultural pursuits were forced to look for openings in the industrial labour market which up to that time had been the undisputed domain of the Indian immigrant[2]. ...

This unfortunate constellation of a large wave of indigenous males looking for employment, and of an industrial labour market which offered no chances of employment, undoubtedly contributed to the series of unprecedented riots, and the intense growth of racial animosity, which—busily fanned by interested parties—has never since subsided, and which is likely now to produce the usual crop of obnoxious restrictionism, although—among other things—the labour market is long again in perfect adjustment, and although at the present, and during the next years, the entry of indigenous people into the labour market will be by no means brisk (See the depression GHD of the 1931 wave which is relevant for the present phase).

3. Naturally the question arises, how do such waves originate? A little reflection shows that population cycles easily can be produced synthetically. Take a species, say a beetle, which lives three years only, and which propagates in its third year of life. Let the survival rate of the first age-group be $\frac{1}{2}$, of the second $\frac{1}{3}$, and assume that each female in the age 2—3 produces, in the average, 6 new living females. (These values are, of course, entirely arbitrary.) Then a population of these beetles which, we may assume, has in a certain year 1 a thousand females

---

[1] Census of India, 1931, Vol. XI, Burma, Part I, Rept. by J.J. Bennison, Rangoon 1933, pp. 73ff.
See also Census of India, 1921, Vol. X, Burma, Part I, Rept. by S.G. Grantham, Rangoon 1923, p. 121ff.
[2] "The economic depression"—says Bennison l.c. p. 77—"has caused a great deal of unemployment, and this has apparently been aggravated by the unusually large proportion of males between 20 and 30."

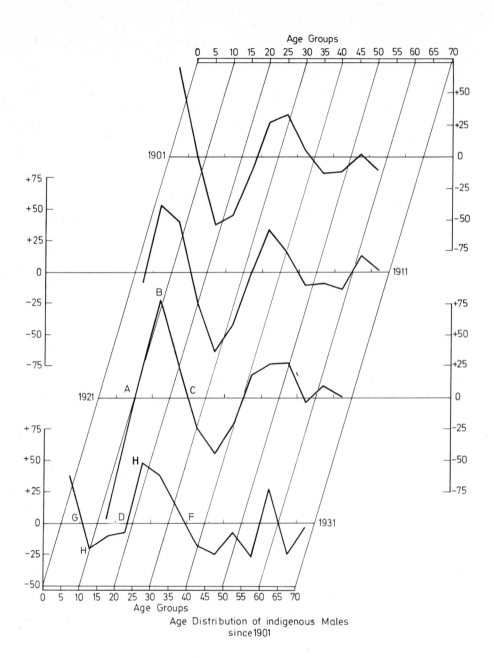

Age Distribution of indigenous Males
since 1901

Fig. 1

218

in each age-group—males are irrelevant in matters of reproduction—will produce waves in the following manner:—

| | II | | | |
Age Group	1	2	3	4
0—1	1000	6000	1999.9	1000
1—2	1000	500	3000	1000
2—3	1000	333.3	166.6	1000
Total	3000	6833.3	5166	3000

So long as the rates of fertility and mortality remain constant, this development will, of course, continue indefinitely in regular three-year cycles. The population will oscillate round a stationary level, as the Net Reproduction Rate R is here assumed to be $=1$.

What determines the amplitude of these waves? Without going into the Mathematics of the process, it can be seen by means of simple Arithmetic, that the amplitudes depend on the assumptions made with regard to the initial distribution of females in the age-groups. If one takes in the year 1 the numbers 1000, 600, 200 respectively, it is easily seen that the amplitudes of the resulting waves become much smaller: and with the numbers 1000, 500, 167 instead, the waves disappear altogether.

For a mathematical analysis interested readers are referred to the Appendix where a system of equations is given from which all the relevant features of the process can be deduced with greatest ease. Here only the following observations will be used:

I. For any given set of fertility and mortality conditions there exists always one characteristic age distribution (1000:500:167 in the above example) which does not give rise to disequilibrium waves; but whenever the population deviates from this equilibrium structure, waves will be set up with amplitudes that depend entirely on the degree of these deviations.

II. It is not necessary that the population should be stationary, as in the above example; waves will also occur, if the Net Reproduction Rate $R > 1$, indicating a growing population, or if $R < 1$, so that the population eventually dies out.

III. It is not necessary that the waves repeat themselves year after year with the same amplitude. The occurrence of dampening, or of magnifying factors depends entirely on the given fertility and mortality rates.

IV. Similarly the wave length depends entirely on those rates, so that any change of the fertility and mortality conditions will produce waves of a different character. Wave length as well as the degree of dampening, and the amplitudes will be modified, if the situation changes.

25. On the Generation and Growth of a Population

E. G. Lewis (1942)

Sankhya 6: 93—96.

1. Some of the factors which complicate the theoretical and practical quantitative study of human populations may be negligible in the treatment of certain lower biological populations. The comparatively short lifetime and the regularity of breeding epochs is one simplification, which, incidentally suggests the interval between breeding epochs as a natural unit of age. It may also be that fertility and mortality rates at ages are more nearly constant in time—in any case, time fluctuations in these rates due to environment changes are more amenable to experimental control in colonies of lower animals or plants than in human populations.

2. Suppose a group y_1 of individuals to be 'born' at some epoch ($t=0$) and consider the population generated by this group. Assume that 'breeding' occurs at regular epochs $t=z, 2z, \ldots NZ, \ldots$ and let us adopt z as the unit of age for individuals. Let f_r be the fertility factor per individual entering the rth age group ($r-1z$ to rz) and let S_r be the survival factor per individual for survival from the rth to the $(r+1)^{th}$ age group. Both f_r and S_r are taken to be constant in time and we assume, further, that for reproductive purposes there is no interaction between different age groups.

Under these somewhat stringent conditions, the relaxation of which can be considered later, the numerical history of the population generated by the original y_1 individuals, will, just after the third breeding epoch (say), be as follows:

	After epoch $t=0$	After epoch $t=Z$	After epoch $t=2Z$	After epoch $t=3Z$
No. aged $0+$	y_1	$f_1 y_1$	$(f_1^2+f_2 s_1)y_1$	$(f_1^3+2f_1 f_2 s_1 + f_3 s_2 s_1 y_1)$
No. aged $Z+$	0	$s_1 y_1$	$f_1 s_1 y_1$	$(s_1 f_1^2 + f_2 s_2^2)y_1$
No. aged $2Z+$	0	0	$s_2 s_1 y_1$	$f_1 s_2 s_1 y_1$
No. aged $3Z+$	0	0	0	$s_3 s_2 s_1 y_1$

If we now suppose that no individual survives the age nZ (i.e. $S_n = 0$), the frequency distribution in the n age groups 0 to Z; Z to $2Z \ldots, n-1 Z$ to nz will, after the Nth breeding epoch be given by the Nth power of the linear substitution.

$$A \equiv (f_1 y_1 + f_2 y_2 + \ldots + f_n y_n, s_1 y_1 \ldots s_{n-1} y_{n-1}), \tag{1}$$

the initial set of frequencies being $y_1 = y_1; y_2 = y_3 = \ldots y_n = 0$.

3. Now consider the n age groups 0 to z, z to $2z$, etc. to be filled by an *arbitrary* set of $y_1, y_2, \ldots, y_n$ individuals and let the fertility and survival factors for age groups be $f_1 f_2 \ldots f_n$ and $s_1, s_2 \ldots s_{n-1}, 0$ as before. Then we may take the arbitrary column matrix

$$\{y\} = \{y_1, y_2 \ldots y_n\}$$

as an initial age distribution corresponding to the epoch $t=0$. The age distribution after the first breeding epoch will now be given by

$$\{y'\} = \{y_1', y_2' \ldots y_n'\}$$

222

where

$$y'_1 = f_1 y_1 + f_2 y_2 + \ldots + f_n y_n$$

$$y'_2 = s_1 y_1$$

$$y'_3 = s_2 y_2$$

$$y'_n = s_{n-1} y_{n-1}$$

which is, of course, the linear substitution A of equation (1). The matrix of this substitution is the square matrix

$$M = \begin{vmatrix} f_1 f_2 \cdots f_{n-1} f_n \\ s_1\, 0 \ldots 0 \quad 0 \\ 0\, s_2 \ldots 0 \quad 0 \\ 0\, 0 \ldots s_{n-1}\, 0 \end{vmatrix}$$

all elements not in the first row or sub-diagonal being zero. The age frequency distribution after the Nth breeding epoch will be given by the Nth power of A or, what is the same thing, by the column matrix $M^N \{y\}$.

Now the characteristic equation of M is easily shown to be

$$\lambda^n - f_1 \lambda^{n-1} - \ldots - f_{n-1} s_1 s_2 \ldots s_{n-2} \lambda - f_n s_1 s_2 \ldots s_{n-1} = 0$$

where, of course $f_r \geq 0$ $(r = 1, 2, \ldots, n)$ and $0 \leq s_r \leq 1$ $(r = 1, 2, \ldots n-1)$, $s_n = 0$. This equation determines the form to which the age distribution of the population envisaged ultimately settles down under the action of the given system of fertility and survival factors.

4. Under certain conditions however the initial age distribution will be repeated exactly after the cycle of n breeding epochs. This follows at once since the matrix M satisfies its own characteristic equation

$$M^n - f_1 M^{n-1} - \ldots - f_{n-1} s_1 s_2 \ldots s_{n-2} M - f_n s_1 \ldots s_{n-1} I = 0$$

where I is the unit matrix of order n.

Now if $M^n - I = 0$ the initial distribution $\{y\}$ will be repeated exactly after n operations of M. We see that the condition for this is

$$f_1 = f_2 \ldots f_{n-1} = 0 \quad \text{and} \quad f_n s_1 s_2 \ldots s_{n-1} = 1.$$

If

$$f_1 = f_2 \ldots f_{n-1} = 0 \quad \text{and} \quad f_n s_1 s_2 \ldots s_{n-1} = k$$

then the distribution after n breeding epochs will be

$$\{k y_1 \ldots, k y_n\}, \quad i.e., \quad k\{y\}.$$

After $2n$ epochs it will be $k^2 \{y\}$ and so on.

223

5. In general, however, the age distribution will display no pure periodicity but will settle down to a definite (stable) distribution depending on the dominant root (root of maximum modulus) of the characteristic equation.

In this equation, namely

$$\lambda^n - f_1 \lambda^{n-1} - f_2 \lambda^{n-2} s_1 \ldots - f_{n-1} s_1, s_2 \ldots s_{n-2} \lambda - f_n s_1, \ldots s_{n-1} = 0$$

the f and s are all positive so that the equation has at most one positive root, and moreover, it always has one. It can be shown that this positive root is also the dominant root[1] (a fact of itself not without interest). The only exception which can arise yields the periodicity mentioned in para 4.

6. Now when the characteristic equation has a single (non repeated) real dominant root l(say) the ratio of an element in the $(m+1)^{th}$ power of M to the corresponding element in the mth power will, for large m, tend to l. The column matrix giving the age distribution, being acted upon successively by the matrix M, will also tend to a form in which an element of $M^{m+1}\{y\}$ will bear to the corresponding element in $M^m\{y\}$ the ratio l.

Now let $Y_1, Y_2, \ldots Y_n$ be a set of frequencies such that

$$\left.\begin{aligned}
\lambda Y_1 &= f_1 Y_1 + f_2 Y_2 \ldots + f_n Y_n \\
\lambda Y_2 &= s_1 Y_1 \\
\lambda Y_3 &= s_2 Y_2 \\
&\vdots \\
\lambda Y_n &= s_{n-1} Y_{n-1}
\end{aligned}\right\} \tag{2}$$

Eliminating $Y_1 Y_2 \ldots Y_n$ we obtain the characteristic equation of M, the matrix of A.

The set $Y_1 Y_2 \ldots Y_n$ constitutes a 'pole' of the substitution, A corresponding to the root λ of the characteristic equation. We have seen that the frequencies in all age groups will ultimately increase in the ratio l, it follows that the ultimate ratios of the n frequencies one to another will be $Y_1, Y_2, \ldots Y_n$ where $Y_1 \ldots Y_n$ is a pole corresponding to the dominant root l. Now from equation (2) with $\lambda = l$ we obtain

$$Y_1 : Y_2 : \ldots Y_n = 1 : \frac{s_1}{l} : \frac{s_1 s_2}{l^2} : \ldots \frac{s_1 s_2 \ldots s_{n-1}}{l^{n-1}} \tag{3}$$

7. We see, therefore, that under the operation of the matrix M the age distribution column matrix $\{y\}$ will in general settle down to a form in which the relative frequencies in the different age groups are determined by the ratios (3) and that after each breeding epoch the numbers in each group (and so of course in the total population) will increase in geometric progression of ratio equal to the positive root of the characteristic equation of M. The population will

---

[1] This theorem and its consequences, which lead to a simple upper bound to the modulus of the roots of any equation, have been investigated by Mr. D. B. Lahiri and the present author in a paper awaiting publication.

increase, just maintain itself, or decrease ultimately according as $l \gtreqless 1$. The number l is related to Kuczynski's net reproduction rate, the analogue of which would here be

$$(f_1 + f_2 s_1 + \ldots + f_n s_1 \ldots s_{n-1})$$

but l appears to be a better index of what can be called the true (ultimate) rate of increase. We have

$$1 = \frac{f_1}{l} + \frac{f_2 s_1}{l^2} + \ldots + \frac{f_n s_1 \ldots s_{n-1}}{l^n}$$

so that a net reproduction rate $\gtreqless 1$ means $l \gtreqless 1$ and *vice versa*, but the magnitudes of the two measures will be different.

8. The question of approximation to the case of pure periodicity discussed in para 4 may be of practical importance for some lower animals with few age groups. The rapidity with which an arbitrary initial age distribution settles down close to the 'ultimate' form depends on the difference between the dominant and sub-dominant root of the characteristic equation.

As for the relaxation of the conditions assumed in para 1, the question of interaction between age groups would in many cases be of no importance but to take account of time changes in f_r and s_r would mean assuming suitable laws of variation for the matrix elements. Such laws lead to much more complicated analysis but are being investigated.

26. On the Use of Matrices in Certain Population Mathematics

P. H. LESLIE (1945)

From *Biometrika* 33. Excerpts are from pages 183—185, 187—193, 199—202.

Leslie's work, rather than that of his predecessors Bernardelli and Lewis, is most commonly cited in the widespread literature using matrices, largely for the reason that Leslie worked out the mathematics and the application with great thoroughness. Some of his elaboration was designed to save arithmetic—for example his transformation of the projection matrix into an equivalent form with unity in the subdiagonal positions. Such devices, like a considerable part of classical numerical analysis, are unnecessary in a computer era.

The case of double roots seemed to require attention, for Leslie could not then know that no case of double roots would ever arise with real data. In the present excerpt we omit this, as well as the spectral decomposition of the matrix, which has not found extensive application.

Among the latent roots or eigen values and the corresponding stable vectors, as Leslie points out, demographic and biological interest is confined to three. The first, of largest absolute value, is positive and represents the ratio of population at the end of a cycle to that at the beginning, when the process has been operating for a considerable period; it provides the component of geometric increase. The second and third roots produce waves of diminishing amplitude having the length of the generation, usually 25 to 30 years. These waves measure the echo effect—after a baby boom they provide for a smaller boom a generation later—on the condition that the age-specific rates of birth and death remain constant.

CONTENTS

1. INTRODUCTION

If we are given the age distribution of a population on a certain date, we may require to know the age distribution of the survivors and descendants of the original population at successive intervals of time, supposing that these individuals are subject to some given age-specific rates of fertility and mortality. In order to simplify the problem as much as possible, it will be assumed that the age-specific rates remain constant over a period of time, and the female population alone will be considered. The initial age distribution may be entirely arbitrary; thus, for instance, it might consist of a group of females confined to only one of the age classes.

The method of computing the female population in one unit's time, given any arbitrary age distribution at time t, may be expressed in the form of $m+1$ linear equations, where m to $m+1$ is the last age group considered in the complete life table distribution, and when the same unit of age is adopted as that of time. If

n_{xt} = the number of females alive in the age group x to $x+1$ at time t,

P_x = the probability that a female aged x to $x+1$ at time t will be alive in the age group $x+1$ to $x+2$ at time $t+1$,

F_x = the number of daughters born in the interval t to $t+1$ per female alive aged x to $x+1$ at time t, who will be alive in the age group 0–1 at time $t+1$,

then, working from an origin of time, the age distribution at the end of one unit's interval will be given by

$$\sum_{x=0}^{m} F_x n_{x0} = n_{01}$$
$$P_0 n_{00} = n_{11}$$
$$P_1 n_{10} = n_{21}$$
$$P_2 n_{20} = n_{31}$$
$$\vdots \qquad \vdots$$
$$P_{m-1} n_{m-1,0} = n_{m1}$$

or, employing matrix notation, $M n_0 = n_1$, where n_0 and n_1 are column vectors giving the age distribution at $t = 0$ and 1 respectively, and the matrix

$$M = \begin{bmatrix} F_0 & F_1 & F_2 & \cdots & & F_k & F_{k+1} & \cdots & F_{m-1} & F_m \\ P_0 & \cdot & \cdot & \cdots & & \cdot & \cdot & \cdots & \cdot & \cdot \\ \cdot & P_1 & \cdot & \cdots & & \cdot & \cdot & \cdots & \cdot & \cdot \\ \cdot & \cdot & P_2 & \cdots & & \cdot & \cdot & \cdots & \cdot & \cdot \\ \cdots & \cdots & \cdots & \cdots & \cdots & \cdots & \cdots & \cdots & \cdots & \cdots \\ \cdot & \cdot & \cdot & \cdots & P_{k-1} & \cdot & \cdot & \cdots & \cdot & \cdot \\ \cdot & \cdot & \cdot & \cdots & \cdot & P_k & \cdot & \cdots & \cdot & \cdot \\ \cdots & \cdots & \cdots & \cdots & \cdots & \cdots & \cdots & \cdots & \cdots & \cdots \\ \cdot & \cdot & \cdot & \cdots & \cdot & \cdot & \cdot & \cdots & P_{m-1} & \cdot \end{bmatrix} \quad 0 < P_x < 1; \ F_x \geqq 0.$$

228

This matrix is square and consists of $m+1$ rows and $m+1$ columns. All the elements are zero, except those in the first row and in the subdiagonal immediately below the principal diagonal. The P_x figures all lie between 0 and 1, while the F_x figures are by definition necessarily positive quantities. Some of the latter, however, may be zero, their number and position depending on the reproductive biology of the species we happen to be considering in any particular case, and on the relative span of the pre- and post-reproductive ages. If $F_m = 0$, the matrix M is singular, since the determinant $|M| = 0$.

Since $Mn_0 = n_1$, and $Mn_1 = M^2_{n_0} = n_2$, etc., the age distribution at time t may be found by pre-multiplying the column vector $\{n_{00} n_{10} n_{20} \ldots n_{m0}\}$, i.e. the age distribution at $t = 0$, by the matrix M^t. Moreover, it will be seen that with the help of the jth column of M^t the age distribution and number of the survivors and descendants of the $n_{j-1,0}$ individuals, who were alive at $t = 0$, can readily be calculated. Thus, $n_{j-1,0}$ times the sum of the elements in the jth column of M^t gives the number of living individuals contributed to the total population at time t by this particular age group.

2. Derivation of the matrix elements

The basic data, from which the numerical elements of this matrix may be derived, are given usually in the form of a life table and a table of age specific fertility rates. To take the P_x figures first; if at $t = 0$ there are n_{x0} females alive in the age group x to $x+1$, the survivors of these will form the $x+1$ to $x+2$ age group in one unit's time, and thus $P_x n_{x0} = n_{x+1,1}$. Then it is usually assumed (e.g. Charles, 1938, p. 79; Glass, 1940, p. 464) that

$$P_x = \frac{L_{x+1}}{L_x},$$

where
$$L_x = \int_x^{x+1} l_x \, dx,$$

or the number alive in the age group x to $x+1$ in the stationary or life table age distribution. This method of computing the survivors in one unit's time would be exact if the distribution of those alive within a particular age group was the same as in the life-table distribution.

The F_x figures are more troublesome, and in the numerical example which will be given later they were obtained from the basic maternal frequency figures (m_x = the number of live daughters born per unit of time to a female aged x to $x+1$) by an argument which ran as follows. Consider the n_{x0} females alive at $t = 0$ in the age group x to $x+1$, and let us suppose that they are concentrated at the midpoint of the group, $x+\frac{1}{2}$. During the interval of time 0–1 some of these individuals are dying off, and at $t = 1$ the $n_{x+1,1}$ survivors can be regarded as concentrated at the age $x+1\frac{1}{2}$. Although these deaths are taking place continuously, we may assume them all to occur around $t = \frac{1}{2}$, so that at this latter time the number of females alive in the age group we are considering changes abruptly from n_{x0} to $n_{x+1,1} = P_x n_{x0}$. Then during the time interval 0–$\frac{1}{2}$ these n_{x0} females will have been exposed to the risk of bearing daughters, and the number of the latter they will have given birth to per female alive will be given by the maternal frequency figure for the ages $x+\frac{1}{2}$ to $x+1$. This figure may be obtained by interpolating in the integral curve of the m_x values, and thus expressing the latter in $\frac{1}{2}$ units of age throughout the reproductive span instead of in single units. The daughters born during the interval of time 0–$\frac{1}{2}$ will be aged $\frac{1}{2}$–1 at $t = 1$, the number of them surviving at this time being determined approximately by multiplying the appropriate $m_{x+\frac{1}{2}}$ figure by the factor $2\int_{\frac{1}{2}}^{1} l_x \, dx$ according to the given life table. Similarly, each of the $P_x n_{x0}$ females during the interval of time $\frac{1}{2}$–1 give birth to $m_{x+1-x+1\frac{1}{2}}$ daughters, the survivors of which form part of the 0–$\frac{1}{2}$ age group at $t = 1$. The survivorship factor is in this case taken to be $2\int_0^{\frac{1}{2}} l_x \, dx$.

Combining these two steps together we obtain a series of F_x figures, which may be defined as the number of daughters alive in the age group 0–1 at $t = 1$ per female alive in the age group x to $x+1$ at $t = 0$. Putting

$$k_1 = 2\int_0^{\frac{1}{2}} l_x\,dx, \quad k_2 = 2\int_{\frac{1}{2}}^1 l_x\,dx,$$

then

$$F_x = (k_2 m_{x+\frac{1}{2}-x+1} + k_1 P_x m_{x+1-x+1\frac{1}{2}}),$$

and

$$\sum_{x=0}^m F_x n_{x0} = n_{01},$$

the total number of daughters alive aged 0–1 at $t = 1$. ...

4. PROPERTIES OF THE BASIC MATRIX

The matrix M is square and of order $m+1$; it is not necessary, however, in what follows to consider this matrix as a whole. For, if $x = k$ is the last age group within which reproduction occurs, F_k is the last F_x figure which is not equal to zero. Then, if the matrix be partitioned symmetrically at this point,

$$M = \begin{bmatrix} A & \cdot \\ B & C \end{bmatrix}.$$

The submatrix A is square; B is of order $(m-k) \times (k+1)$; C again is square consisting of $m-k$ rows and columns, the only numerical elements being in the subdiagonal immediately below the principal diagonal. The remaining submatrix is of order $(k+1) \times (m-k)$ and consists only of zero elements. Then in forming the series of matrices M^2, M^3, M^4, etc.,

$$M^l = \begin{bmatrix} A^l & \cdot \\ f(ABC) & C^l \end{bmatrix}.$$

The submatrix C is, however, of such a type that $C^{m-k} = 0$, so that M^l, $t \geqq m-k$, will have all its last $m-k$ columns consisting of zero elements. This is merely an expression of the obvious fact that individuals alive in the post-reproductive ages contribute nothing to the population after they themselves are dead. It is the submatrix A which is principally of interest, and in the mathematical discussion which follows, attention is focused almost entirely on it and on age distributions confined to the prereproductive and reproductive age groups.

The matrix A is of order $(k+1) \times (k+1)$, where $x = k$ is the last age group in which reproduction occurs, and written in full,

$$A = \begin{bmatrix}
F_0 & F_1 & F_2 & F_3 & \cdots & F_{k-1} & F_k \\
P_0 & \cdot & \cdot & & \cdots & & \cdot \\
\cdot & P_1 & \cdot & \cdot & \cdots & \cdot & \cdot \\
\cdot & \cdot & P_2 & \cdot & \cdots & \cdot & \cdot \\
\multicolumn{7}{c}{\cdots\cdots\cdots\cdots\cdots\cdots\cdots\cdots} \\
\cdot & \cdot & \cdot & \cdot & \cdots & P_{k-1} & \cdot
\end{bmatrix}.$$

This matrix is non-singular, since the determinant $|A| = (-1)^{k+2}(P_0 P_1 P_2 \ldots P_{k-1} F_k)$. There exists, therefore, a reciprocal matrix of the form

$$A^{-1} = \begin{bmatrix}
\cdot & P_0^{-1} & \cdot & \cdot & \cdots & \cdot \\
\cdot & \cdot & P_1^{-1} & \cdot & \cdots & \cdot \\
\cdot & \cdot & \cdot & P_2^{-1} & \cdots & \cdot \\
\multicolumn{6}{c}{\cdots\cdots\cdots\cdots\cdots\cdots\cdots} \\
\cdot & \cdot & \cdot & \cdot & \cdots & P_{k-1}^{-1} \\
F_k^{-1} & -(P_0 F_k)^{-1} F_0 & -(P_1 F_k)^{-1} F_1 & -(P_2 F_k)^{-1} F_2 & \cdots & -(P_{k-1} F_k)^{-1} F_{k-1}
\end{bmatrix}.$$

Thus, given an initial age distribution n_{x0} ($x = 0, 1, 2, 3, \ldots, k$) at $t = 0$, in addition to the forward series of operations $A n_0$, $A^2 n_0$, $A^3 n_0$, ..., etc., there is also a backward series $A^{-1} n_0$,

$A^{-2}n_0$, $A^{-3}n_0$, ..., etc. There is, however, a fundamental difference between these; for, whereas the forward series can be carried on for as long as we like, given any initial age distribution, the backward series can only be performed so long as n_{xt} remains ≥ 0, since a negative number of individuals in an age group is meaningless. Apart from this limitation, it is possible to foresee that the reciprocal matrix might be of some use in the solution of certain types of problem.

5. TRANSFORMATION OF THE CO-ORDINATE SYSTEM

Hitherto an age distribution n_{xt} has been regarded as a matrix consisting of a single column of elements. For simplicity in notation, this column vector will now be termed the vector ξ and different ξ's will be distinguished by different subscripts (ξ_a, ξ_x, etc.). We may picture an age distribution as a vector having a certain magnitude and related to a definite direction in a vector space, the space of the ξ's. The different age distributions which may arise in the case of any particular population will be assumed to be ξ's all radiating from a common origin. The numerical elements of a ξ vector are thus taken to be the co-ordinates of a point in multi-dimensional space referred to a general Cartesian co-ordinate system, in which the reference axes may make any angles with one another. At this point in the argument another type of vector will be introduced, which in matrix notation will be written as a row vector, and which will be termed the vector η. There is an intimate relationship between this new type and the old, for, associated with each vector ξ_a, there is a uniquely determined vector η_a, and vice versa. The inner or scalar product, $\eta_a \xi_a$, is the square of the length of the vector ξ_a. Either we may picture each of these vectors as associated with a different kind of vector space, the space of the ξ's and the dual space of the η's, which are not entirely disconnected but related in a special way; or, alternatively, we may regard them as two different kinds of vector associated with the same vector space. The relationship between η and ξ is precisely the same as that between covariant and contravariant vectors in differential geometry.

If we pass from our original co-ordinate system to a new frame of reference, and the variables η and ξ undergo the non-singular linear transformations,

$$\eta = \phi H, \quad \xi = H^{-1}\psi, \quad |H| \neq 0,$$

it can be seen that since the variables are contragredient, $\eta\xi = \phi\psi$, so that the square of the length of a vector remains invariant. Moreover, since the result of operating on a vector ξ_a with the matrix A is, in general, another vector ξ_b, where ξ_a and ξ_b are both referred to the original co-ordinate system, it follows that in the new frame of reference which is defined by the linear transformations given above, the relationship

$$A\xi_a = \xi_b$$

becomes

$$HAH^{-1}\psi_a = \psi_b,$$

or

$$B\psi_a = \psi_b.$$

Thus, in the new frame of reference the matrix $B = HAH^{-1}$ operating on the vector ψ_a is equivalent to the matrix A operating on the vector ξ_a in the original frame.

It is convenient, for the purposes of studying the matrix A and of performing any numerical computations with it, to transform the variables η and ξ in the above way, choosing the matrix H so as to make $B = HAH^{-1}$ as simple as possible. For $B^t = (HAH^{-1})^t = HA^tH^{-1}$, and since A is non-singular, by the reversal law, $(HAH^{-1})^{-1} = HA^{-1}H^{-1}$. Thus, if $f(A)$ is a rational integral function of A, $f(B) = f(HAH^{-1}) = Hf(A)H^{-1}$; and the properties of matrix functions $f(A)$ can be studied by means of the simpler forms $f(B)$. Moreover, the matrices A and B have the same characteristic equation and, therefore, the same latent roots. For $B - \lambda I = H(A - \lambda I)H^{-1}$ and, forming the determinants of both sides,

$$|B - \lambda I| = |H||A - \lambda I||H|^{-1},$$

231

so that the characteristic equation is

$$|A - \lambda I| = |B - \lambda I| = 0.$$

If, in the present case, the transforming matrix is taken to be

$$H = \begin{bmatrix} (P_0 P_1 P_2 \dots P_{k-1}) & \cdot & \cdot & \cdots & \cdot & \cdot \\ \cdot & (P_1 P_2 P_3 \dots P_{k-1}) & \cdot & \cdots & \cdot & \cdot \\ \cdot & \cdot & (P_2 P_3 \dots P_{k-1}) & \cdots & \cdot & \cdot \\ \hdashline \cdot & \cdot & \cdot & \cdots & (P_{k-2} P_{k-1}) & \cdot & \cdot \\ \cdot & \cdot & \cdot & \cdots & \cdot & P_{k-1} & \cdot \\ \cdot & \cdot & \cdot & \cdots & \cdot & \cdot & 1 \end{bmatrix}$$

in which, it is to be noted, the only numerical elements lie in the principal diagonal and are derived entirely from the life table, then

$$B = HAH^{-1} = \begin{bmatrix} F_0 & P_0 F_1 & P_0 P_1 F_2 & P_0 P_1 P_2 F_3 & \cdots & (P_0 P_1 P_2 \dots P_{k-1}) F_k \\ 1 & \cdot & \cdot & \cdot & \cdots & \cdot \\ \cdot & 1 & \cdot & \cdot & \cdots & \cdot \\ \cdot & \cdot & 1 & \cdot & \cdots & \cdot \\ \cdot & \cdot & \cdot & 1 & \cdots & \cdot \\ \hdashline \cdot & \cdot & \cdot & \cdot & \cdots & 1 & \cdot \end{bmatrix}.$$

Comparing this matrix B with the original form A, it can be seen that the latter has been simplified to the extent that the original P_x figures in the principal subdiagonal are now replaced by a series of units, and the matrix A has been reduced to the rational canonical form $B = HAH^{-1}$ (see Turnbull & Aitken, 1932, chap. v). In this way any computations with the matrix A are made easier, and we may work henceforward in terms of ϕ and ψ vectors together with the matrix B, instead of with the original η and ξ vectors, and the matrix A. Any results obtained in this new system of co-ordinates may be transformed back again to the original system whenever necessary. It is evident that by suitably enlarging H the original matrix M may be transformed in a similar way.

This linear transformation of the original co-ordinate system is equivalent biologically to the transformation of the original population we were considering into a new and completely imaginary type which, although intimately connected with the old, has certain quite different properties. Thus, it can be seen from the transformed matrix B that the individuals in this new population, instead of dying off according to age as the original ones did, live until the whole span of life is completed, when they all die simultaneously. This is indicated by the P_x figures being now all equal to unity; an individual alive in the age group x to $x+1$ at $t = 0$ is certain of being alive at $t = 1$, excepting in the last age group of all where none of the individuals will be alive in one unit's time. Accompanying this somewhat radical change in the life table, there is a compensatory adjustment made in the rates of fertility so that the new population has the same inherent power of natural increase (r) as that of the old. This follows from the fact that the latent roots of the matrices A and B are the same, and, as will be shown later, the dominant latent root is closely related to the value of r obtained by the usual methods of computation. Insomuch as the transformation is reversible and $A = H^{-1}BH$, it can be seen that by changing H we could transform the canonical form B, if we wished, into another matrix in which the P_x subdiagonal might be a specified set of figures derived from some other form of life table. But, for our present purposes, the canonical form B, in which all the P_x figures are units, offers advantages over any other matrix of a similar type owing to the greater ease with which it can be handled. ...

232

7. The stable age distribution

The result of operating on an age distributon ψ_x with the matrix B is, in general, a different distribution ψ_y. But, in the special case when the relation between the two distributions is such that

$$B\psi_a = \lambda\psi_a,$$

where λ is an algebraic number, then ψ_a may be said to be a stable age distribution appropriate to the matrix B. For the sake of brevity it will be referred to as a stable ψ. Similarly for initial row vectors, if

$$\phi_a B = \lambda\phi_a,$$

then ϕ_a is said to be a stable ϕ.

The matrix equation defining a stable ψ may be written as $k+1$ linear equations, of which the ith is

$$\sum_{j=1}^{k+1} b_{ij} n_j - \lambda n_i = 0,$$

where n_i ($i = 1, 2, ..., k+1$) are the co-ordinates of the stable ψ, and b_{ij} the element in the ith row and jth column of B. Eliminating the n_i from this system of equations, we obtain the characteristic equation of B, namely,

$$|B - \lambda I| = 0;$$

and, expanding this determinant in powers of λ, we have in the present case,

$$\lambda^{k+1} - F_0\lambda^k - P_0 F_1\lambda^{k-1} - P_0 P_1 F_2\lambda^{k-2} - ... - (P_0 P_1 ... P_{k-2}) F_{k-1}\lambda - (P_0 P_1 ... P_{k-1}) F_k = 0.$$

The $k+1$ roots λ_a of this equation are the latent roots of B, and corresponding to each distinct λ_a there is a pair of stable vectors, ϕ_a and ψ_a, determined except for an arbitrary scalar factor.

Once a latent root λ_a has been determined, it is a comparatively simple matter to find the appropriate stable ψ_a and ϕ_a vectors. Thus, it is easily shown that the stable ψ_a is the column vector $\{\lambda_a^k \lambda_a^{k-1} \lambda_a^{k-2} ... \lambda_a 1\}$. A short method of estimating ϕ_a is the following. Suppose, to take a simple case, that

$$B = \begin{bmatrix} a & b & c & d \\ 1 & . & . & . \\ . & 1 & . & . \\ . & . & 1 & . \end{bmatrix}$$

and let y_x ($x = 1, 2, 3, 4$) be the elements of the stable ϕ_a appropriate to the root λ_a. Then

$$\phi_a B = [ay_1 + y_2 \quad by_1 + y_3 \quad cy_1 + y_4 \quad dy_1]$$
$$= [\lambda_a y_1 \quad \lambda_a y_2 \quad \lambda_a y_3 \quad \lambda_a y_4].$$

By equating similar elements and putting $y_1 = 1$, $y_4 = d/\lambda_a$, $y_3 = \dfrac{c + y_4}{\lambda_a}$, etc., it is easy to see how the required row vector can be built up. Having in this way obtained the stable ψ and ϕ vectors for the matrix B, they may be transformed to the appropriate stable ξ and η for the matrix A by means of the relations

$$\eta = \phi H, \quad \xi = H^{-1}\psi.$$

The characteristic equation of the matrix B, when expanded, is of degree $k+1$ in λ, and once B has been obtained this equation can immediately be written down, since the numerical coefficients of λ^k, λ^{k-1}, λ^{k-2}, etc., are merely the elements of the first row taken with a negative sign. Since there is only one change of sign in this equation, only one of the latent roots will be real and positive. Excluding the rather special case when the first row of B has only a single non-zero element, and taking the more usual type of matrix which will be met with, namely, that for a species breeding continuously over a large proportion of its total life span, it will be found that the modulus of this root (λ_1) is greater than any of the others,

$$|\lambda_1| > |\lambda_2| > |\lambda_3| > ... > |\lambda_{k+1}|,$$

233

the remaining roots being either negative or complex.

This dominant latent root λ_1, which will be $\gtreqless 1$ according as to whether the sum of the elements in the first row of B is $\gtreqless 1$, is the one which is principally of interest. Since it is real and positive, it is the only root which will give rise to a stable ψ or ξ vector consisting of real and positive elements. It is this stable ξ_1 associated with the dominant root λ_1 which is ordinarily referred to as the stable age distribution appropriate to the given age specific rates of fertility and mortality. Since

$$A'\xi_1 = \lambda_1^t \xi_1,$$

it can be seen that the latent root λ_1 of the matrix A and the value of r obtained in the usual way from

$$\int_0^\infty e^{-rx} l_x m_x \, dx = 1,$$

are related by

$$\log_e \lambda_1 = r. \ \dots$$

8. Properties of the stable vectors

Before proceeding further it is necessary to mention briefly the reasons why the methods given above for the computation of the stable ψ and ϕ vectors were adopted, apart from their simplicity in practice. If the $k+1$ distinct roots of the characteristic equation are known, we may form a set of $k+1$ matrices $f(\lambda_a)$ by inserting in turn the numerical value of each root in the matrix $[B - \lambda_a I]$. The adjoint of $f(\lambda_a)$ is

$$F(\lambda_a) = \prod_{b \neq a} [B - \lambda_b I] \quad \text{and} \quad f(\lambda_a) \, F(\lambda_a) = 0.$$

It may be shown that the stable ψ_a appropriate to the root λ_a can be taken proportional to any column, and the stable ϕ_a proportional to any row of the matrix $F(\lambda_a)$ (see e.g. Frazer, Duncan & Collar, 1938, chap. III). Moreover, $F(\lambda_a)$ is a matrix product of the type $\psi\phi$, where the ψ vector is given by the first column and the ϕ vector by the last row of $F(\lambda_a)$, each divided by the square root of the element in the bottom left-hand corner; and the trace of the matrix is equal to the scalar product $\phi\psi$. Now $[B - \lambda_a I]$ is a square matrix of order $k+1$ with only zero elements below and to the left of the principal subdiagonal, which itself consists of units. The product of k such matrices, which gives $F(\lambda_a)$, will have therefore a unit in the bottom left-hand corner. Since the stable ϕ_a and ψ_a vectors obtained by the methods suggested in § 7 have respectively their first and last elements $= 1$, it follows that

$$\psi_a \phi_a = F(\lambda_a), \quad \phi_a \psi_a = \text{trace } F(\lambda_a).$$

The stable vectors may now be normalized. If the scalar product, $\phi_a \psi_a = z^2$, say, then

$$\frac{\phi_a}{|z|} \frac{\psi_a}{|z|} = 1.$$

From now on it will be assumed that the stable vectors appropriate to each of the latent roots have been normalized in this way.

These vectors have the following important properties:

(1) The $k+1$ stable ψ are linearly independent. There is thus no such relationship, with non-zero coefficients c, as

$$c_1 \psi_1 + c_2 \psi_2 + c_3 \psi_3 + \dots + c_{k+1} \psi_{k+1} = 0.$$

(2) The scalar product of a stable ψ, ψ_a with the associated vector of another stable ψ, ψ_b is zero, i.e.

$$\phi_b \psi_a = 0 \quad (a \neq b).$$

The normalized stable ψ thus form a set of $k+1$ independent and mutually orthogonal vectors of unit length.

(3) Any arbitrary ψ—ψ_x say—can be expanded in terms of the stable ψ, thus

$$\psi_x = c_1 \psi_1 + c_2 \psi_2 + c_3 \psi_3 + \dots + c_{k+1} \psi_{k+1},$$

where the coefficients c may be either real or complex. Similarly an arbitrary vector ϕ_x can be expanded in terms of the stable ϕ. ...

13. THE APPROACH TO THE STABLE AGE DISTRIBUTION

A stable age distribution appropriate to the matrix B has been defined mathematically by the equation
$$B\psi = \lambda\psi,$$
and it has already been shown that since only one latent root of B is real and positive, only one of the stable ψ will consist of real and positive elements. But, in addition to this Malthusian age distribution, it is also of some interest to inquire whether any significance can be attached to the remaining stable ψ associated with the negative and complex roots of the characteristic equation.

Any age distribution ψ_x, the elements of which are necessarily ≥ 0, may be expressed as a vector of deviates from the stable ψ_1 associated with the dominant latent root, and we may therefore write the expansion of ψ_x in terms of the stable ψ as
$$(\psi_x - c_1\psi_1) = c_2\psi_2 + c_3\psi_3 + \ldots + c_{k+1}\psi_{k+1} = \psi_d,$$
where the coefficients c are given by the vector $c = U\psi_x$. Thus, the way in which a particular type of age distribution will approach the stable form may be studied by means of the vector ψ_d.

Among the terms occurring in the right-hand side of this expression there will be, corresponding to each negative root, a single term $c_a\psi_a$ which will consist of real elements alternately positive and negative in sign. (Even if the normalized ψ_a is imaginary this term will consist of real numbers, since in this case c_a will also become imaginary.) Moreover, corresponding to every pair of complex roots there will be a pair of terms $(c_m\psi_m + c_n\psi_n)$ which taken together will also give a single vector with real elements. This follows from the fact that c_m is the conjugate complex of c_n owing to the way in which the matrix U is constructed. Then, apart from the scalar c_1 which must necessarily be > 0, some of the remaining coefficients $c_2, c_3, \ldots, c_{k+1}$ in the expansion of ψ_d may be zero. The first and most obvious case is when they are all zero, and the age distribution ψ_x is therefore already of the stable form. But, if either
$$\psi_d = c_a\psi_a,$$
where ψ_a corresponds to a negative latent root, or
$$\psi_d = c_m\psi_m + c_n\psi_n,$$
where ψ_m and ψ_n are associated with a conjugate pair of complex roots, then it follows that the age distribution ψ_x will, as time goes on, approach the stable form in a particular way defined by either
$$B^t\psi_d = c_a\lambda^t\psi_a \quad \text{or} \quad B^t\psi_d = c_m\lambda^t\psi_m + c_n\overline{\lambda}^t\psi_n,$$
in which λ^t for a pair of complex roots $u + iv$ with modulus r may be written in the form of $r^t(\cos\theta t \pm i\sin\theta t)$. Thus, the negative and complex latent roots of B serve to determine a number of age distributions which are of some interest owing to the fact that they will approach the Malthusian form in what may be termed a stable fashion.

Since $|\lambda_1| > |\lambda_2| > |\lambda_3| > \ldots > |\lambda_{k+1}|$, the vector of deviates ψ_d will tend towards zero as $t \to \infty$ whenever $\lambda_1 \lessgtr 1$. Thus, in the case of a stationary population, any ψ_x will converge to the stable form of age distribution. But if $\lambda_1 > 1$, there is a possibility of one or more of the remaining roots having a modulus ≥ 1, e.g. $|\lambda_2| \geq 1$. In the latter case there may be certain age distributions with $c_2 \neq 0$ for which the amplitude of the deviations from the stable form tend either to increase ($|\lambda_2| > 1$), or to remain constant ($|\lambda_2| = 1$). From the practical point of view, however, we may still say that a population with such an age distribution approaches or becomes approximately equal to the stable population, since λ_1^t is much greater than λ_2^t when t is large.

14. Special case of the matrix with only a single non-zero F_x element

The interesting case of the matrix A having only a single non-zero element in the first row has been illustrated in a numerical example by Bernardelli (1941).* This author has also used a matrix notation in the mathematical appendix to his paper, and the form of his basic matrix is the same as that referred to here as M or A. It is not clear, however, from the definitions which he gives whether he regards the elements in the first row of his matrix as being constituted by the maternal frequency figures (m_x) themselves, or by a series of values similar to those defined here as the F_x figures. He refers to them merely as the specific fertility rates for female births.

In discussing the causes of population waves, Bernardelli describes a hypothetical species, such as a beetle, which lives for only three years and which propagates in the third year of life. He assumes, for the sake of argument, that—to employ the terminology used here—$P_0 = \frac{1}{2}$ and $P_1 = \frac{1}{3}$, and that 'each female in the age 2–3 produces, on the average, 6 new living females'. Assuming for the moment that he is here defining a F_x figure, we may write this system of mortality and fertility rates as

$$ A = \begin{bmatrix} 0 & 0 & 6 \\ \frac{1}{2} & 0 & 0 \\ 0 & \frac{1}{3} & 0 \end{bmatrix}, \quad B = HAH^{-1} = \begin{bmatrix} 0 & 0 & 1 \\ 1 & 0 & 0 \\ 0 & 1 & 0 \end{bmatrix}. $$

The characteristic equation expanded in terms of λ is $\lambda^3 - 1 = 0$; and the latent roots are therefore $1, -\frac{1}{2} \pm \frac{\sqrt{3}}{2}i$, all three being of equal modulus. The matrix A has the interesting properties
$$ A^2 = A^{-1}, \quad A^3 = I, $$
so that any initial age distribution repeats itself regularly every three years. Thus, as Bernardelli shows, a population of 3000 females distributed equally among the three age groups becomes a total population of 6833 at $t = 1$; of 5166 at $t = 2$; and again 3000 distributed equally among the age groups at $t = 3$. Unless a population has already an initial age distribution in the ratio of $\{6 : 3 : 1\}$, no approach will be made to the stable form associated with the real latent root, and the vector of deviates ξ_d will continue to oscillate with a stable amplitude, which will in part depend on the form of the initial distribution. Although this numerical example refers specifically to a stationary population, it is evident that a similar type of argument may be developed in the case when $|\lambda| > 1$ and $A^3 = \lambda^3 I$.

We have assumed here that his definition of the fertility rate refers to a F_x figure. But, if we were to interpret the words quoted above as referring to a maternal frequency figure, namely that every female alive between the ages 2–3 produces on the average 6 daughters per annum, then the results become entirely different. For, deriving the appropriate F_x figures by the method described in § 2, the matrix is now

$$ A = \begin{bmatrix} 0 & 1 & 3 \\ \frac{1}{2} & 0 & 0 \\ 0 & \frac{1}{3} & 0 \end{bmatrix}, \quad B = HAH^{-1} = \begin{bmatrix} 0 & \frac{1}{2} & \frac{1}{2} \\ 1 & 0 & 0 \\ 0 & 1 & 0 \end{bmatrix}, $$

and the latent roots are $1, -\frac{1}{4} \pm \frac{1}{4}i$. The modulus of the pair of complex roots is $1/\sqrt{2}$, which is < 1, so that every age distribution will now converge to the stable form associated with the

* At this point I should like to acknowledge the gift of a reprint of this paper, which was received by the Bureau of Animal Population at a time when I was in the middle of this work, and when I was just beginning to appreciate the interesting results which could be obtained from the use of matrices and vectors: also a personal communication from Dr Bernardelli, received early in 1942, at a time when it was difficult to reply owing to the developments of the war situation in Burma. Although the problems we were immediately interested in differed somewhat, this paper did much to stimulate the theoretical development given here, and it is with great pleasure that I acknowledge the debt which I owe to him.

real root. Thus, to take the same example as before, 3000 females distributed equally among the age groups will tend towards a total population of 4000 distributed in the ratio of $\{6:3:1\}$, and it was found that this age distribution would be achieved at approximately $t = 23$. During the approach to this stable form periodic waves are apparent both in the age distribution and in the total number of individuals, but these oscillations are now damped, in contrast with the results obtained with the first type of matrix.

This simple illustration serves to emphasize the importance which must be attached to the way in which the basic data are defined and to the marked difference which exists between what are termed here the m_x and F_x figures. Nevertheless, apart from the question of the precise way in which the definition of the fertility rates is to be interpreted in this example, the first type of matrix with only a single element in the first row does correspond to the reproductive biology of certain species. Thus, in the case of many insect types the individuals pass the major portion of their life span in various immature phases and end their lives in a short and highly concentrated spell of breeding. The properties of this matrix suggest that any stability of age structure will be exceptional in a population of this type, and that even if the matrix remains constant we should expect quite violent oscillations to occur in the total number of individuals.

15. Numerical comparison with the usual methods of computation

From the practical point of view it will not always be necessary to estimate the actual values of all the stable vectors and of the associated matrices which are based on them. Naturally, much will depend on the type of information which is required in any particular case. In order to compute, for instance, the matrices U, Q and G, it is necessary first of all to determine all the latent roots of the basic matrix. The ease with which these may be found depends very greatly upon the order of the matrix. Thus, in the numerical example for the brown rat used previously in §3, the unit of age and time is one month and the resulting square matrix A is of order 21. To determine all the 21 roots of the characteristic equation would be a formidable undertaking. It might be sufficient in this case to estimate the positive real root and the stable vector associated with it. On the other hand, it is possible to reduce the size of the matrix by taking a larger unit of age, and in some types of problem, where extreme accuracy is not essential, a unit say three times as great might be equally satisfactory, which would reduce the matrix for the rat population to the order of 7×7. It is not too difficult to find all the roots of a seventh degree equation by means of the root-squaring method (Whittaker & Robinson, 1932, p. 106). But the reduction of the matrix in this way will generally lead to a value of the positive real root which is not the same as that obtained from the larger matrix, and it is therefore necessary to see by how much these values may differ owing to the adoption of a larger unit of time.

Another important point which must be considered is the following. By expressing the age specific fertility and mortality rates in the form of a matrix and regarding an age distribution as a vector, an element of discontinuity is introduced into what is ordinarily taken to be a continuous system. Instead of the differential and integral calculus, matrix algebra is used, a step which leads to a great economy in the use of symbols and consequently to equations which are more easily handled. Moreover, many quite complicated arithmetical problems can be solved with great ease by manipulating the matrix which represents the given system of age specific rates. But the question then arises whether these advantages may not be offset by a greater degree of inaccuracy in the results as compared with those obtained from the previous methods of computation. It is not easy, however, to settle this point satisfactorily. In the way the usual equations of population mathematics are solved, a similar element of discontinuity is introduced by the use of age grouping. Thus, in the case of a human population, if we were estimating the inherent rate of increase in the ordinary way, we should not expect to obtain the same value of r from the data grouped in five year intervals of age as that from the data grouped in one year intervals. The estimates of the seminvariants would not be precisely the same in both cases. Nevertheless, the estimate from the data grouped in five year intervals is usually considered to be sufficiently accurate for all ordinary purposes, and there is little doubt that if we merely require the inherent rate of increase and the stable age distribution, these methods of computation are perfectly satisfactory when applied to human data.

REFERENCES

BERNARDELLI, H. (1941). Population waves. *J. Burma Res. Soc.* **31**, no. 1, 1–18.

CHARLES, ENID (1938). The effect of present trends in fertility and mortality upon the future population of Great Britain and upon its age composition. Chap. II in *Political Arithmetic*, ed. Lancelot Hogben. London: Allen and Unwin.

FRAZER, R. A., DUNCAN, W. J. & COLLAR, A. R. (1938). *Elementary Matrices.* Camb. Univ. Press.

GLASS, D. V. (1940). *Population Policies and Movements in Europe.* Oxford: Clarendon Press.

TURNBULL, H. W. & AITKEN, A. C. (1932). *An Introduction to the Theory of Canonical Matrices.* London and Glasgow: Blackie and Son.

WHITTAKER, E. T. & ROBINSON, G. (1932). *The Calculus of Observations,* 2nd ed. London and Glasgow: Blackie and Son.

27. Matrix Representation of Changes in the Active Population

Léon Tabah (1968)

From Population *23*, pp. 437—454, 468—470. Translated by David Smith.

We have omitted the second half of Tabah's paper, which projects both rural-urban migration and regional or occupational changes. Part of the Annex is included to indicate the output format.

Whether employed by economists, sociologists or demographers a projection obeys the same principles and uses the same language: it is a hypothetical and deductive operation intended to permit the emergence of new structures, by applying hypotheses of mobility to the most recent known structure. A distribution of the active population by occupation, for example, can be projected if one lays out a table of assumptions of entry, exit and interprofessional mobility that apply to the different elements of the initial distribution. From this new structure one then passes to others by chaining the probabilities of entry, exit and migration supposed for the successive periods. *A single structure and combinations of probabilities of change suffice for the construction of a projection.*

In this class of operations the choice of assumptions and their translation to a quantitative form assume major importance, for it is these which dictate the results, the calculations being purely arithmetical. This is not to say that the operational form is of little interest. On the contrary, the form appears increasingly deserving of attention and care, for it must be convenient enough to lend itself to the integration of an ever larger number of variables, while offering a clear picture of relationships between the assumptions, and facilitating calculation by modern means such as the computer.

Recourse to linear transformations, and more particularly to matrix notation, says nothing about the principle underlying the projections itself. It does not, for example, guarantee the greater validity of the assumptions. However, it is particularly suited to a dynamic description of structures in demography, as earlier in economics and sociology.

The procedure is the following: one lays out a table of the situation at a moment of time, often as revealed in censuses, which is then placed in a column vector at time zero. To go from this cross-section to a longitudinal statement, that is to animate this initial structure, the vector is premultiplied by a square matrix. The matrix contains the same number of rows and columns as the vector and in its interior, arranged in an appropriate fashion, are the probabilities of entry, exit and mobility; these being generally obtained from official registers and surveys. The matrix fixes the movement that one proposes for the future. It serves in this way as a means of communication between the structures, or as an "interchange" in the manner of cloverleafs and other figures that permit the distribution of automobiles in different directions on an expressway.

In every projection there exist factors independent of human will (passage from one age group to the next, for example), and others that depend on it (passage in and out of the labor force, migration, etc.). The first factors can be described as "servitude". In the example of aging, one is interested in a process that is continuous and irreversible. This is the type of projection first elaborated in demography especially by the efforts of Leslie[1] and extended in an interesting fashion by N. Keyfitz.[2]

---

[1] Leslie, P. H.: "On the use of matrices in certain population mathematics." Biometrika **33**, p. 184.
[2] Numerous articles on the subject have been published by this author, see in particular: Keyfitz, N: "Utilisation des machines électroniques pour les calculs démographiques." Population **19**: 673—682 (1964); "The population projection as a matrix operator." Demography **1**: 56—73 (1964).

The projection matrix is here very simple. From the second row onward it contains non-zero elements only in the sub-diagonal: these are the probabilities of passage from one age-group to the next. The first row contains combinations of probabilities of fertility at different ages and of survival for infants born in the projection interval.

The second type of projection has been used above all by sociologists in calculations of social mobility, and by those concerned with anticipating political opinion. These groups also were first to apply Markov processes, the theoretical basis of all projections, to the social sciences.

We propose to describe here two projection models that belong to both types at the same time: they involve economically active populations classified by sex and age, and in which one estimates either migration between rural and urban zones, or such migration combined with interregional displacements.

1. Projection of the Active Population with Internal Migration by Sex and Age

Let us project a population distributed by sex and age, supposing it to be subject not only to a process of entry and exit between participation and non-participation, but also and simultaneously to movements of internal migration. For convenience we will suppose that the internal displacements take place between two sectors only: rural and urban for example.

Let us designate by $_0n_{x;m;a;u}$ the number of males of age x, active and urban at the initial moment of the projection. To symbolize the female population it is sufficient to substitute the index f for the index m. The inactive population is represented by substituting i for a, and the rural population by substituting r for u. Time is registered by means of the index placed before the letter n.

Let us likewise designate by $_0N_{m;a;u}$ the sub-vector for active urban males at the initial timepoint. The other sub-vectors are represented easily by resorting to the indices f, i, and r as appropriate.

If only three age groups are retained, the layout of the projection is of the type shown in Table 1. States that are possible at time 0 are given by the column headings of the matrix and final states are given in the margins of the rows. Several ways of presenting the matrix product fairly conveniently are possible, one of which we have chosen here.

[For quinquennial age groups] the vector which characterises the initial structure comprises 136 elements (2 for sex, 2 for urban or rural residence, 2 for active and inactive populations, and 17 for age). It is subdivided into 8 sub-vectors, that bring together all the characteristics except age.

Each of the 64 sub-matrices (8 possible states initially, not counting age, which are multiplied by the same 8 states at the end of the period) can be represented by the symbol M_i^j, in which i symbolizes the final state and j the initial state, with age not taken into account.

241

Table 1. Projection of the Active Population by Sex and Age, with Migration between Rural and Urban Zones. (To make the Table clearer only non-zero elements are shown.)

Initial

Final State	#	Active urban males			Inactive urban males			Active urban females			Inactive urban females		
		1	2	3	4	5	6	7	8	9	10	11	12
Active urban males	1		M_1^1			M_1^2			M_1^3			M_1^4	
	2	m_2^1			m_2^4								
	3	m_3^2			m_3^5								
Inactive urban males	4							m_4^7	m_4^8	m_4^9	m_4^{10}	m_4^{11}	m_4^{12}
	5	m_5^1	M_2^1		m_5^4	M_2^2			M_2^3			M_2^4	
	6	m_6^2			m_6^5								
Active urban females	7		M_3^1			M_3^2			M_3^3			M_3^4	
	8							m_8^7			m_8^{10}		
	9								m_9^8			m_9^{11}	
Inactive urban females	10							m_{10}^7	m_{10}^8	m_{10}^9	m_{10}^{10}	m_{10}^{11}	m_{10}^{12}
	11		M_4^1			M_4^2		m_{11}^7	M_4^3		m_{11}^{10}	M_4^4	
	12								m_{12}^8			m_{12}^{11}	
Active rural males	13		M_5^1			M_5^2			M_5^3			M_5^4	
	14	m_{14}^1			m_{14}^4								
	15	m_{15}^2			m_{15}^5								
Inactive rural males	16							m_{16}^7	m_{16}^8	m_{16}^9	m_{16}^{10}	m_{16}^{11}	m_{16}^{12}
	17	m_{17}^1	M_6^1		m_{17}^4	M_6^2			M_6^3			M_6^4	
	18	m_{18}^2			m_{18}^5								
Active rural females	19		M_7^1			M_7^2			M_7^3			M_7^4	
	20							m_{20}^7			m_{20}^{10}		
	21								m_{21}^8			m_{21}^{11}	
Inactive rural females	22							m_{22}^7	m_{22}^8	m_{22}^9	m_{22}^{10}	m_{22}^{11}	m_{22}^{12}
	23		M_8^1			M_8^2		m_{23}^7	M_8^3		m_{23}^{10}	M_8^4	
	24								m_{24}^8			m_{24}^{11}	

State

Active rural males	Inactive rural males	Active rural females	Inactive rural females	Vector describing the initial distribution	Sub-vectors
13 14 15	16 17 18	19 20 21	22 23 24		
$m_2^{13}\ M_1^5$ m_3^{14}	$m_2^{16}\ M_1^6$ m_3^{17}	M_1^7	M_1^8	$_0n_{1;m;a;u}$ $_0n_{2;m;a;u}$ $_0n_{3;m;a;u}$	$_0N_{m;a;u}$
$m_5^{13}\ M_2^5$ m_6^{14}	$m_5^{16}\ M_2^6$ m_6^{17}	$m_4^{19}\ m_4^{20}\ m_4^{21}$ M_2^7	$m_4^{22}\ m_4^{23}\ m_4^{24}$ M_2^8	$_0n_{1;m;i;u}$ $_0n_{2;m;i;u}$ $_0n_{3;m;i;u}$	$_0N_{m;i;u}$
M_3^5	M_3^6	$m_8^{19}\ M_3^7$ m_2^{20}	$m_8^{22}\ M_3^8$ m_9^{23}	$_0n_{1;f;a;u}$ $_0n_{2;f;a;u}$ $_0n_{3;f;a;u}$	$_0N_{f;a;u}$
M_4^5	M_4^6	$m_{10}^{19}\ m_{10}^{20}\ m_{10}^{21}$ $m_{11}^{19}\ M_4^7$ m_{12}^{20}	$m_{10}^{22}\ m_{10}^{23}\ m_{10}^{24}$ $m_{11}^{22}\ M_4^8$ m_{12}^{23}	$_0n_{1;f;i;u}$ $_0n_{2;f;i;u}$ $_0n_{3;f;i;u}$	$_0N_{f;i;u}$
$m_{14}^{13}\ M_5^5$ m_{15}^{14}	$m_{14}^{16}\ M_5^6$ m_{15}^{17}	M_5^7	M_5^8	$_0n_{1;m;a;r}$ $_0n_{2;m;a;r}$ $_0n_{3;m;a;r}$	$_0N_{m;a;r}$
$m_{17}^{13}\ M_6^5$ m_{18}^{14}	$m_{17}^{16}\ M_6^6$ m_{18}^{17}	$m_{16}^{19}\ m_{16}^{20}\ m_{16}^{21}$ M_6^7	$m_{16}^{22}\ m_{16}^{23}\ m_{16}^{24}$ M_6^8	$_0n_{1;m;i;r}$ $_0n_{2;m;i;r}$ $_0n_{3;m;i;r}$	$_0N_{m;i;r}$
M_7^5	M_7^6	$m_{20}^{19}\ M_7^7$ m_{21}^{20}	$m_{20}^{22}\ M_7^8$ m_{21}^{23}	$_0n_{1;f;a;r}$ $_0n_{2;f;a;r}$ $_0n_{3;f;a;r}$	$_0N_{f;a;r}$
M_8^5	M_8^6	$m_{22}^{19}\ m_{22}^{20}\ m_{22}^{21}$ $m_{23}^{19}\ M_8^7$ m_{24}^{20}	$m_{22}^{22}\ m_{22}^{23}\ m_{22}^{24}$ $m_{23}^{22}\ M_8^8$ m_{24}^{23}	$_0n_{1;f;i;r}$ $_0n_{2;f;i;r}$ $_0n_{3;f;i;r}$	$_0N_{f;i;r}$

The components of the matrix are easily explained:

—rows 1, 7, 13, and 19 represent the net probabilities of entry into the first age group and [simultaneously] the active population, which may be masculine or feminine, urban or rural. If one supposes that entry into the active population takes place after the initial age or group of ages, all the elements of these rows will be zero.

—by assumption, new population is provided exclusively by females, who may be active or inactive, urban or rural. Only elements in the first row of each of these sub-matrices can be non-zero. They are to be multiplied by elements in the sub-vectors for females. This is the case for the sub-matrices $M_2^3, M_2^4, M_2^7, M_2^8, M_4^3$, etc. In other words, it is supposed that new population is generated by females, and that they give birth to individuals not in the labor force;

—elements of the sub-matrices situated on the horizontals of a given sex but which are multiplied by elements of the vector corresponding to the opposite sex are necessarily zero, since passage from one sex to the other is clearly impossible. This is the case for the sub-matrices $M_1^3, M_1^4, M_1^7, M_1^8, M_2^3$, and M_2^4 beginning with their second row, etc.

Makeup of the Elements of the Projection Matrix

Assumptions concerning migration: We will make the assumption that during each period of projection emigrants are subject to the laws of fertility and mortality of the zone from which they come. It is thus only at later periods that they adopt the laws of the populations into which they are received. It is very probable that this assumption does not correspond exactly to reality, but we are constrained to accept it because available data generally don't permit us to formulate more satisfactory assumptions, as by following for example cohorts of migrants whose behavior is made to change through time. The assumption we use has the merit of not further complicating calculations that are already fairly cumbersome.

We will likewise suppose that several moves in one direction or another can be made by an individual in each projection period. But the rates of migration reflect the net balance of these moves for a given age group. Stated differently, if at the end of the projection period a person is found in the place he was originally, it is as if no migration had taken place; if he is found in another place, it is as if he had passed directly from his place of origin to his place of residence at the end of the projection period.

Moreover, the observed data do not in general permit knowing the migration flows in each direction. For example if one looks at urban and rural sectors only the positive balances by age in favor of the urban sector are available: the excess of migrants by age in the rural to urban direction relative to those in the opposite direction. The calculation of this positive balance is most often carried out by following cohorts between successive censuses. One therefore has available only information on the net migration flows.

Finally, it is useful to keep clearly in mind the fact that migration confounds two types of movement: persons who actually move from rural to urban areas or vice versa, and those who migrate "in place" in the sense that their area of residence crosses the definitional boundary separating urban and rural areas.

We will suppose that children born during a projection period do not migrate independently of their mothers in the course of this period, so that we do not distinguish migrations of mothers and migrations of these children.

Assumptions Concerning Entry Into and Exit From the Active Population

Normally, to calculate the probabilities of entry and exit with respect to activity it is necessary to ask the following question in the course of a census or survey carried out at time t: "Were you in the labor force or not in it during the period i to t?" One then finds the ratio for a specified cohort between the number who left the active population during the period, in the absence of mortality, and the number in the labor force at the start of the period. One obtains in this way the probability of exit from activity for a given period and age interval; an inverse calculation provides the rate of entry into activity.

Unfortunately, existing data do not permit carrying out such calculations and one must be content with information derived from censuses, which are in general restricted to the proportions active and inactive at different ages.

Let a_x be the rate of activity at age x. It is easily shown, making certain approximations, that the probabilities of "migration" between activity and inactivity are given as a function of a_x by:
—probability of entry into activity:

$$\theta_{i,a} = \frac{a_{x+1} - a_x}{1 - a_x}$$

—probability of exit from activity:

$$\theta_{a,i} = \frac{a_x - a_{x+1}}{a_x}$$

—probability of remaining inactive:

$$\theta_{i,i} = \frac{1 - a_{x+1}}{1 - a_x}$$

—probability of remaining active:

$$\theta_{a,a} = \frac{a_{x+1}}{a_x}$$

These four probabilities are the prospective quotients for entry or exit from activity, similar to the prospective quotients for survival $p_{x,x+4}$ or the prospective

245

quotients for migration $\mu_{x,x+4}$ that we will define later. When the calculation is carried out in intervals of five years instead of one year, the formulas are the same apart from the interval width. We will designate by $a_{x,x+4}$ the proportion active between ages x and $x+4$.

The interpretation of the elements of the first line of the sub-matrices is as follows (see Table 1):

— m_2^1 and m_3^2 are the probabilities of survival for active urban males, combined with the probabilities of not exiting from activity and not emigrating to the rural sector;

— m_2^4 and m_3^5 are the probabilities of survival for inactive urban males, combined with the probabilities of passage from inactivity to activity and of not emigrating to the rural sector;

— m_2^{13} and m_3^{17} are the probabilities of survival for active rural males, combined with the probabilities of not exiting from activity and of migrating to the urban sector;

— m_2^{16} and m_3^{17} are the probabilities of survival for inactive rural males, combined with the probabilities of entering into activity and of emigrating to the urban sector.

The number of active urban males at the end of the period is decomposable into a sum of four products, as can be seen by multiplying the first row of the sub-matrix by the vector:

$$_1N_{m;a;u} = M_{10}^1 N_{m;a;u} + M_{10}^2 N_{m;i;u} + M_{10}^5 N_{m;a;r} + M_{10}^6 N_{m;i;r}$$

active urban	inactive urban	active rural	inactive rural
remaining	becoming	becoming	becoming
active urban	active urban	active urban	active urban

An example of the calculation corresponding to this equation is seen in Table 2.

The number of inactive urban males at the end of the period is obtained by means of a sum of eight products, as is indicated in the multiplication of the second row of the sub-matrices by the vector:

$$_1N_{m;i;u} = M_{20}^1 N_{m;a;u} + M_{20}^2 N_{m;i;u} + M_{20}^3 N_{f;a;u} + M_{20}^4 N_{f;i;u}$$

active urban	inactive urban	surviving male	surviving male
becoming	remaining	infants of	infants of
inactive urban	inactive urban	non-migrant	non-migrant
		active urban	inactive urban
		females	females

$$+ \quad M_{20}^5 N_{m;a;r} + M_{20}^6 N_{m;i;r} + M_{20}^7 N_{f;a;r} + M_{20}^8 N_{f;i;r}$$

active rural	inactive rural	surviving male	surviving male
becoming	becoming	infants of	infants of
active urban	active urban	emigrating	emigrating
		active rural	inactive rural
		females	females

Multiplying the other rows of the sub-matrices by the vector, one easily finds the components of the other elements at the end of the period.

This manner of projecting the active population has the great advantage of permitting the calculation not only of the population according to its different characterisitics at the end of each period, but also the movement in the intervals, as can be seen in the application to Mexico. It is of interest to estimate not only the number of active urban males from 1965 to 1970, for example, but also how this group decomposes in 1965 into the proportion who have remained urban and active in the 1960—1965 period and the proportion of new arrivals into the labor stock, whether urban or rural, male or female, and at whatever ages. Similar questions present themselves for each of the other groups of the population.

Calculation of the Elements of the Matrix

Let us return to the population of Mexico in 1960, separated into five year age groups by sex, urban or rural residence and activity or inactivity, and see how to establish the formulas for advancing the structure in five year steps.

We begin by indicating the meanings of the symbols that will intervene:

—$_{60}n_{x,x+4;f;a;u}$ is the number of females ages x to $x+4$ in 1960 who are active and urban;

—$_{60}n_{x,x+4;f;a;r}$ the number of active rural females in 1960 at ages x to $x+4$;

—$_{60}n_{x,x+4;f;i;u}$ the number of inactive urban females in 1960 at ages x to $x+4$;

—$_{60}n_{x,x+4;f;i;r}$ the number of inactive rural females in 1960 at ages x to $x+4$;

—$_{60}\phi_{x,x+4;u}$ and $_{65}\phi_{x,x+4;u}$ the rates of fertility at ages x to $x+4$ in urban areas in 1960 and 1965. For lack of data we have not distinguished fertility differentials according to activity or inactivity; we would not however run into any difficulty if this distinction were introduced;

—$_{60}\phi_{x,x+4;r}$ and $_{65}\phi_{x,x+4;r}$ the same rates in rural areas;

—$P_{x,x+4;f;u}$ and $P_{x,x+4;f;r}$ the expected proportions surviving in urban and rural areas between 1960 and 1964, among females at ages x to $x+4$ on January 1, 1960, again without distinguishing mortality differentials according to activity or inactivity, though it is possible to do so;

—$P_{n;f;u}$ and $P_{n;f;r}$ the expected proportions surviving among daughters born between 1960 and 1964 in urban and rural areas, without distinguishing mortality differentials according to whether the births occur to active or inactive females;

—$\mu_{x,x+4;f;u,r}$ the probability of migration from urban to rural zones in the 1960 to 1964 period, in the absence of mortality, for females aged x to $x+4$ on January 1, 1960, not taking into account migration differentials according to activity or inactivity;

—$\mu_{x,x+4;f;r,u}$ the same probabilities for rural to urban migration;

—$\theta_{x,x+4;f;i,a;u}$ the probabilities for inactive urban females ages x to $x+4$ to enter into activity before ages $x+5$ to $x+9$;

—$\theta_{x,x+4;f;i,i;u}$ the probabilities for the same women of remaining inactive;

—$\theta_{x,x+4;f;i,a;r}$ and $\theta_{x,x+4;f;i,i;r}$ the same probabilities as above, but for rural women;

—$\theta_{x,x+4;f;a,i;u}$ the probabilities for active urban females at ages x to $x+4$ of becoming inactive before ages $x+5$ to $x+9$;

—$\theta_{x,x+4;f;a,a;u}$ the probabilities for these same women of remaining in the active population;

—$\theta_{x,x+4;f;a,i;r}$ and $\theta_{x,x+4;f;a,a;r}$ the same probabilities as above, but for rural women;

—k the proportion of births that are male;

—finally, all of these expressions remain the same when they relate to males except that the index m replaces f.

We can now calculate each of the elements of the sub-matrices:

Sub-matrix M_1^1: *active urban males remaining active and urban.*

The elements of the sub-diagonal m_1^2 and m_3^2 represent combinations of probabilities of survival in urban areas, probabilities that those in the active urban population remain active, and probabilities that they do not emigrate, that is:

$$P_{x,x+4;m;u} \cdot \theta_{x,x+4;m;a,a;u} \cdot \mu_{x,x+4;m;u,u}$$

Sub-matrix M_1^2: *inactive urban males entering the active urban population.*

The elements of the sub-diagonal m_2^4 and m_3^5 now represent combinations of probabilities of survival in urban areas, probabilities that inactive urbanites enter into activity, and probabilities that they do not emigrate from the urban sector, that is:

$$P_{x,x+4;m;u} \cdot \theta_{x,x+4;m;i,a;u} \cdot \mu_{x,x+4;m;u,u}$$

The elements of the two other sub-matrices of the first row are easily obtained:
Sub-matrix M_1^5: *active rural males migrating and remaing active.*

$$P_{x,x+4;m;r} \cdot \theta_{x,x+4;m;a,a;r} \cdot \mu_{x,x+4;m;r,u}$$

Sub-matrix M_1^6: *inactive rural males emigrating and becoming active.*

$$P_{x,x+4;m;r} \cdot \theta_{x,x+4;m;i,a;r} \cdot \mu_{x,x+4;m;r,u}$$

Similarly, for the first two sub-matrices of the second row:
Sub-matrix M_2^1: *active urban males becoming inactive but remaining urban.*

$$P_{x,x+4;m;u} \cdot \theta_{x,x+4;m;a,i;u} \cdot \mu_{x,x+4;m;u,u}$$

Sub-matrix M_2^2: *inactive urban males remaining inactive and urban.*

$$P_{x,x+4;m;u} \cdot \theta_{x,x+4;m;i,i;u} \cdot \mu_{x,x+4;m;u,u}$$

Sub-matrices M_2^3 *and* M_2^4: *surviving male births to active or inactive urban females who do not emigrate.*

248

The elements of the two sub-matrices that follow, M_2^3 and M_2^4, bring in births during the interval of the projection and thus serve to maintain the population. They are calculated in the following way if one does not suppose a fertility differential according to whether females are active or inactive.

The number of births of both sexes to urban women between the ages x and $x+4$ during the year 1960 can be estimated as:

$$_{60}n_{x,x+4;f;u} \cdot {}_{60}\phi_{x,x+4;u}$$

The figure is only an approximation, as it is the number of urban women between ages x and $x+4$ on June 30, 1960 that should be multiplied by:

$$_{60}\phi_{x,x+4;u}$$

Among these births, those to urban females who did not emigrate during the entire period of projection number, in the absence of mortality:

$$_{60}n_{x,x+4;f;u} \cdot \mu_{x,x+4;f;u,u} \cdot {}_{60}\phi_{x,x+4;u}$$

Births at the end of the period, that is to say in 1964, can be estimated for this group of women as:

$$_{60}n_{x,x+4;f;u} \cdot P_{x,x+4;f;u} \cdot \mu_{x,x+4;f;u,u} \cdot {}_{65}\phi_{x+5,x+9;u}$$

Again the figure is only an approximation, as it is the number of urban women who did not emigrate and have come to form the age group from $x+4$ to $x+8$ on June 30, 1964 that should be multiplied by the rate $_{60}\phi_{x+4,x+8;u}$ to find births for the group during the year 1964.

Mean annual births between 1960 and 1964 can be estimated by the arithmetic mean of the two expressions above:

$$\frac{1}{2} \, {}_{60}n_{x,x+4;f;u} \cdot \mu_{x,x+4;f;u,u} \cdot \left({}_{60}\phi_{x,x+4;u} + {}_{65}\phi_{x+5,x+9;u} \cdot P_{x,x+4;f;u} \right)$$

To reach the total number of births during the five years of the projection for this same group of women it suffices to multiply this quantity by 5. To then obtain male births alone the product must be multiplied by k, and finally to estimate boys surviving to January 1, 1965 it suffices to apply the expected male survival rate for urban areas $P_{n;m;u}$, and we have finally:

$$\frac{1}{2} \, {}_{60}n_{x,x+4;f;u} \cdot 5k \, P_{n;m;u} \cdot \mu_{x,x+4;f;u,u}$$
$$\cdot \left({}_{60}\phi_{x,x+4;u} + {}_{65}\phi_{x+5,x+9;u} \cdot P_{x,x+4;f;u} \right)$$

In these expressions the coefficients of $_{60}n_{x,x+4;f;u}$ form the elements of the first row of the sub-matrix M_2^3.

249

The sum of these expressions over all fertile age groups ($x = 15, \ldots, 45$) represents the complete number of surviving boys ages 0 to 4 on January 1, 1965.

We have supposed throughout this calculation, as was noted earlier, that children born in the projection period cannot migrate independently of their mothers, so that we identify migration of mothers with that of their children. If the independence of these two migratory movements were supposed, it would be necessary to replace the probability of the mothers' [remaining urban] $\mu_{x,x+4;u,u}$ by that for births $\mu_{n;m;u,u}$ and we would have:

$$\frac{1}{2}\,_{60}n_{x,x+4;f;u}\cdot 5k\,P_{n;m;u}\cdot\mu_{n;m;u,u}$$
$$\cdot\left(_{60}\phi_{x,x+4;u} + \,_{65}\phi_{x+5,x+9;u}\cdot P_{x,x+4;f;u}\right)$$

Sub-matrix M_2^5: active rural males becoming inactive and emigrating.

$$P_{x,x+4;m;r}\cdot\theta_{x,x+4;m;a,i}\cdot\mu_{x,x+4;m;r,u}$$

Sub-matrix M_2^6: inactive rural males remaining inactive and emigrating.

$$P_{x,x+4;m;r}\cdot\theta_{x,x+4;m;i,i}\cdot\mu_{x,x+4;m;r,u}$$

Sub-matrices M_2^7 and M_2^8: surviving male infants of emigrating rural women, active or inactive.

This expression is the same as for M_2^3 and M_2^4, with the substitution of the index r for the index u in the functions of mortality and fertility and for the first subscript u of the migration function, that is:

$$\frac{1}{2}\,5k\,P_{n;m;r}\cdot\mu_{x,x+4;f;r,u}\cdot\left(_{60}\phi_{x,x+4;r} + \,_{65}\phi_{x+5,x+9;r}\cdot P_{x,x+4;f;r}\right)$$

The reader can easily find the other elements of the matrix himself, by relying on similar reasoning.

We have reproduced in Tables 2 and 3 two examples of the calculation: Table 2 indicates how the number of active urban males at the end of the period is obtained, adding active urban males remaining active and not emigrating (column 4), inactive urban males becoming active and not emigrating (column 8), inactive rural males becoming active and urban (column 14), and finally, active rural males remaining active and becoming urban (column 19). To carry out the calculations we make use of sub-matrices M_1^1, M_1^2, M_1^5 and M_1^6 in Table 1.

In Table 3 the calculation of surviving female infants to rural or rural-emigrant women, active or inactive, is indicated. The upper part of the table corresponds to sub-matrices M_4^7 and M_4^8, and the lower part to sub-matrices M_8^7 and M_8^8. It has been supposed in these calculations that the migration rates between rural and urban zones are net rates. As the migration flows at all ages are more important in the rural-urban direction than in the urban-rural direction, all the elements in the lower left quarter of the matrix are zero. ...

Table 2. Example of the Calculation of Active Urban Males in 1965 (numbers in thousands).

Age group in 1960	Number of active urban males in 1960	Expected proportion surviving	Proba-bility of remaining active	"Permanent" active urban males $(1) \times (2) \times (3)$	Number of inactive urban males in 1960	Expected proportion surviving	Proba-bility of becoming active
	(1)	(2)	(3)	(4)	(5)	(6)	(7)
0— 4 years	0	0.9772	1	0	1712.7	0.9772	0
5— 9	0	0.9945	1	0	1322.2	0.9945	0.070
10—14	73.8	0.9923	1	0	980.1	0.9923	0.602
15—19	543.6	0.9874	1	73.23	319.2	0.9874	0.756
20—24	668.5	0.9839	1	526.75	66.1	0.9839	0.222
25—29	568.3	0.9809	1	657.74	42.8	0.9809	0.142
30—34	506.8	0.9770	1	557.44	32.3	0.9770	0.166
35—39	426.6	0.9717	1	495.14	22.5	0.9717	0.200
40—44	355.0	0.9627	1	414.53	14.8	0.9627	0
45—49	294.0	0.9488	0.990	341.76	12.2	0.9488	0
50—54	251.6	0.9309	0.958	276.16	13.2	0.9309	0
55—59	196.7	0.9028	0.934	224.37	19.5	0.9028	0
60—64	128.3	0.8582	0.918	165.86	22.6	0.8582	0
65—69	88.5	0.7953	0.833	101.08	24.9	0.7953	0
70—74	50.1	0.7066	0	58.63	27.0	0.7066	0
75—79	0	0.5904	0	0	96.5	0.5904	0
80 +	0	0.3774	0	0	0	0.3774	0

Table 2 (continued)

Age group in 1960	Inactive becoming active urban males $(5) \times (6) \times (7)$	Total non-emigrant active urban males in 1965 $(4) + (8)$	Number of inactive rural males in 1960	Expected proportion surviving	Proba-bility of becoming active	Proba-bility of emigrating	Inactive rural becoming active urban males $(10) \times (11) \times (12) \times (13)$
	(8)	(9)	(10)	(11)	(12)	(13)	(14)
0— 4 years	0	0	1746.7	0.9418	0	0.070	0
5— 9	0	0	1398.9	0.9811	0.220	0.085	0
10—14	92.05	92.05	889.4	0.9833	0.821	0.090	25.66
15—19	585.48	658.71	127.4	0.9803	0.429	0.090	64.62
20—24	238.28	775.03	58.0	0.9748	0.125	0.085	4.82
25—29	14.44	672.18	43.1	0.9671	0.143	0.082	0.60
30—34	5.96	563.40	31.0	0.9584	0.167	0.080	0.49
35—39	5.24	500.38	22.1	0.9496	0	0.075	0.40
40—44	4.37	418.90	17.5	0.9395	0	0.070	0
45—49	0	341.76	14.6	0.9268	0	0.067	0
50—54	0	276.16	12.8	0.9094	0	0.064	0
55—59	0	224.37	12.8	0.8846	0	0.060	0
60—64	0	165.86	13.7	0.8463	0	0.058	0
65—69	0	101.08	13.0	0.7914	0	0.056	0
70—74	0	68.63	12.6	0.7167	0	0.055	0
75—79	0	0	52.6	0.6177	0	0	0
80 +	0	0	52.2	0.4212	0	0	0

Table 2 (continued)

Age group in 1960	Number of active rural males in 1960	Expected proportion surviving	Probability of remaining active	Probability of emigrating	Active rural becoming active urban males $(15) \times (16) \times (17) \times (18)$	Rural males entering the active urban pop. by 1965 $(14) + (19)$	Total active urban males in 1965 $(9) + (20)$
	(15)	(16)	(17)	(18)	(19)	(20)	(21)
0— 4 years	0	0.9418	1	0.070	0	0	0
5— 9	0	0.9811	1	0.085	0	0	0
10—14	250.9	0.9833	1	0.090	0	25.66	117.71
15—19	782.6	0.9803	1	0.090	22.20	86.82	745.53
20—24	667.0	0.9748	1	0.085	69.05	73.87	848.90
25—29	572.6	0.9671	1	0.082	55.27	55.87	728.05
30—34	486.4	0.9584	1	0.080	45.41	45.90	609.30
35—39	420.1	0.9496	1	0.075	37.29	37.69	538.07
40—44	333.4	0.9395	1	0.070	29.92	29.92	448.82
45—49	277.4	0.9268	1	0.067	21.93	21.93	363.69
50—54	243.0	0.9094	0.989	0.064	17.26	17.26	293.42
55—59	200.9	0.8846	0.979	0.060	13.99	13.99	238.36
60—64	157.9	0.8463	0.967	0.058	10.44	10.44	176.30
65—69	105.4	0.7914	0.955	0.056	7.49	7.49	108.57
70—74	71.1	0.7167	0	0.055	4.46	4.46	63.09
75—79	0	0.6177	0	0	0	0	0
80+	0	0.4212	0	0	0	0	0

Table 3.   Calculation of Surviving Female Infants to Women Emigrating to Urban Areas or Remaining Rural.

Age group	Fertility rate in 1960	Fertility rate in 1965	Expected proportion surviving	Product of (3) by (2) of the row following	Sum $(1) + (4)$	Probability of emigrating from rural to urban zones
	(1)	(2)	(3)	(4)	(5)	(6)
10—14 years	0	0	0.9854	12.3175	12.3175	0.088
15—19	12.5	12.5	0.9811	34.2404	46.7404	0.090
20—24	34.9	34.9	0.9754	31.7980	66.6980	0.090
25—29	32.6	32.6	0.9710	29.5184	62.1184	0.088
30—34	30.4	30.4	0.9664	22.4205	52.8205	0.085
35—39	23.2	23.2	0.9614	8.2680	31.4680	0.080
40—44	8.6	8.6	0.9550	2.7695	11.3695	0.075
45—49	2.9	2.9	0.9464		2.9000	0.071
Total						

Table 3 (continued)

Age group	Product $(5) \times (6) \times$ 1.0762* (7)	Number of rural women in 1960 $_{60}n_{x,x+4:f:r}$ (8)	Surviving female infants to emigrant rural women $(7) \times (8)$ (9)	Proba-bility of remaining rural (10)	Product $(5) \times (10) \times$ 1.0762* (11)	Surviving female infants to women remaining rural $(8) \times (11)$ (12)
10—14 years	1.1665	1053.8	12.3	0.912	12.0896	127.4
15—19	4.5272	843.1	38.2	0.910	45.7748	385.9
20—24	6.4602	707.9	45.7	0.910	65.3202	462.4
25—29	5.8830	623.4	36.7	0.912	60.9689	380.1
30—34	4.8319	517.9	25.0	0.915	52.0136	269.4
35—39	2.7093	440.5	11.9	0.920	31.1566	137.2
40—44	0.9177	348.6	3.2	0.925	11.3182	39.5
45—49	0.2216	283.4	0.6	0.929	2.8994	8.2
Total			173.6			1810.1

* 1.0762 is the product $\frac{1}{2}5(1-k)P_{n:f:r}$, with $k = 0.4878$ and $P_{n:f:r} = 0.8825$.

The complete results for the period 1960—1965 are indicated in the annex. One can see there the numbers entering and leaving the active and inactive categories in different age groups according to sex and according to whether they are "permanent" (i.e., non migrants) or migrant. ...

Having arrived at this degree of complexity recourse to a computer proves indispensible, especially if one desires to vary the functions that are being assumed, or if he wishes to gage the effects of an error in one of the functions on a selected element of the resulting vector.

The same methodology could be applied for projecting the occupational structure if one possessed a table of interprofessional migration and rates of entry of young people into the various professions. For this it would suffice to ask in censuses questions similar to those which Mexico has adopted for residence. Unfortunately no country prepares this highly interesting information, and it cannot be drawn from surveys due to insufficient numbers of observations.

Annex

Projection of the Active Population of Mexico, 1960—1965

Composition of the Active Urban Male Population in 1965, According to its Distribution in 1960.

Age group in 1965	"Permanent" active urban	Inactive urban	Total urban	Active rural	Inactive rural	Total rural	Total in 1965
			Distribution in 1960				Total in 1965
0— 4 years	0	0	0	0	0	0	0
5— 9	0	0	0	0	0	0	0
10—14	0	92.0	92.0	0	25.7	25.7	117.7
15—19	73.2	585.5	658.7	22.2	64.6	86.8	745.5
20—24	536.8	238.3	775.1	69.0	4.8	73.8	848.9
25—29	657.7	14.4	672.1	55.3	0.6	55.9	728.0
30—34	557.4	6.0	563.4	45.4	0.5	45.9	609.3
35—39	495.1	5.2	500.4	37.3	0.4	37.7	538.1
40—44	414.5	4.4	418.9	29.9	0	29.9	448.8
45—49	341.8	0	341.8	21.9	0	21.9	363.7
50—54	276.2	0	276.2	17.3	0	17.3	293.4
55—59	224.4	0	224.4	14.0	0	14.0	238.4
60—64	165.9	0	165.9	10.4	0	10.4	176.3
65—69	101.1	0	101.1	7.5	0	7.5	108.6
70—74	58.6	0	58.6	4.5	0	4.5	63.1
75—79	0	0	0	0	0	0	0
80 +	0	0	0	0	0	0	0
Total	3902.7	945.8	4848.6	334.7	96.6	431.3	5279.8

Composition of the Inactive Urban Male Population in 1965, According to its Distribution in 1960.

Age group in 1965	Active urban	"Permanent" inactive urban	Total urban	Active rural	Inactive rural	Total rural	Total in 1965
			Distribution in 1960				Total in 1965
0— 4 years	0	1899.8	1899.8	0	180.5	180.5	2080.3
5— 9	0	1673.7	1673.7	0	115.1	115.1	1788.8
10—14	0	1222.9	1222.9	0	90.9	90.9	1313.8
15—19	0	386.1	386.1	0	14.1	14.1	400.2
20—24	0	76.6	76.6	0	6.4	6.4	83.0
25—29	0	50.5	50.5	0	4.2	4.2	54.7
30—34	0	36.0	36.0	0	2.9	2.9	38.9
35—39	0	26.3	26.3	0	2.0	2.0	28.3
40—44	0	17.5	17.5	0	1.6	1.6	19.1
45—49	0	14.2	14.2	0	1.2	1.2	15.4
50—54	2.8	11.6	14.4	0	0.9	0.9	15.3
55—59	9.8	12.3	22.1	0.1	0.7	0.8	22.9
60—64	11.7	17.6	29.3	0.2	0.7	0.9	30.2
65—69	9.3	19.4	28.7	0.3	0.7	1.0	29.7
70—74	11.8	19.8	31.6	0.2	0.6	0.8	32.4
75—79	35.4	19.1	54.5	2.8	0.5	3.3	57.8
80 +	0	57.0	57.0	0	32.5	32.5	89.5
Total	80.8	5560.4	5641.2	3.6	455.5	459.1	6100.3

28. Weak Ergodicity

ALVARO LOPEZ (1961)

From *Problems in Stable Population Theory*, pp. 4—5, 66—68. Princeton: Office of Population Research.

Lopez' proof of his ergodicity theorem is omitted. Its application is limited to matrices that are irreducible and primitive, whose meanings we discuss in the introduction to Parlett (1970, paper 29 below).

It is a common-sense fact that age structure of a closed population at a given point in time is determined once we know 1) its age structure at an arbitrary point in the past and 2) the age-schedules of fertility and mortality at every moment in the intervening period. We know from stable population theory that the effect of the first determinant is washed out when sufficient time is allowed to elapse between the two referred moments, if fertility and mortality conditions are unchanging in time.

It has been conjectured by Coale in 1957 (Ref. 8) that the same factors that cause the transient effects to disappear from a population with unchanging fertility and mortality rates would also operate for *any* time path of these rates. That is, the age distribution of a closed population tends to "forget" its past shape and to be determined exclusively by the recent history of fertility and mortality. This property will be called *weak ergodicity*. ...

It is essential to distinguish between *weak* and *strong* ergodicity, the latter being present when besides the tendency to forget the past, the age structure approaches a stable limiting shape. Strong ergodicity is of course characteristic of Lotka's model but it is not to be expected when fertility and mortality rates are not constant through time. To make clearer what we are talking about we can say that the age distribution is weak ergodic if two arbitrarily chosen populations submitted after an initial moment to the same conditions of age specific fertility and mortality will in the long run differ less and less in their age structures, even when these are changing themselves through time without converging to a limit.

... We can put the theorem on weak ergodicity as a simple implication of the following statement: Under the relevant assumptions, the product of n transition matrices $A^{(t)}$ is a matrix whose rows tend to become proportional as $n \to \infty$.

This statement is helpful to clarify the sense in which the term "weak ergodicity," as applied here to the chain of transition matrices $A^{(t)}$, differs from the meaning given to it by John Hajnal, who seems to have coined the term (see Refs. 1, 2). For one thing, Hajnal deals in the referred papers with stochastic matrices only. Reading those papers will show how relevant this is in order to obtain his results. On the other hand, Hajnal defines weak ergodicity as the property possessed by a product of a chain of stochastic matrices of having rows which tend to become identical when the number of factor matrices is increased indefinitely, even if the rows approach no limit. We have on our part used the term weak ergodicity in nonhomogeneous chains to denote the tendency for the rows in the product matrix to become proportional, not necessarily identical.

The reformulation of the weak ergodic theorem in terms of the ultimate proportionality of the rows in the product matrix makes it very clear why it is necessary to have non-cyclical fertility with respect to age. If we formed the sequence

$\prod\limits_{t=0}^{n} A^{(t)}$ for successive values of n under conditions of cyclical fertility with respect

to age, no matrix in this sequence would be strictly positive and ... successive products would have zeros in positions changing in a cyclical way (see Kemeny and Snell, Ref. 3, pp. 5—7, 99). It is then impossible for the rows to approach proportionality, so weak ergodicity will be absent.

Other demographic implications of weak ergodicity of the age structure are the following:

Consider two populations differing initially in an arbitrary way as to their age structure, neither of which is headed towards extinction. If they are subsequently submitted to coincident regimes of age specific fertility and mortality rates, they will ultimately take on practically coincident age structures, hence they will have birth rates, death rates and rates of natural increase which can be considered equal for all practical purposes.

Also, if V^0 and W^0 are the initial population row vectors with components $V_i^{(0)}$ and $W_i^{(0)}, \ldots$ the number of females of age i at time t will be given by the asymptotic expression:

$$V_i^{(t)} \rightarrow a_i^{(t)} \sum_{j=0}^{\beta} V_j^{(0)} k_j$$

$$W_i^{(t)} \rightarrow a_i^{(t)} \sum_{j=0}^{\beta} W_j^{(0)} k_j \qquad (2)$$

as $t \rightarrow \infty. \ldots$

From (2) it follows that:

$$\frac{V_i^{(t)}}{W_i^{(t)}} \rightarrow \frac{\sum\limits_{j=0}^{\beta} V_j^{(0)} k_j}{\sum\limits_{j=0}^{\beta} W_j^{(0)} k_j} \qquad (3)$$

as $t \rightarrow \infty$.

That is, the ratio of the population sizes at all ages, and hence the ratio of the total population sizes, tends toward a limit which depends on the initial condition (V_j^0, W_j^0) as well as on the characteristics of fertility and mortality history (k_j). Thus if we imagine that from an initial moment in time onwards all populations in the world had coincident regimes of age specific fertility and mortality, changing together in time for all populations; and if we could exclude migratory movements, then with the passage of time all populations would coincide in their age structure, in their birth and death rates and in their rate of natural increase. The relative sizes of the different populations would vary for some time but would ultimately tend to adopt a fixed pattern.

References

1. Hajnal, J.: The ergodic properties of nonhomogeneous finite Markov chains. Proceedings of the Cambridge Philosophical Society **52**, pp. 67 ff. (1956).
2. Hajnal, J.: Weak ergodicity in nonhomogeneous Markov chains. Proceedings of the Cambridge Philosophical Society **54**, pp. 233 ff. (1958).
3. Kemeny, J.G., Snell, J.L.: Finite Markov Chains. D. Van Nostrand and Company 1960.
8. Coale, A.J.: How the age distribution of a human population is determined. Cold Spring Harbor Symposia on Quantitative Biology **22**, pp. 83 ff. (1957).

29. Ergodic Properties of Population I: The One Sex Model

BERESFORD PARLETT (1970)

From *Theoretical Population Biology* 1. Excerpts are from pages 191—202.

The requirements Parlett introduces for stability of non-negative matrices, irreducibility and primitivity, have an intuitive base. Irreducibility translates as the restriction that the projection matrix be for a single population, as in the continuous form of the renewal equation. (Matrix techniques for extracting stable roots and vectors do not apply to non-interacting populations; nor in general would such populations have the same roots.)

Primitivity is associated with the persistence of oscillations in births, the oscillations being damped if at least two fertile age groups form circuits lacking common divisors. For example, the circuit {birth to age 1, age 1 to birth of offspring} which has the matrix form $\begin{bmatrix} 0 & 1 \\ 1 & 0 \end{bmatrix}$ is of length 2, and the circuit {birth to age 1, age 1 to age 2, age 2 to age 3, age 3 to birth} is of length 4, an integer multiple of the other. These two circuits and the combined circuit have the matrix forms

$$\begin{bmatrix} 0 & 1 \\ 1 & 0 \end{bmatrix} \qquad \begin{bmatrix} 0 & 0 & 0 & 1 \\ 1 & 0 & 0 & 0 \\ 0 & 1 & 0 & 0 \\ 0 & 0 & 1 & 0 \end{bmatrix} \qquad \begin{bmatrix} 0 & 1 & 0 & 1 \\ 1 & 0 & 0 & 0 \\ 0 & 1 & 0 & 0 \\ 0 & 0 & 1 & 0 \end{bmatrix}.$$

Where there is only one fertile age or where circuits are all integer multiples of some number, as in the cases here, oscillations that derive from the age distribution of the initial population are perpetuated. In contrast, lines of descent from the separate age cohorts in the initial population become blended with time if at least two circuits exist that do not have a common factor, and the distinctiveness that gives rise to oscillations is lost. Note that in continuous analysis there exist an infinity of circuit lengths, hence oscillations become damped in every case.

We have omitted Parlett's appendix, a proof of Fact 10. In Fact 7 note that the indices of the positive elements of the first row of the projection matrix sum to the circuit lengths.

The references to Perron and Frobenius are Perron (1907) and Frobenius (1912).

Introduction

The present paper develops stable population theory from certain general propositions of mathematics. It takes advantage of the large body of analysis developed during this century concerning matrices whose elements are either zero or positive. This mathematical literature is scattered through many sources and the reorganization of its relevant parts into a coherent form bearing on population is a worthwhile undertaking. Once the mathematics is presented the reader will see that the propositions of stable population theory follow immediately. The suggested new organization of the subject seems an advance on existing demographic literature in being both more general and simpler. It assumes only such elementary knowledge of matrices as their multiplication, partitioning, ranks, and determinants. ...

Notation

If v denotes a column vector with real or complex elements then v^* will denote the row vector obtained by transposing v and taking the conjugate of any complex elements. We try to reserve capital roman letters for matrices. The determinant of A is denoted by $\det(A)$.

Let A be a $n \times n$ matrix and x a vector. Put $y = Ax$ ($y_i = a_{i1}x_1 + a_{i2}x_2 + \cdots + a_{in}x_n$, $i = 1,..., n$). The explicit relation of y to x is complicated and we get no feeling for how A acts on the set of all x's. The eigenvectors of A are just those vectors on which the action of A is as simple as it could be, namely y is a multiple of x. The multiple is the eigenvalue. It is a major result of matrix theory that for "most" matrices a knowledge of the eigenvectors and eigenvalues captures the action of A completely. It follows that the eigenvalues are those numbers λ which make $\det(A - \lambda I)$ vanish, I being the identity matrix.

Now A also acts on row vectors by multiplication on the left. Analogously the eigenvalues and their row eigenvectors also capture the action of A except in certain special cases.

These quantities are to be regarded as the hidden foundations of the geometric structure of A.

Population Projection

Suppose that a population is partitioned into age groups so that at time t it may be represented by a column vector $p(t)$. (When it is not necessary to allude to t explicitly we shall write simply p.) The j-th element, written p_j, denotes the number of members of the population in the j-th age group.

In discrete models the age groups all span an equal time interval Δt and the population is only considered at intervals of Δt. The model is determined by the way in which $p(t)$ is transformed or "projected" into $p(t + \Delta t)$. We shall ignore migration and study the simple but nontrivial[1] model

$$p(t + \Delta t) = L(\Delta t)\, p(t)$$

where L is the population projection matrix, based on a set of observed or hypothetical rates of birth and death. The notation $L(\Delta t)$ indicates that the survival and fertility rates do *not* change with time.

In real life populations, the rates do change with time. If there is no migration, we may still write $p(t + \Delta t) = B(\Delta t, t, p, ...)\, p(t)$ but B is then too complicated to be useful. B certainly depends on both Δt and t. Moreover the rates may depend on $p(t)$ itself and on previous states $p(\tau)$, $\tau < t$. In other words a particular model arises out of the restrictions placed on B, not out of writing the transformation in the linear-looking form Bp.

In our model, for any time $t = \nu\Delta t$, $\nu = 1, 2, ...$, $p(t)$ is completely determined by $p(0)$, the starting population, and $L(\Delta t)$. A question of interest is to know the rate at which the population grows, and the ratios of the age groups after a long time has elapsed (ν large). At first glance, it is surprising that these quantities are independent of the initial age distribution $p(0)$ and depend only on the regime of mortality and fertility embodied in L.

How should these quantities be calculated? Good programming procedures will become apparent in the context of the next few sections.

The analysis of $p(\nu\Delta t) = L^{\nu}p(0)$ is a mathematicial task which is almost completely accomplished. In the next few sections we will bring together the various parts of the theory together with some discussion. The object of such a summary is economy—to relieve others from continually reproving known results. Mathematicians strip abstract arguments down to their essentials. Any "demographic" proof of a property of a mathematicial model must inevitably be a reformulation of the mathematical argument. To make the line of argument stand out more clearly in what follows proofs in excess of a few lines will be omitted.

ANALYSIS OF L

Three key integers are associated with out model: the first and last age groups, α and β, in which reproduction is possible, and the total number of age groups γ. All three depend on the interval of time Δt over which projection takes place.

---

[1] Nontrivial means that the first row of L is not zero.

We assume that the form of L is familiar to the reader, and need here say only that the j-th element of its i-th row, l_{ij}, is the contribution to the population in the i-th age group at time t due to the j-th age group at time $t - \Delta t$. See Keyfitz (1968), p. 31. It is convenient to rename the element of L by partitioning; that is we write

$$L = \begin{pmatrix} P & 0 \\ M & N \end{pmatrix}$$

where the first row of P contains the fertility rates of the various age groups, its elements immediately below the diagonal are the survival rates of the age groups $1,...,\beta - 1$. M is zero except for its $(1, \beta)$ element and N is zero except for the elements immediately below the diagonal. By our definition of P we have $p_{1\beta} > 0$.

The zero submatrix in the North East corner makes L a reduced matrix. In different words L is lower triangular by blocks and it is easy to verify that all powers of L have the same form. In fact

$$L^\nu = \begin{pmatrix} P^\nu & 0 \\ X_\nu & N^\nu \end{pmatrix}$$

where $X^\nu = \sum_{j=0}^{\nu-1} N^j M P^{\nu-j-1}$. The matrix N is called *nilpotent* because it enjoys the property that $N^\nu = 0$ for all sufficiently large values of ν; in our case for $\nu \geqslant \gamma - \beta$. This corresponds to the demographic fact that the initial population above the reproductive age has no effect on the subsequent population after $\gamma - \beta$ projections of length Δt.

The matrix X_ν involves linear combinations of powers of P less than ν. Thus X_ν will behave like P^ν but less dramatically; if P^ν grows, or decays, as ν increases then so does X_ν.

We shall see that the important information about L is all contained in P. For example, the age groups beyond reproduction do not contribute to the intrinsic growth rate of the whole population. Below we write down the known facts about L and P, arranging them in a natural order. It will be seen that most of stable population theory is contained in this list. Some of the facts are deep, others not.

FACT 1. *The eigenvalues of P and the eigenvalues of N together constitute the eigenvalues of L.*

Proof. Because L is reduced $\det(L - \lambda I) = \det(P - \lambda I) \cdot \det(N - \lambda I)$ for any λ. ∎

FACT 2. *No eigenvalue of P is zero. All eigenvalues of N are zero.*

Proof. The product of P's eigenvalues $= \det(P) = p_{21} p_{32} \cdots p_{\beta,\beta-1} p_{1\beta} > 0$. Furthermore, $N^{\gamma-\beta} = 0$ which implies that for any eigenvalue μ of N we have $\mu^{\gamma-\beta} = 0$. Hence $\mu = 0$. ∎

Before confining our attention to P we indicate how its eigenvectors are related to those of L. Corresponding to any eigenvalue λ of P there are two vectors, x and y^*, called the column and row eigenvectors and they satisfy the equations $Px = x\lambda = \lambda x$, $y^*P = \lambda y^* = y^*\lambda$.

FACT 3. *The eigenvectors of P determine the eigenvectors of L. Specifically $L\binom{x}{\tilde{x}} = \binom{x}{\tilde{x}} \lambda$, $(y^*, \tilde{y}^*)L = \lambda(y^*, \tilde{y}^*)$ where $\tilde{x}_1 = m_{1\beta} x_\beta / \lambda$, and $\tilde{x}_j = n_{j,j-1} \tilde{x}_{j-1}/\lambda$, $j = 2,..., \gamma - \beta$; $\tilde{y}_k = 0$, $k = 1,..., \gamma - \beta$.*

Proof. The equation $(\lambda I - N) \tilde{x} = Mx$ can be solved explicitly row by row. Moreover since λ cannot be an eigenvalue of N the relation $\tilde{y}^*(\lambda I - N) = 0$ implies that $\tilde{y} = 0$. ∎

Fact 3 was restated by Lopez (1961) as the fact that if there is a stable age distribution up to the end of reproduction then so there is for the whole population.

Of course 0 is an eigenvalue of L and its corresponding eigenvectors are called null vectors. Let the j-th column of the identity matrix I be called e_j. The dimension e_j is to be understood from its context.

FACT 4. $Le_\gamma = 0$, $(z^*, e_1^*)L = 0$ *where* $z_1 = -m_{1\beta}/p_{1\beta}$, $z_{j+1} = -p_{1j} z_1/p_{j+1,j}$, $j = 1,..., \beta - 1$. *There are no other linearly independent null vectors.*

Proof. $L\binom{w}{\tilde{w}} = 0$ implies that $Pw = 0$, $Mw + N\tilde{w} = 0$. Since P is non-singular $w = 0$ and, by its form $N\tilde{w} = 0$ implies that $\tilde{w} = e_{\gamma-\beta}$. Secondly $(z^*, \tilde{z}^*)L = 0$ implies (i) $\tilde{z}^*N = 0$ and (ii) $z^*P + \tilde{z}^*M = 0$. By its form (i) implies $\tilde{z}^* = e_1^*$ and $e_1^*M = m_{1\beta} e_\beta^*$. Solving (ii) column by column gives the result. The uniqueness follows from the fact that the only freedom of choice in the above solution lay in picking $\tilde{w} = \theta e_{\gamma-\beta}$ and $\tilde{z}^* = \phi e_1^*$, for any nonzero θ, ϕ. ∎

PROPERTIES OF P

It is (quite well) known that P satisfies the hypotheses of the Perron–Frobenius theory of nonnegative matrices. What we want to do here is to set down the argument so that recent results in demographic theory such as the Coale–Lopez weak ergodicity theorem are seen to be very close to the original strong ergodicity theorem.

It is usual to say that any matrix A is *reducible* if there exists a permutation matrix S such that SAS^{-1} has the same form as L; that is SAS^{-1} is block-triangular. If no such S exists then A is *irreducible* (or indecomposable). This is probably the briefest definition of the term. However it appears at first to be a little contrived. Let us look at the situation more intuitively.

Consider again our matrix L. Suppose we wish to know what happens to the first β components of some vector v when multiplied by L. We observe that the result is independent of the value of the remaining elements $v_{\beta+1}, \ldots, v_{\gamma}$. A matrix A is *irreducible* if there is no set of vector elements which is independent of the remainder as far as multiplication by A is concerned.

The actual values of the non-zero elements of A are irrelevant to its reducibility. This is in strong contrast to A's eigenvalues. Those properties of A which depend only on whether elements are zero or not are often called graph-theoretic. If the nonzero elements are replaced by 1's then the new matrix $\tilde{A}$ is the so-called incidence matrix of a graph G and some properties of A are reflected most simply in G. For an introduction to Graph Theory see Berge (1962).

Consider for a moment the elements of the product of three square matrices A, B, and C. A typical element

$$(ABC)_{il} = \sum_{j=1}^{\beta} \sum_{k=1}^{\beta} a_{ij} b_{jk} c_{kl} .$$

A little reflection shows that there is a nonzero contribution to the sum on the right when and only when there is a sequence of index pairs (i, j), (j, k), (k, l) such that $a_{ij} \neq 0$, $b_{jk} \neq 0$, $c_{kl} \neq 0$. Such a sequence is called a *path* (from i to l) of length 3. If A, B, C are nonnegative then $(ABC)_{il}$ is positive if, and only if, there is at least one path from i to l.

Often we wish to put $B = A$, $C = A$ and study powers of A. An alternative definition of irreducibility is that there exists a path (of some length) connecting any pair of row indices. This is a third definition of irreducibility. In other words, as powers of A are formed there is no position (i, j) which remains permanently zero. Of course it still can happen that when one position is zero another is not and vice versa.

FACT 5. *P is nonnegative and irreducible.*

We shall not prove this fact here. Sykes [8] points out that it is necessary only that $p_{1\beta}$ does not vanish; the other elements in row 1 may be ignored. For primitivity (see Fact 7) more is required. It suffices to suppose that $p_{1\beta}$ is the only positive element in row 1. Replacing each positive element in P by 1 we obtain a permutation matrix and it is a standard result (Marcus and Minc, 1964) that such matrices are irreducible. Each member of age group β at time t contributes $p_{1\beta}$ new living offspring at time $t + \Delta t$.

264

By clever arguments Perron and Frobenius established the following consequence of Fact 5.

FACT 6. *P has an eigenvalue λ_1 (called the Perron root) which is simple, positive, and is not exceeded in magnitude by any other eigenvalue of P. The row and column eigenvectors u, v* may be chosen positive (i.e., all the elements are strictly positive).*

Proof. See Gartmacher (1959).

There may be some other eigenvalues (necessarily negative or complex) equal in absolute value to λ_1. The number of eigenvalues of maximal magnitude is called the *index of imprimitivity* of P. If this index is 1, then P is said to be *primitive*. An alternative definition states that a nonnegative irreducible matrix is primitive if some power of it is positive (that is all its elements are actually positive). This usage is not to be confused with the term *positive definite* used for symmetric matrices and quadratic forms.

The equivalence of the two definitions of primitivity is given, for example, in Marcus and Minc (1964).

DEFINITION. The *index of primitivity* of a primitive matrix A is the smallest integer q such that $A^q > 0$.

Note the difference between index of imprimitivity and index of primitivity.

FACT 7. *P is primitive if and only if the greatest common divisor of the indices of the positive elements in the first row is 1.*

The problem of determining this index goes back to Frobenius although in rather different language. A proof of Fact 7 is given by Rosenblatt (1957). Most of the literature on this problem is in journals concerned with number theory and graph theory.[2] Fact 7 has been reproved by Lopez (1961).

The connection with graph theory comes in strongly here. A *circuit* is a path in which the first and last indices are the same. It is called *elementary* if no other index appears twice.

Let $\alpha = \alpha_1 , \alpha_2 ,..., \alpha_k = \beta$ be the indices of the age groups with positive reproduction. Then it happens that the only elementary circuits in A have lengths $\alpha_1 ,..., \alpha_k$. (See Heap and Lynn, 1964, p. 122). Rosenblatt shows that A is primitive if and only if the lengths of all circuits in A (and therefore at least two) are relatively prime. Moreover the characteristic polynomial of P is given by

$$\det(\lambda I - P) = \lambda^\beta - p_{11}\lambda^{\beta-1} - \cdots - \left(\prod_{i=1}^{k-1} p_{i+1,i}\right) p_{1k}\lambda^{\beta-k} - \cdots - \det P.$$

Thus the nonzero exponents are

$$\beta, \beta - \alpha_1 , \beta - \alpha_2 ,..., \beta - \alpha_{k-1} , 0.$$

265

The differences between these exponents are

$$\alpha_1, \alpha_2 - \alpha_1, \alpha_3 - \alpha_2, ..., \beta - \alpha_{k-1}.$$

Since the greatest common divisor of both sets of integers is the same as for the $\{\alpha_i\}$ we have an alternative characterization for primitivity. In fact Dulmage and Mendelsohn (1964) have a paper entitled "Gaps in the exponent set of primitive matrices" in which they describe the situation in terms of the last set above.

INDEX OF PRIMITIVITY

Let us assume that the indices $\{\alpha_i\}$ of the age groups with positive reproduction are indeed relatively prime. It follows that there are at least two of them. It is natural to ask what is Index(P), the smallest q such that $P^q > 0$? From the previous discussion this will be the smallest m such that there is a path in P of exact length m connecting each pair of row indices. Among the paths are the circuits.

FACT 8. *The length l of any circuit can be written in the form $l = \sum_{j=1}^{\beta} v_j \alpha_j$, where the v_j are nonnegative.*

Proof. See Heap and Lynn (1964), Appendix.

Now Frobenius introduced $g[\alpha_1, ..., \alpha_k]$ the greatest integer which *cannot* be expressed in the form of l above. It is related to the index of primitivity by

FACT 9. $g[\alpha_1, ..., \alpha_k] = \text{Index}(P) - \beta.$

Proof. See Heap and Lynn (1964), Theorem 3.1.

In general there are no closed formulas for g and it is the study of g which has stimulated most of the papers cited here. Efforts have been directed to concocting algorithms for computing g and in finding upper and lower bounds for g.

In the case of human populations the $\{\alpha_i\}$ are the consecutive integers between α and β, inclusive. See Sykes (1969) for an interesting discussion of the case when the α_i are not consecutive.

FACT 10. *For the human population model*

$$\text{Index}(P) = \alpha \left[\frac{\beta - 2}{\beta - \alpha} \right] + \beta,$$

where $[x]$ denotes the greatest integer $\leqslant x$.

[2] See Heap and Lynn (1964) and (1965); also Karlin (1966).

Proof. An apparently new and elementary proof is given in the Appendix. Actually this theorem is a special case of results in Section 3 of Dulmage and Mendelsohn (1964). Let us illustrate this in the case $\Delta t = 1$ year, $\alpha = 30$, $\beta = 40$. Consider the ancestors of all people who had their 39-th birthday this year. We must go back 130 years before we find ancestors of all ages between 1 and 40 at that time. In other words, every age group this year will contribute to the population aged 39 in 130 years time.

It is important to notice that a stronger, and important result has been established. Let $\{P_j\}$ be a set of nonnegative matrices with the same pattern of nonzero elements as P. Thus each P_j has the same graph as P.

FACT 11. *The product of any* Index(P) *or more nonnegative matrices with the same graph as P is positive.*

Proof. Each path of length l corresponds to a positive element in the product of any l matrices from $\{P_j\}$.

It follows that much of the argument in Lopez (1961) concerning projection matrices varying in time is, in essence, the same argument required to establish the primitivity of P.

ASYMPTOTIC STRUCTURE OF P^ν

Fact 6, the Perron–Frobenius theorem, coupled with Fact 7 shows that models of *human* populations with more than one reproductive age group have P matrices which are primitive. From this follows

FACT 12. *Let* $Pu = u\lambda_1$, $v^*P = \lambda_1 v^*$ *where u, v are chosen positive and normalized by the condition $v^*u = 1$. Then*

$$P^\nu = \lambda_1{}^\nu(uv^* + E_\nu),$$

$E_\nu \to 0$ *as* $\nu \to \infty$. *Here λ_1 is the Perron root of P.*

Proof. See Gartmacher (1959), p. 81; Karlin (1966), p. 479.

Note that uv^* is a neat way to write a matrix of rank one; u is a column vector and each column of uv^* is a multiple of it. Recall that v^* is to be read v transpose since v is real.

The strong ergodic theorem follows immediately. Let $\hat{p}(0)$ denote the vector of the first β age groups of any viable initial population. It is nonnegative and hence

$$v^*\hat{p}(0) = \sum_{i=1}^{\beta} v_i p_i(0) \equiv \xi > 0.$$

267

Thus as $\nu \to \infty$

$$P^\nu \hat{p}(0) \to u(\lambda_1{}^\nu \xi).$$

Returning to the projection matrix L and using Fact 3 gives

$$L^\nu p(0) \to \binom{u}{\tilde{u}} \lambda_1{}^\nu [(v^*, \tilde{v}^*)\, p(0)].$$

Thus the population distribution by age group is, in the limit, proportional to $\binom{u}{\tilde{u}}$ and thus independent of $p(0)$.

Fact 12 suggests that uv^* could be approximated, once λ_1 is known, by computing $(\lambda_1^{-1}P)^\nu$ for suitable ν. We wish to point out here that this is completely unnecessary. Once λ_1 is known u and v^* can be written down *explicitly* in terms of λ_1, of the survival rates and of the fertility rates. Actually a little more is true. The eigenvectors corresponding to *any* eigenvalue λ of P are readily found.

FACT 13. *Let* $Pu = u\lambda$, $v^*P = \lambda v^*$ *and write* $s_j = p_{j,j-1}$, $f_j = p_{1j}$ *then* $u_1 = $ *arbitrary* > 0, $u_j = u_{j-1}s_j/\lambda$, $j = 2,...,\beta$; $v_1 = $ *arbitrary* > 0, $v_j = (\lambda v_{j-1} - v_1 f_{j-1})/s_j$, $j = 2,...,\beta$.

Proof. The equations $(P - \lambda I)u = 0$ can be solved sequentially from the second equation down in terms of the arbitrary constant u_1. Similarly $v^*(P - \lambda I) = 0$ can be solved sequentially beginning with column 1 in terms of the arbitrary constant v_1. ∎

Recall that the arbitrary choices above correspond to the fact that an eigenvector is really only an eigendirection, the vector is determined only to within a scalar multiple.

INTRINSIC GROWTH RATE

From Fact 12 we see that the population increases by a factor λ_1 (the Perron root) in a time period of Δt. Moreover λ_1 itself depends on our choice of Δt as well as on the population itself. We would prefer a growth measure independent of Δt.

One way to do this is to consider what happens as $\Delta t \to 0$. Clearly $\lambda_1(\Delta t) \to 1$. Now we must make some assumption about the nature of the population which we happen to have sampled with a particular frequency Δt. The simplest assumption mathematically is that the derivative λ_1' is constant. Yet this is not at all realistic for Δt values as great as a year or two.

The next simplest assumption, made by Lotka, is that the derivative of λ_1 is proportional to λ_1 for every Δt. He calls this ratio r the *intrinsic growth rate*.

Thus

$$\lambda'(\tau)/\lambda(\tau) = r.$$

Integrating this from $\tau = 0$ to $\tau = \Delta t$ yields

$$\ln \lambda_1(\Delta t) = r\Delta t,$$

which permits us to determine r from one sample of λ_1.

THE WEAK ERGODIC THEOREM

Suppose that we now consider models in which the fertility and survival rates vary in time: Then

$$p(t + \Delta t) = L(\Delta t, t)\, p(t)$$

and L has the same triangular structure as before. However we now have, e.g.,

$$p(4\Delta t) = L(3\Delta t)\, L(2\Delta t)\, L(\Delta t)\, p(0).$$

The significant part of the product is the top left corner $P(3\Delta t)\, P(2\Delta t)\, P(\Delta t)$ because the product of any $\gamma - \beta$ matrices $N(j\Delta t)$ will vanish. Each matrix $P(j\Delta t)$ is nonnegative and irreducible. With the same assumption that the *indices* of the reproductive age groups remain constant and relatively prime we have Fact 11, namely if $q = \mathrm{Index}(P)$ then

$$P_{j+q}P_{j+q-1} \cdots P_{j+2}P_{j+1} > 0$$

for any $j \geqslant 0$. Here P_k is written instead of $P(k\Delta t)$.

In this more general situation there will be no result with all the detail of Fact 12. No eigenvalues or eigenvectors are relevant and we aim to show simply that the influence of $p(0)$ on $p(t)$ vanishes in the limit as $t \to \infty$.

The usual approach to this theorem is to take two different initial populations $p(0)$ and $\bar{p}(0)$ and then show that $p(t)$ and $\bar{p}(t)$ become proportional as $t \to \infty$. The demographic proofs of this are quite lengthy. See [6] and [12].

The central idea is as follows. Let x and y be any two positive vectors. Define μ and ν, the extreme values of the ratios, by

$$\mu \leqslant \frac{x_i}{y_i} \leqslant \nu, \quad \text{all} \quad i.$$

More precisely split the indices into three nonempty sets: denote by i those

269

indices for which $x_i = \mu y_i$, by k those indices for which $x_k = \nu y_k$, and by j the rest. The j-set might be empty. For *any* positive matrix Q we have

$$(Qx)_l = \sum q_{li} x_i + \sum q_{lj} x_j + \sum q_{lk} x_k$$

$$\geqslant \mu \sum q_{li} y_i + \mu \sum q_{lj} y_j + \nu \sum q_{lk} y_k$$

$$= \mu(Qy)_l + (\nu - \mu) \sum q_{lk} y_k \,.$$

Note that it is the middle term which gives the inequality. Similarly, with $\delta = \nu - \mu$,

$$(Qx)_l \leqslant \nu(Qy)_l - \delta \sum q_{li} y_i \,.$$

With $\phi_l = (\sum q_{lk} y_k)/(Qy)_l$, $\psi_l = (\sum q_{li} y_i)/(Qy)_l$ these two relations can be stated as

LEMMA. $\mu + \delta \phi_l \leqslant (Qx)_l/(Qy)_l \leqslant \nu - \delta \psi_l \,.$

Lopez essentially takes the $P(j \varDelta t)$ matrices Index(P) at a time and gets positive matrices to substitute for Q above. He takes x and y as population vectors with different original distributions. By repeated application of the Lemma we have, as $t \rightarrow \infty$, $\min_i p_i(t)/\bar{p}_i(t) \rightarrow \mu^*$ and $\max_i p_i(t)/\bar{p}_i(t) \rightarrow \nu^*$. Thus certainly $\delta(t)\phi_l(t) \rightarrow 0$, $\delta(t)\psi_l(t) \rightarrow 0$ for each $l = 1,...,\beta$. The ergodic theorem would be proved if only $\mu^* = \nu^*$. This would be true if there were some constant θ bounding ϕ_l and ψ_l away from 0, for then convergence of $\nu(t)$ to ν^* would force $\delta(t) = \nu(t) - \mu(t)$ to converge to 0.

The assumption made by Lopez and MacFarland is that the nonzero elements of $P(t)$ are bounded[†] away from 0 and ∞. We do not wish to pursue here the weakest conditions for the ergodic theorem to hold.

An alternative formulation which is quite illuminating is as follows. Instead of taking x and y as population vectors in our fundamental inequality above let us take them as two arbitrary columns of the matrix

$$F_\nu = P(\nu \varDelta t) \cdots P(\varDelta t) P(0).$$

With the same bounds on the elements of $P(t)$ assumed by Lopez (1961) and MacFarland (1969) we then see that the column of F_ν become proportional to each other as $\nu \rightarrow \infty$. It is *not* true, in general, that F_ν converges to a matrix of rank one. What the above argument shows is that F_ν approaches the *set* of matrices of rank one.

† Uniformly in t.

270

FACT 14. *Assume that the nonzero elements of $P(t)$ are bounded away from 0 and ∞ uniformly in t. For each ν there exists a positive rank one matrix $u_\nu v_\nu^*$ such that $F_\nu = u_\nu v_\nu^* + G_\nu$ and $\| G_\nu \|/\|u_\nu v_\nu^* \| \to 0$ as $\nu \to \infty$ for any matrix norm.*

For a given initial population $p(0)$ we have as $\nu \to \infty$

$$p(t) = p(\nu \Delta t) = F_\nu p(0)$$

$$= u_\nu(v_\nu^* p(0)) + G_\nu p(0).$$

Thus the population distribution is given by u_ν with a correction proportional to $G_\nu p(0)$ which dies out as $\nu \to \infty$. There are some details which we have omitted. For example our argument above gives Fact 14 only for $\nu = kq$ where q is the primitivity index of each $P(t)$. The extension to all sufficiently large integers ν is a straightforward technical matter.

REFERENCES

DULMAGE, A. L. AND MENDELSOHN, N. S. 1964. Gaps in the exponent set of primitive matrices, *Illinois J. Math.* 8, 642–656.

HEAP, B. R. AND LYNN, M. S. 1964. The index of primitivity of a non-negative matrix, *Numer. Math.* 6, 120–144.

HEAP, B. R. AND LYNN, M. S. 1964. A Graph-theoretic algorithm for the solution of a linear diophantine problem of Frobenius, *Numer. Math.* 6, 346–354.

HEAP, B. R. AND LYNN, M. S. 1965. On a linear diophantine problem of Frobenius: An improved algorithm, *Numer. Math.* 7, 226–231.

KEYFITZ, N. 1968. "Introduction to the Mathematics of Population," Addison-Wesley Publ. Co., Reading, Mass.

McFARLAND, D. D. 1969. On the theory of stable populations. A new and elementary proof of the theorems under weaker assumptions, *Demography* 6, 303–322.

ROSENBLATT, D. 1957. On the graphs and asymptotic forms of finite Boolean relation matrices and stochastic matrices, *Naval Res. Logist. Quart.* 4, 151–167.

SYKES, Z. M. 1969. On discrete stable population theory, *Biometrics 25*.

MARCUS, M. AND MINC, H. 1964. "A Survey of Matrix Theory and Matrix Inequalities," Allyn and Bacon, Inc., Boston, Mass.

GARTMACHER, F. R. 1959. "Theory of Matrices," Vol. 2, Chelsea, New York.

KARLIN, —. 1966. "A First Course in Stochastic Processes," Academic Press, Inc., New York.

LOPEZ, A. 1961. "Problems in Stable Population Theory," Office of Population Research, Princeton, N. J.

BERGE, C. 1962. "The Theory of Graphs," Methuen, London.

271

Parameterization and Curve Fitting

An early attempt to describe demographic observations in a mathematical formula was contributed by Abraham DeMoivre, whose hypothesis (1725, p. 4) "consists in supposing that the number of lives existing at any age is proportional to the number of years intercepted between the age given and the extremity of old age," i.e.

$$l_x = l_0 \left(1 - \frac{x}{\omega} \right),$$

from which we have for the force of mortality

$$\mu_x = \frac{1}{\omega - x}.$$

DeMoivre chose the upper limit $\omega = 86$ for its quite good fit to Halley's (1693) life table, and qualified his analysis by setting a lower age limit of 12. His assumption that l_x is a linear function of age was used for short age intervals by Joshua Milne (1815) to develop his q_x formula and it remains in use in that context. As an approximation to the full life table it has long been obsolete. DeMoivre (1733) was also first to develop the normal distribution, discussed below in connection with fertility.

We begin this chapter with an article by Benjamin Gompertz (1825) who used the better mortality statistics of his time to suggest the expressions

$$\mu_x = B c^x,$$
$$l_x = l_0 g^{(c^x - 1)},$$

where B, c, and g are constants; in effect a law that mortality increases exponentially with age. Gompertz was innovative in that he reasoned from a law of mortality in finding l_x, where a century earlier DeMoivre had not attempted to justify his implied mortality function. Gompertz is as much remembered for having reasoned well: his Law is a better description of the upper ages than DeMoivre's hypothesis

273

that populations collapse towards a specific age limit. From Gompertz are derived Farr's (1864) and indirectly Reed and Merrell's (1939) excellent approximations to $_nq_x$.

An improvement on Gompertz' Law was contributed by William Makeham (1860, 1867), whose 1867 article is included here as paper 31. In a careful examination of causes of death (as they were then listed), Makeham found that overall mortality levels could be better represented if a constant term were added to μ_x to account for causes of mortality not dependent on age, a possibility Gompertz also had noted. This gives

$$\mu_x = A + Bc^x,$$
$$l_x = l_0 s^x g^{(c^x - 1)}.$$

For ages beyond childhood and youth Makeham's Law combines intuitive plausibility and a close fit to observed mortality.

Other equations have been suggested that give a slightly better fit to mortality rates overall but have spurious inflection points or are otherwise not intuitive. For these the reader may consult Hugh M. Wolfenden (1942, pp. 79—85; 1954, pp. 164—167). A further contribution to the interpretation of Gompertz' and Makeham's Laws and their generalization will be found in Brillinger (1961).

The lack of a satisfactory analytic expression for mortality that includes both infancy and adulthood has driven workers in the field to the alternative approach provided by model life tables, which average observed mortality schedules by regression, usually on life expectancy at birth or at age ten where the strong impact of infant mortality is not felt. An early attempt was published by the United Nations (1955); but by far the best known is that of Ansley Coale and Paul Demeny (1966). From their examination of historical mortality patterns in countries having reliable data, Coale and Demeny were able to isolate four general patterns and to construct regressions of the forms

$$_nq_x = a + b \mathring{e}_{10},$$
$$\ln {_nq_x} = a + b \mathring{e}_{10}$$

for the separate age groups within each of the four families of tables. Their blending of the regressions and general methodology are given as paper 32. How the model tables are applied to countries with poor demographic data is discussed in paper 33, from *United Nations Manual IV* (1967) which Coale and Demeny prepared. An application of factor analysis to the problem of sorting out existing life tables and making the models will be found in Ledermann and Breas (1959).

William Brass (Brass and Coale 1968) has suggested an ingenious combination of analytic curves and model tables that provides greater flexibility in table use, which we include here as paper 34. Brass takes a selected model table as the standard, say $l_x^{(s)}$, and adjusts it to an observed set of rates by a regression of the form

$$\ln\left[(1 - l_x)/l_x\right] = \alpha + \beta \ln\left[(1 - l_x^{(s)})/l_x^{(s)}\right].$$

The equation preserves the extreme values $l_0 = 1$, $l_\omega = 0$, and allows model tables to be used in cases where observed rates are incomplete and not quite of the (usually European) pattern the models reflect. The Brass technique also has much potential value for forecasting, since in many instances the constants α and β show fairly clear trends over time (Brass 1974, pp. 546—551).

The early attention and degree of success achieved in mapping and analyzing mortality patterns contrasts sharply with the very meager efforts that were made to understand marriage and fertility. Most of the 19th century elapsed between Nicander's publication of fertility rates and their use by Milne (1815), and Richard Böckh's (1886, p. 30) introduction of the Net Reproduction Rate, defined as the integral

$$R_0 = \int_\alpha^\beta p(a)m(a)\,da\,,$$

where $p(a)$ is the probability that an individual survives from birth to age a and $m(a)$ is the expected number of offspring of the same sex born to him (her) in the interval a to $a + da$. The integration is across all fertile ages and hence gives expected progeny to an individual just born. [Richard R. Kuczynski, at one time a student of Böckh's and an assistant in the Berlin statistical office, in several of his works drew attention to the implications of the rate for population survival and growth; e.g. (Kuczynski 1928, pp. 41—42): "The pertinent question is not: is there an excess of births over deaths? but rather: are natality and mortality such that a generation which would be permanently subject to them would during its lifetime, that is until it has died out, produce sufficient children to replace that generation? If, for instance, 1,000 newly born produce in the course of their lives exactly 1,000 children, the population after the death of the older 1,000 will remain unaltered ... and if natality and mortality remain permanently the same, the population will always exactly hold its own. If more than 1,000 children are produced by a generation of 1,000 newly born, the population will increase; if less than 1,000 are produced, the population will decrease and finally die out."] Neither the net reproduction rate nor the simpler age-specific fertility rates were in use in England when Cannan carried out his population projections in 1895.

With Lotka's contributions to stable population theory good approximations to the net maternity function and related measures became immediately valuable, and several attempts to fit equations to them were made. The earliest employ Pearson distributions (Elderton and Johnson 1969, pp. 35—109; Kendall and Stuart 1969, Vol. 1, pp. 148—154), which are solutions to the differential equation

$$\frac{d\ln f(x)}{dx} = \frac{(x-a)}{b_0 + b_1 x + b_2 x^2}\,,$$

most but not all having a single mode at $x = a$ and making smooth contact with the x-axis at the extremities. They include the normal distribution, investigated by Dublin and Lotka (1925); the Pearson Type I by S.J. Pretorius (1930) and Lotka (1933); and Pearson Types I and III by S.D. Wicksell (1931). (For the normal, $b_1 = b_2 = 0$; for the Type III, $b_2 = 0$.)

The normal, Type III and an exponential introduced by Hadwiger (1940) are analyzed in detail in Keyfitz (1968, pp. 141—169). We include here as paper 35 Wicksell's remarks about the normal and Pearson curves.

None of the equations we mention has been defined to express either marriage patterns or birth interval and family size distributions, and the most recent work in the field has been directed toward providing this essential base. A large measure of success has been achieved by Ansley Coale (Coale 1971; Coale and McNeil 1972; Coale and Trussell 1974), in work directed toward the development of model nuptiality and fertility tables. The work combines four elements. Drawing on evidence from a number of populations, Coale (1971) found that the age-specific risk of marriage among those who ever marry could be closely approximated by a double exponential (Gompertz) distribution, differing between populations in origin (age at which marriages essentially begin) and the relative intensity of the marriage process once underway. For fertility by duration of marriage, Coale suggested taking as a standard that of a natural fertility population (i.e., one in which birth limitation is not intentionally practiced), reducible by an exponential decay function to represent fertility patterns imposed by family planning practice.

In later work extending Griffith Feeney's (1972) suggestion that the age-specific probability of marriage could be represented as the convolution of a normal distribution to represent age at eligibility with an exponential distribution representing delays between eligibility and marriage, Coale and McNeil found that as further exponentials are introduced the distribution takes as its limiting value a Makeham curve of the form (Coale and McNeil 1972, p. 744)

$$\bar{g}(x) = \frac{\lambda}{\Gamma(\alpha/\lambda)} \exp\{-\alpha(x-\mu) - \exp[-\lambda(x-\mu)]\},$$

where Γ is the gamma function, $\mu = a + \dfrac{\Gamma'(\alpha/\lambda)}{\lambda\,\Gamma(\alpha/\lambda)}$, and a is the mean of $\bar{g}(x)$.

The distribution has the important properties that it closely matches the probability distribution $g(x)$ generated by the Coale risk function and can be approximated to a high degree of accuracy using 1 to 3 exponentials. The distribution is thus both empirical and intuitive. The process of fitting the curve is greatly simplified by taking advantage of the similarities between different populations to establish a standard distribution. This distribution, fitted by two constants, and the distribution for fertility by duration of marriage, fitted by one constant, form the base for the Coale and Trussell (1974) model fertility tables. Paper 36 is from this article. [For another model expressing age at first marriage, also with an intuitive base, see Hernes (1972). The various models about equally reflect observed marriage patterns.]

The future growth of populations was first treated mathematically by Pierre-François Verhulst (1838), in response to Malthus' argument that populations would tend to grow exponentially until checked by resource limits. By implication, long range projections might be formed without detailed reference to current age structure, fertility or mortality, an assumption compatible with the lack of fertility information in Verhulst's time. His suggestion, paper 37, was to represent population growth by the logistic

$$r_t = b\left(1 - \frac{P_t}{k}\right),$$

$$P_t = \frac{k}{1 + e^{a - bt}},$$

which makes the intrinsic growth rate r_t a linearly decreasing function of population size P_t, k being the asymptotic population. The equation was independently applied to U.S. population growth by Raymond Pearl and Lowell Reed (1920), whose article here follows Verhulst's. Pearl and Reed's satisfaction with the logistic was not warranted when they wrote; in particular they failed to notice that the good fit of their equation reflected extraordinary contributions of technology, immigration and territorial expansion to U.S. population growth, making the correspondence of r_t and P_t rather more fortuitous than imperative. (As it turned out, the curve was 0.3% under the 1930 census but over by 3.5% at the next count. The higher figure translates as a 57% overestimate of intercensal growth. It may be compared with Whelpton's fine componant projection for 1940, 0.2% above the census, and his 6.0% and 49% underestimates for 1950 population and 1940—1950 intercensal growth respectively. In this case the better method was not a better guarantee of success.) Pearl and Reed toward the end of the paper take care in suggesting limitations of the logistic, and the reader will find there some of the reasons why they should not have used it; others were provided by Lancelot Hogben (1931, pp. 176—184) and William Feller (1940).

In another work that deserves mention, Thomas Edmonds (1852) applied Gompertz' Law to population growth by defining the rate of population increase as:

$$\frac{dP}{dt} = Bc^t,$$

with the distinction that $c < 1$ in order that the rate of growth would decline rather than increase over time. (In Verhulst, this corresponds to the damping factor $\phi(p)$.) Edmonds' equations are those later reapplied to mortality by Farr (1864), and he is almost certainly Farr's immediate source. The equations have not been remembered in population projections.

The chapter concludes with an examination by Otis Dudley Duncan (1958) of human spatial measurement and efforts that have been made to fit curves to city size distributions, using either Felix Auerbach's (1913) rank-size rule or Mario Saibante's (1928) power function. These analyses confront measurement problems more complex than we have treated elsewhere in the book.

Readers should be aware that the equations presented here are fitted to real data in different ways: DeMoivre's and those given by Duncan are solved by regression; the early fertility functions by moments; and Gompertz and Verhulst used simultaneous equations with selected data points. [Here, note that we might not follow the same approach. Curves in which constants enter non-linearly are accessible to least squares fitting by computer, using iterative methods; neither moment nor selected-point fittings are of equivalent quality.]

The method of least squares was first suggested by Carl Friedrich Gauss in personal communications and by Adrien Marie Legendre (1805) in print; fitting by moments is due principally to Pearson (1893, 1948).

30. On the Nature of the Function Expressive of the Law of Human Mortality

<small>Benjamin Gompertz (1825)</small>

From *Philosophical Transactions* 27. Excerpts are from pages 513—519.

We omit Gompertz' method of fitting, which the reader will find in Makeham, below. Also omitted are Gompertz' mortality tables, and his discussion of life expectancy and annuities under his Law. The hyperbolic logarithms in Art. 5 are the natural logs; we are not able to follow completely his integration, which is by Newton's method of fluxions.

Article 1. In continuation of Art. 2. of my paper on the valuation of life contingencies, published in the Philosophical Transactions of this learned Society, in which I observed the near agreement with a geometrical series for a short period of time, which must pervade the series which expresses the number of living at ages in arithmetical progression, proceeding by small intervals of time, whatever the law of mortality may be, provided the intervals be not greater than certain limits: I now call the reader's attention to a law observable in the tables of mortality, for equal intervals of long periods; and adopting the notation of my former paper, considering L_x to express the number of living at the age x, and using λ for the characteristic of the common logarithm; that is, denoting by $\lambda(L_x)$ the common logarithm of the number of persons living at the age of x, whatever x may be, I observe that if $\lambda(L_n)-\lambda(L_{n+m})$, $\lambda(L_{n+m})-\lambda(L_{n+2m})$ $\lambda(L_{n+2m})-\lambda(L_{n+3m})$, etc. be all the same; that is to say, if the differences of the logarithms of the living at the ages $n, n+m; n+m, n+2m; n+2m, n+3m$; etc. be constant, then will the numbers of living corresponding to those ages form a geometrical progression; this being the fundamental principle of logarithms.

Art. 2. This law of geometrical progression pervades, in an approximate degree, large portions of different tables of mortality; during which portions the number of persons living at a series of ages in arithmetical progression, will be nearly in geometrical progression; thus, if we refer to the mortality of Deparcieux, in Mr. Baily's life annuities, we shall have the logarithm of the living at the ages 15, 25, 35, 45, and 55 respectively, 2.9285; 2.88874; 2.84136; 2.79379; 2.72099, for $\lambda(L_{15})$; $\lambda(L_{25})$; $\lambda(L_{35})$; etc. and we find $\lambda(L_{25})-\lambda(L_{35})=.04738$, $\lambda(L_{35})-\lambda(L_{45})=$ .04757, and consequently these being nearly equal (and considering that for small portions of time the geometrical progression takes place very nearly) we observe that in those tables the numbers of living in each yearly increase of age are from 25 to 45 nearly, in geometrical progression. If we refer to Mr. Milne's table of Carlisle, we shall find that according to that table of mortality, the number of living at each successive year, from 92 up to 99, forms very nearly a geometrical progression, whose common ratio is $\frac{3}{4}$; thus setting out with 75 for the number of living at 92, and diminishing continually by $\frac{1}{4}$, we have to the nearest integer 75, 56, 42, 32, 24, 18, 13, 10, for the living at the respective ages 92, 93, 94, 95, 96, 97, 98, 99, which in no part differs from the table by $\frac{1}{37}$th part of the living at 92. ...

Such a law of mortality would indeed make it appear that there was no positive limit to a person's age; but it would be easy, even in the case of the hypothesis, to show that a very limited age might be assumed to which it would be extremely improbable that any one should have been known to attain.

For if the mortality were, from the age of 92, such that $\frac{1}{4}$ of the persons living at the commencement of each year were to die during that year, which I have observed is nearly the mortality given in the Carlisle tables between the ages 92 and 99,[1] it would be above one million to one that out of three millions of persons, whom history might name to have reached the age of 92, not one would have

---

[1] If from the Northampton tables we take the numbers of living at the age of 88 to be 83, and diminish continually by $\frac{1}{4}$ for the living, at each successive age, we should have at the ages 88, 89, 90, 91, 92, the number of living 83; 61.3; 45.9; 34.4; 25.8; almost the same as in the Northampton table.

attained to the age of 192, notwithstanding the value of life annuities of all ages above 92 would be of the same value. And though the limit to the possible duration of life is a subject not likely ever to be determined, even should it exist, still it appears interesting to dwell on a consequence which would follow, should the mortality of old age be as above described. For, it would follow that the non-appearance on the page of history of a single circumstance of a person having arrived at a certain limited age, would not be the least proof of a limit of the age of man; and further, that neither profane history nor modern experience could contradict the possibility of the great age of the patriarchs of the scripture. And that if any argument can be adduced to prove the necessary termination of life, it does not appear likely that the materials for such can in strict logic be gathered from the relation of history, not even should we be enabled to prove (which is extremely likely to be the state of nature) that beyond a certain period the life of man is continually becoming worse.

Art. 4. It is possible that death may be the consequence of two generally co-existing causes; the one, chance, without previous disposition to death or deterioration; the other, a deterioration, or an increased inability to withstand destruction. If, for instance, there be a number of diseases to which the young and old were equally liable, and likewise which should be equally destructive whether the patient be young or old, it is evident that the deaths among the young and old by such diseases would be exactly in proportion of the number of young to the old; provided those numbers were sufficiently great for chance to have its play; and the intensity of mortality might then be said to be constant; and were there no other diseases but such as those, life of all ages would be of equal value, and the number of living and dying from a certain number living at a given earlier age, would decrease in geometrical progression, as the age increased by equal intervals of time; but if mankind be continually gaining seeds of indisposition, or in other words, an increased liability to death (which appears not to be an unlikely supposition with respect to a great part of life, though the contrary appears to take place at certain periods) it would follow that the number of living out of a given number of persons at a given age, at equal successive increments of age, would decrease in a greater ratio than the geometrical progression, and then the chances against the knowledge of any one having arrived to certain defined terms of old age might increase in a much faster progression, notwithstanding there might still be no limit to the age of man.

Art. 5. If the average exhaustions of a man's power to avoid death were such that at the end of equal infinitely small intervals of time, he lost equal portions of his remaining power to oppose destruction which he had at the commencement of those intervals, then at the age x his power to avoid death, or the intensity of his mortality might be denoted by aq^x, a and q being constant quantities; and if L_x be the number of living at the age x, we shall have $aL_x \times q^x \cdot \dot{x}$ for the fluxion of the number of deaths $= -(L_x)'; \therefore abq^x = -\dfrac{L_x}{L_x}, \therefore abq^x = -\text{hyp.log of } b \times \text{hyp.log}$ of L_x, and putting the common logarithm of $\dfrac{1}{b} \times$ square of the hyperbolic logarithm of $10 = c$, we have $c \cdot q^x = $ common logarithm of $\dfrac{L_x}{d}$; d being a constant quantity, and therefore L_x or the number of persons living at the age of $x = d \cdot g^{q^x}$; g being put for the number whose common logarithm is c. The reader should be aware that I mean g^{q^x} to represent g raised to the power q^x and not g^q raised to the x power; which latter I should have expressed by $(g^q)^x$, and which would evidently be equal to g^{qx}. I take this opportunity to make this observation, as algebraists are sometimes not sufficiently precise in their notation of exponentials.

This equation between the number of the living, and the age, becomes deserving of attention, not in consequence of its hypothetical deduction, which in fact is congruous with many natural effects, as for instance, the exhaustions of the receiver of an air pump by strokes repeated at equal intervals of time, but it is deserving of attention, because it appears corroborated during a long portion of life by experience; as I derive the same equation from various published tables of mortality during a long period of man's life, which experience therefore proves that the hypothesis approximates to the law of mortality during the same portion of life; and in fact the hypothesis itself was derived from an analysis of the experience here alluded to.

31. On the Law of Mortality

WILLIAM M. MAKEHAM (1867)

From *Journal of the Institute of Actuaries* 13. Excerpts are from pages 335—340, 348.

We omit Makeham's opening comments on life table functions, some of the mortality tables he examined, and his appendix, in which he develops some of his formulas. Makeham's F_x is the force of mortality.

The formula for F_x, according to Mr. Gompertz's theory, is Bq^x. For this I propose to substitute $A + Bq^x$, where A is the sum of certain partial forces which we assume to be, in the aggregate, of equal amount at all ages. The quantity Bq^x may also consist of the aggregate of several forces of a similar nature. So that we may put

$$F_x = (a + a' + a'' + \ldots) + (b + b' + b'' + \ldots)q^x,$$

where $a + a' + a'' + \ldots = A$, and $b + b' + b'' + \ldots = B$.

I do not profess to be able to separate the whole category of diseases into the two classes specified—viz., diseases depending for their intensity solely upon the gradual diminution of the vital power, and those which depend upon other causes, the nature of which we do not at present understand. I apprehend that medical science is not sufficiently advanced to render such a desideratum possible of attainment at present. I propose only at present to show that there are certain diseases—and those too of a well-defined and strictly homogeneous character— which follow Mr. Gompertz's law far more closely than the aggregate mortality from all diseases taken together. I shall then have given sufficient reason for the substitution of the form $Bq^x + \phi(x)$ for the force of mortality in lieu of Bq^x: the proof that the terms included in $\phi(x)$ form, *in the aggregate*, a constant quantity, I shall leave until we come to the examination of data more satisfactory than the returns of population and the public registers of deaths.

The two following tables are taken from the supplement before referred to. They give, first, the number of annual deaths (from all causes) to 1,000,000 living; and secondly, the number of annual deaths from certain specified causes to the same number living. The causes of death, as well as the ages for which they are given, have of course been selected as the most favourable exponents of the law of geometrical progression; but is will be observed that the former embrace all the principal vital organs of the body, and the latter include the whole of the period from early manhood to the confines of extreme old age.

The column headed "total force of mortality" should form a geometrical progression if Gompertz's law were applicable thereto. That it does not, however, form such a progression, is evident by inspection; the rate of increase in the earlier terms being less than 50 per cent., and gradually increasing until it exceeds 100 per cent. A similar result is found in all the known tables when the law is applied to the *total* force of mortality, the remedy for which (in constructing mortality tables by Mr. Gompertz's formula) is usually sought in a change of the constants of the formula after certain intervals. It is this gradual but constant variation of the rate of increase *in one direction*, and the fact of its being uniformly found in *all* tables, that show unmistakeably that if the law itself be true, its application stands imperatively in need of some modification.

The modification which I have suggested, viz., there are certain partial forces of mortality (how many I do not pretend to say) which increase in intensity with the age in a constant geometrical ratio, while there are also certain other partial forces which do not so increase, may be tested by an examination of the six columns which follow that of the *total* force above referred to. The tendency to a geometrical progression is more or less apparent in all of them; the average rate

Male Life, 1851—60

Ages	Total Force of Mortality	Partial Forces of Mortality					
		Lungs	Heart	Kidneys	Stomach and Liver	Brain	Sum of five preceding Columns
25—34	9,574	772	514	174	464	638	2,562
35—44	12,481	1,524	1,002	292	890	1,180	4,888
45—54	17,956	3,092	1,898	471	1,664	1,990	9,115
55—64	30,855	6,616	4,130	937	3,032	4,097	18,812
65—74	65,332	13,416	8,714	2,453	4,837	9,831	39,251

Female Life, 1851—60

Ages	Total Force of Mortality	Partial Forces of Mortality					
		Lungs	Heart	Kidneys	Stomach and Liver	Brain	Sum of five preceding Columns
25—34	9,925	582	603	109	570	532	2,395
35—44	12,147	1,049	1,118	151	937	872	4,127
45—54	15,198	2,062	2,064	212	1,608	1,681	7,627
55—64	27,007	5,027	4,558	317	2,967	3,818	16,687
65—74	58,656	11,016	8,916	485	4,692	8,905	34,014

of increase being such that the force of mortality somewhat more than doubles itself in 10 years.

It should be observed that, in addition to the diseases of the particular organs specified, other diseases of a kindred nature are also included under each of the above five partial forces. Possibly if more detailed information were accessible, we might be able to trace the geometrical character during a still more extended period of life. This, at least, I find to be the case in reference to one particular disease, viz., bronchitis, which in the preceding tables is included in the class of "lung diseases." Now it so happens that the deaths from bronchitis alone, for a long series of years, are given in the 26th Annual Report of the Registrar-General, from which the materials for the following table are taken. The number living is supposed to be 100,000, instead of 1,000,000 as in the two preceding tables.

In the preceding examination of the results of the Registrar-General's returns of deaths, I have confined myself to the object of proving that Gompertz's law is traced much more distinctly in the deaths arising from certain specified diseases, than in the deaths arising from all causes together. If I have succeeded in this object (and I think it can scarcely be denied that I *have* succeeded), I have justified the introduction of an additional term in the formula representing the total force of mortality; but I have as yet brought forward nothing to show that such additional term is a constant in respect of the age, and varying only with the peculiar characteristics which distinguish different sets of observations from each other.

285

Ages	1848 to 1854 (7 Years)		1855 to 1857 (3 Years)		1858 to 1863 (6 Years)	
	Males	Females	Males	Females	Males	Females
15—25	8	9	9	9	9	9
25—35	17	16	21	22	22	21
35—45	42	34	55	45	59	50
45—55	107	85	133	112	151	126
55—65	259	218	333	316	379	351
65—75	589	525	801	697	876	834
75—85	1,027	906	1,463	1,325	1,614	1,479

The several observations, however, which I now proceed to examine, if they do not enable us (like the former) to test particular terms of the function referred to, yet they will nevertheless afford a very satisfactory criterion of the complete expression. Not only, therefore, do they form by themselves (on account of their unquestionable accuracy and trustworthiness) ample evidence of the truth of the supposed law of mortality, but they also supply the deficiency, above adverted to, in the preceding investigation, as regards the requisite proof of the constancy of the term representing the aggregate of the remaining partial forces of mortality.

Commencing with the very valuable observations on the "Peerage Families" (both sexes), I find, by dividing the entire period of life into intervals of 14 years—neglecting, however, the first—the following results:—

$$
\begin{aligned}
\log L_{14} &= 0.99034 \\
&\qquad\qquad -0.05068 \\
\log L_{28} &= 0.93966 \qquad\qquad\quad -0.00716 \\
&\qquad\qquad -0.05784 \\
\log L_{42} &= 0.88182 \qquad\qquad\quad -0.02559 \\
&\qquad\qquad -0.08343 \\
\log L_{56} &= 0.79839 \qquad\qquad\quad -0.11395 \\
&\qquad\qquad -0.19738 \\
\log L_{70} &= 0.60101 \qquad\qquad\quad -0.41273 \\
&\qquad\qquad -0.61011 \\
\log L_{84} &= \bar{1}.99090
\end{aligned}
$$

The tendency to a geometrical progression in the four terms of the second order of differences is sufficiently apparent. In order, however, to show this more distinctly, I have devised the following method of correcting the series $\log L_x$ so that the four terms in question shall form a perfect geometrical progression.

If the series consist of *five* terms, and consequently the second order of differences of *three*, the latter may be converted into a pure geometrical progression by substituting for the original series another of the following form, viz.,

$$\log L_0 + p, \log L_n - p, \log L_{2n} + p, \log L_{3n} - p, \log L_{4n} + p,$$

where p is derived from the equation

$$4p = \frac{(\Delta_n^2)^2 - \Delta_0^2 \times \Delta_{2n}^2}{\Delta_0^2 + 2\Delta_n^2 + \Delta_{2n}^2}.^1$$

---

[1] The differences are those of the function $\log L_x$.

This method, it is true, changes the value of the radix of the table, but I see no necessity for making a distinction between that and other terms of the series; for in comparing the terms of the *altered* with those of the *original* series, the object is to ascertain their bearing with respect to the original series *generally*, and not to any one term in particular. Secondly, by the method adopted, the first differences (which are the logarithms of the probabilities of living n years) are increased or diminished by an uniform quantity; whereas by omitting the correction in $\log L_0$, the first term of the first order of differences would be increased or diminished by one-half of the quantity introduced into the remaining terms. Lastly, the equation for p would be of the second order, instead of the simple one given above.

Again, if the series consist of *six* terms—in which case there will be *four* terms in the second order of differences—the required effect may be produced by substituting for $\log L_x$ the series

$$\log L_0 + (v-w),\ log\ L_n - (v-w),\ \log L_{2n} + v,\ \log L_{3n} - v,$$

$$\log L_{4n} + (v+w),\ \log L_{5n} - (v+w),$$

v and w being determined from the equations

$$2w = \frac{AC - B^2}{A + 2B + C} \quad \text{and} \quad 8v = \frac{A'C' - B'^2}{A' + 2B' + C'},$$

where

$$A = \Delta_0^2 + \Delta_n^2, \quad B = \Delta_n^2 + \Delta_{2n}^2, \quad C = \Delta_{2n}^2 + \Delta_{3n}^2,$$

and

$$A' = \Delta_0^3 + 4w, \quad B' = \Delta_n^3, \quad C' = \Delta_{2n}^3 - 4w.$$

Here again, by involving the corrections symmetrically, we obtain for the unknown quantities simple instead of complicated quadratic equations.

Applying these formulae to the [preceding series], we have $2w = 0.002651$, and $8v = 0.007725$. The transformed series therefore becomes—

$\log L'_{14} = 0.989980$	-0.049960		
$\log L'_{28} = 0.940020$	-0.057234	-0.007274	$\log = \bar{3}.86177$
$\log L'_{42} = 0.882786$	-0.085362	-0.028128	$\log = \bar{2}.44914$ $\quad 0.58737$
$\log L'_{56} = 0.797424$	-0.194122	-0.108760	$\log = \bar{1}.03647$ $\quad 0.58733$
$\log L'_{70} = 0.603302$	-0.614694	-0.420572	$\log = \bar{1}.62384$ $\quad 0.58737$
$\log L'_{84} = \bar{1}.988608$			

The logarithms of the third series, and their differences, show that the transformed series fulfils the required conditions.

I have now to show that this result has been attained without a greater alteration of the original series than is warranted by the probable errors of the latter. In the following table the first column contains the age, the second the natural numbers corresponding to the original series $\log L_x$, the third gives the decrement (deduced from the original data) of the year immediately following, while the fourth and fifth contain respectively the transformed series (denoted by L'_x)[2] and the amount by which it differs from the original series in the second column.

x	L_x	D_x	L'_x	$L_x - L'_x$
14	9780.0	41.4	9771.9	+ 8.1
28	8702.8	69.8	8710.0	− 7.2
42	7617.6	78.3	7634.6	−17.0
56	6286.2	150.3	6272.3	+13.9
70	3990.3	220.3	4011.5	−21.2
84	979.3	152.2	974.1	+ 5.2
		712.3		72.6

Comparing columns 3 and 5 together, term by term, we find that in one instance only (viz. at age 42) does the alteration made in the numbers living exceed one-fifth [0.2171] of the corresponding yearly decrement; while from the sums of the same columns it appears that the *average* alteration is little more than one-tenth of the *average* decrement. We may, therefore, say the limit of the variation of the two series (cols. 2 and 4) is about one-fifth of a year. ...

We have seen that the expression for the number living at any given age in a normally-constituted increasing population—in which the yearly births as well as the yearly deaths (and consequently also the excess of the former over the latter) are proportional to the existing population—is of the same form as that representing the numbers living at the given age in a stationary population, and also the numbers living in a table of mortality. But the function $L_x v^x$ is also of the same form as the latter, for L_x contains a factor s^x which combines with that introduced by the interest of money. Hence it follows that to determine the number living between given ages in a population normally constituted— whether increasing or stationary—as well as the expectation of life and the value of annuities, the summation of a function of one form only (viz., $dg^{q^x} s^x$) is required.

---

[2] Hitherto the accent has been used to distinguish the "partial" from the "total" forces of mortality, but as we have now done with this branch of the subject, no confusion will be caused by using it to denote (as it will be used henceforth) the *corrected* values of the function to which it is applied.

32. Calculation of Model Tables

ANSLEY J. COALE and PAUL DEMENY (1966)

From *Regional Model Life Tables and Stable Populations*, Chapter 2, pp. 11—14, 20, 23—26. Princeton: Princeton University Press.

We have omitted sections of the chapter on the correlation of $_nq_x$ to $_nq_y$ and $\mathring{e}_{10}$, which was high in all cases, and on the calculation of stable populations. The regression coefficients for the $_nq_x$ on $\mathring{e}_{10}$ and the separation factors k, α are also left out here, since these are less important than the methods underlying their derivation.

CALCULATION OF FOUR FAMILIES
OF MODEL LIFE TABLES

The model tables presented here are based on a tendency that we noted for the life tables based on accurate data to cluster around four different lines, representing distinct age patterns of mortality in certain geographical regions.

The four families of life tables were isolated as the result of working with a preliminary one-parameter family of model tables designed to represent the entire collection of 326 life tables. The following method was used to construct the preliminary model tables: all of the q_x values were ordered, from highest to lowest, at each age. The values were then plotted as a function of order for each age, and occasional erratic fluctuations removed by hand smoothing. Preliminary model tables were then formed by putting together mortality rates with the same rank. This system of construction has the virtue of simplicity, and of symmetrical treatment of mortality rates at all ages. That is to say, no particular rate or rates are singled out as the basis for estimating others. The usefulness of the preliminary set of tables is questionable, however, because the 326 tables are so widely scattered about the "line" these model tables formed, as we could see by examining two-dimensional scatter diagrams.

The next step was to examine deviations of individual life tables from the age pattern of mortality in the preliminary model tables. The pattern of deviations was measured by the difference between q_x in the given table and q_x in a model life table with the same general level of mortality. The "comparison" model table was selected by first noting the expectation of life at birth in the model life tables with the same $_nq_x$ as the given life table, at $x = 0, 1, 5,$ $\ldots , 75$. The median model table (the table with the median $e_0{}^\circ$) was the "comparison" table. The graph showing $_nq_x - (_nq_x)_{\text{model}}$ displays, for ready visual comprehension, the age pattern of mortality relative to the age pattern in the model tables, in a way that does not depend on the over-all level of mortality. Graphs of this sort were constructed for all 326 life tables. The following general observations emerged from an examination of the patterns:

1. Some life tables have large variations from the model patterns. Large deviations were especially frequent under age 10 and over age 60.

2. The biggest deviations were found in life tables where the quality of the underlying data is suspect: life tables for Western Europe prior to 1850; certain Eastern European life tables—for Russia in 1897 and 1926, for Bulgaria and Greece; and life tables in underdeveloped countries in Asia, Africa, and Latin America. In most of these tables, age-reporting in the censuses, and doubtless in the mortality register, is inaccurate. The completeness of death registration—especially for infants—is uncertain.

3. Where data are known to be accurate, deviations are usually moderate. Statistics that are comparatively free of age misstatement and omission are found in some European countries since 1870, in 20th-century Canada, the United States, Australia, and New Zealand. Unfortunately, these areas have a rather narrow range of cultural diversity and doubtless represent a narrow sample of age patterns of mortality. It is of special interest to note, then, that the life tables of Taiwan and the more recent Japanese life tables—both based on accurate data—conform as well as do European tables to the preliminary model tables.

Two features of the deviations found within life tables based on accurate data led to the construction of four separate families of model tables: (a) The fact that the pattern of deviations is often similar among life tables expressing the mortality of the same population at different times, and (b) the fact that several groups of geographically linked populations exhibited similar patterns of deviations.

Figure 1 shows patterns of deviation in certain Scandinavian life tables; in tables from Germany, Austria, Czechoslovakia, Poland, and North Italy; and in tables from South Italy, Spain, and Portugal. Within each group the similarity of deviations is easily visible, although from group to group the patterns differ.

On the basis of an examination of the patterns of deviations in the 326 life tables for each sex, correlation matrices were calculated for various groups of life tables—zero-order correlations among 19 variables: $\log {}_n q_x$ at various ages ($x = 0, 1, 5, 10, \ldots, 75$), and expectation of life at age 0 and age 10. We experimented with

291

nine principal sets of correlations for each sex: (1) tables before 1870; (2) tables for Russia and certain Balkan areas; (3) tables for selected Central European areas; (4) tables for Scandinavian countries; (5) tables for Spain, Portugal, and Southern Italy; (6) tables for Switzerland; (7) tables for countries with reliable data not included in (3), (4), (5), or (6); (8) tables reflecting mortality when there is an unusually high incidence of tuberculosis; and (9) modern tables based on relatively inaccurate data—primarily from Asia, Africa, and Latin America.

The nature of the life tables underlying each of the "regional" model tables is as follows:

1. Tables underlying "East" model tables. The life tables of Austria, Germany (including tables in 1878 and the 1890's for Bavaria and Prussia), Czechoslovakia, North and Central Italy, Hungary, and Poland show deviations from the preliminary model life tables characterized by high mortality rates in infancy, and increasingly high rates over age 50 (Figure 1, central panel). Switzerland's life tables show deviations very similar to this group until 1920, although the early Swiss life tables have a less conspicuous positive deviation in infancy. After 1920, the Swiss life tables show zero or negative deviations in infant mortality. Hungarian life tables exhibit substantial deviations in an age pattern indicating an extraordinary incidence of tuberculosis. Inclusion of the Hungarian life tables lowers the correlation coefficients from age 5 to 35, but has little other effect, and they were omitted from the calculation of the "East" model tables, as were the Swiss life tables. The tables in the "East" group include 13 from Germany (of which 3 are from Prussia or Bavaria), 5 from Austria, 3 from Poland, 4 from Czechoslovakia, and 6 from North or Central Italy.

2. Tables underlying "North" model tables. The life tables of Norway, Sweden until 1920, and Iceland deviate from the preliminary model tables in having low infant mortality rates, and rates that are lower than the model rates by an increasing margin at ages beyond 45 or 50. Later Swedish life tables do not have this characteristic pattern (Figure 1, left panel). In the life tables of all three "North" countries from 1890 or 1900 to 1940, there are deviations in the mortality rates from age 5 to 35 or 40 indicating the effect of an unusual incidence of tuberculosis. Model tables in-

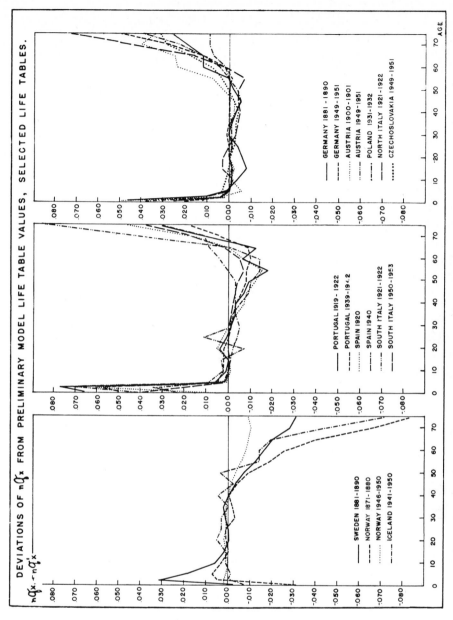

FIGURE 1. Deviations of mortality rates in three regional collections of life tables ($_nq_x$) from mortality rates in a preliminary model life table ($_nq_x'$). Selected tables for females.

293

corporating this experience would be suitable only for populations with a high endemicity of tuberculosis. Consequently, the "North" model tables are based on Swedish mortality from 1851 to 1890 (4 tables), Norwegian mortality from 1856 to 1880 and from 1946 to 1955 (4 tables), and the Iceland life table for 1941–1950.

3. *Tables underlying "South" model tables.* The life tables of Spain, Portugal, and Southern Italy have high mortality under age 5, low mortality from about age 40 to 60, and high mortality over age 65, relative to the preliminary model tables (Figure 1, right panel). Early tables—prior to 1912—for all Italy had these same characteristics. The "South" model life tables were based on 5 tables for all Italy (1876 to 1910), 8 tables for Portugal (1919 to 1958), 1 table for Sicily (1951), 3 for South Italy (1921 to 1957), and 5 for Spain (1910 to 1940).

4. *Tables underlying "West" model life tables.* The "West" model life tables were based on mortality experience recorded in populations known to have relatively good vital statistics, and not showing a persistent systematic pattern of deviations from the preliminary model tables. In other words, the tables underlying the "West" models are a residual collection after the "East," "South," and "North" tables have been removed. (The Swiss tables were also omitted, having a recent pattern resembling the "West" up to age 45 or 50, and the "East" above age 50.) The tables underlying the "West" models include 7 from Australia (1881 to 1955), 4 from Belgium (1880 to 1949), 7 from Canada (1926 to 1959), 11 from Denmark (1895 to 1955), 11 from England and Wales (1871 to 1959), 1 from Estonia (1933), 4 from Finland (1881 to 1955), 16 from France (1871 to 1959), 5 from Ireland (1925 to 1952), 3 from Israel (1949 to 1959), 6 from Japan (1949 to 1960), 1 from Latvia (1936), 1 from Luxemburg (1947), 10 from the Netherlands (1870 to 1955), 12 from New Zealand (1906 to 1959), 2 from Northern Ireland (1925 to 1959), 7 from Scotland (1891 to 1959), 5 from Sweden (1931 to 1959), 3 from Taiwan (1921 to 1959), 4 from the Union of South Africa white population (1920 to 1947), and 10 from the United States (1901 to 1958). Tables before 1870 were eliminated because most of those examined (from France, the Netherlands, and England and Wales) had irregular patterns that appeared to arise from faulty data. ...

The principal steps in the calculation of the four sets of model life tables were as follows:

1. Intercorrelation matrices for $_nq_x$ and $\log_{10}(_nq_x)$ were calculated for the "North," "South," "East," and "West" data. [$e_0{}^\circ$ and $e_{10}{}^\circ$ were left untransformed (no logarithms taken) in the second sets of intercorrelations.]

2. Least-square linear regressions of $_nq_x$ and of $\log _nq_x$ on $e_{10}{}^\circ$ were fitted for both sexes in all four "regions." The regression coefficients are given in Table XI.

3. The values of $_nq_x$ estimated from the logarithmic regression are always above those from the regression of untransformed mortality rates at the high and low extremes of observed life expectancies, and the logarithmic regression values are always lower in the middle range. In other words, the two regression lines always intersect twice within the range of observations. In constructing the model life tables, $_nq_x$ values were taken from the simple regression at all points to the left (i.e., at points with lower life expectancy) of the first intersection of the regression lines; and to the right of the second intersection, $_nq_x$ values were taken from the logarithmic regression. Between the two intersections, the mean of the $_nq_x$ values from the two regressions was used.

4. From various values of the independent variable ($e_{10}{}^\circ$), $_nq_x$'s at ages 0, 1, 5, 10, . . . , 75 were calculated. From each such set of $_nq_x$'s, $l_1, l_5, l_{10}, \ldots, l_{80}$ were computed with l_0 taken as 100,000.

5. $_nL_x$ and $e_x{}^\circ$ were estimated on the use of the following formulae:

$$_1L_0 = k_0 l_0 + (1 - k_0)l_1$$
$$_4L_1 = k_1 l_1 + (4 - k_1)l_5$$
$$_5L_x = 2.5(l_x + l_{x+5}), \; x = 5, 10, \ldots, 75$$
$$e_{80}{}^\circ = 3.725 + 0.0000625 l_{80}$$
$$T_{80} = e_{80}{}^\circ l_{80}$$
$$T_x = \sum_{x}^{75} L_x + T_{80}$$
$$e_x{}^\circ = \frac{T_x}{l_x}$$

6. Age-specific mortality rates $(_nm_x)$ were calculated from the formula $_nm_x = {_nd_x}/{_nL_x}$ where $_nd_x = l_x - l_{x+n}$

7. Five-year survival rates for projecting five-year age groups $(_5P_x)$ were calculated by the formula $_5P_x = {_5L_{x+5}}/{_5L_x}$, $x = 0$, 5, . . . , 70.

The first survival rate is the proportion surviving to the end of a five-year time interval of persons born during the interval, estimated as $_5L_0/5l_0$. The last survival rate is of persons over 75 at the beginning of an interval (and over 80 at the end), estimated as T_{80}/T_{75}.

8. Both male and female tables are calculated by regression of $_nq_x$ on $e_{10}°$. The values of $e_{10}°$ that were used as the independent variable in constructing the female tables were chosen so as to give even 2.5-year intervals of $e_0°$ from 20 to 77.5 years. The values of $e_{10}°$ for males were chosen so as to correspond with the female $e_{10}°$'s in a way that preserves the typical relation of $e_{10}°$ for males and females at each level of mortality within each family of life tables. The relationship posited was as follows:

$$(e_{10}°)_m - \overline{(e_{10}°)}_m = \frac{\sigma_m}{\sigma_f} [(e_{10}°)_f - \overline{(e_{10}°)}_f]$$

where σ_m and σ_f are the standard deviations of expectation of life at age 10 for males and females. This expression is the equation for the straight line with a slope intermediate between the regression of $(e_{10}°)_m$ on $(e_{10}°)_f$, and the inverse of the regression of $(e_{10}°)_f$ on $(e_{10}°)_m$. The correlation between $e_{10}°$ for the two sexes is more than .99 in all instances, so that the two regression lines are almost identical. ...

THE USE OF TWO REGRESSIONS IN CALCULATING $_nq_x$ IN MODEL TABLES

There are two considerations that make questionable the use of regression equations in constructing model life tables: (1) A regression of a particular functional form (linear, quadratic, logarithmic, etc.) often fits the data better over some parts of the range of observations than others, and often represents the data poorly at extremes, and may in particular provide an implausible extension or extrapolation of relationships beyond the range of observations; (2) "least squares" regression equations of any given functional form establish two relations between any pair of variables, as each is considered the independent variable, and the other dependent. In an estimation of one variable from a known value of another, the use of one variable as dependent and the other as independent is clearly appropriate and there is a logical reason for employing one regression rather than the other. But there is no logical basis for selecting one mortality rate or another as independent when trying to estimate a set of interrelationships among mortality rates at different ages. In the original United Nations model life tables, the selection of $_1q_0$ as the sole purely independent variable, and the calculation of a chain of regressions, with each mortality rate first being a dependent and then an independent variable, created a tendency (because of the well-known "regression toward the mean") to incorporate less of the observed range of variability at higher ages than at lower ages. A regression chain starting at the highest ages and ending with $_1q_0$ would have been equally plausible, and would have produced somewhat different model tables.

The use of two linear regressions (logarithmic and non-logarithmic) and the selection of $e_{10}°$ as the index of mortality (the independent variable in calculating mortality rates for every age group) result in a procedure that lacks elegance, but takes account of the considerations stated in the preceding paragraph.

The expectation of life at age 10 is an approximate general index of the level of mortality, an index that is not strongly dependent on the mortality rates for any one age group. (In contrast, $e_0°$ is strongly influenced by the value of $_1q_0$.) Correlations with $e_{10}°$ are near to unity at most ages in all "regions." The choice of $e_{10}°$ as the basis for regression estimates is not an arbitrary fixing of any one mortality rate in the resultant model life tables. All mortality rates are separately subject to "regression toward the mean." Hence if an $e_{10}°$ two standard deviations above the mean is chosen as the basis for estimating a model life table, *all* of the mortality rates will be somewhat less than two standard deviations below their mean value; but the sequence of mortality rates is not ever more compressed toward the mean as age advances. ...

The facts that led us to combine estimates from regressions of $_nq_x$ and log $_nq_x$ on $e_{10}°$ were as follows:

1. The two sets of correlation coefficients were comparable in magnitude. There were 48 instances where the non-logarithmic correlation exceeded the logarithmic by at least .003; 55 where the logarithmic coefficients were larger by this margin; and 33 instances where the difference was less than .003. (Table XII). However, there is a different relative effect on the two kinds of correlation coefficients of "scatter" in mortality rates at high levels of mortality on the one hand, and low levels of mortality on the other. ...

2. The logarithmic regression line at the highest observed expectations of life is closer to the observations, and represents a more plausible extension of the observations, than the non-logarithmic regression line. This closer fit is caused partly by the sensitivity of the logarithmic regression to relative rather than absolute deviations, and partly by the fact that declining linear exponential functions are necessarily asymptotic to zero, and never give negative values. In contrast, the non-logarithmic regression line falls *below* the observed mortality rates for the higher values of $e_{10}°$, and sometimes indicates negative mortality estimates near the upper limits of observed values of $e_{10}°$—a patent absurdity.

3. At the other end—high mortality, low expectation of life—of the range of observed values, the non-logarithmic regression line is usually closer to empirical mortality rates, and, in this range, also represents a more plausible extension of observed relations. The exponential necessarily becomes steeper at lower expectations of life, and would yield absurdly high estimated mortality rates if $e_{10}°$ were assumed to have a low value.

Through the middle range—between the two intersections of the lines—sometimes one line is closer, and sometimes the other to the observed mortality rates. To have a uniform rule, and minimize discontinuity in the sequence of model tables, we estimated mortality rates from the non-logarithmic regression up to the first intersection, used the mean of the two regression estimates between the two intersections, and used the estimates from the logarithmic

regression at expectations of life at age 10 above—to the right of—
the second intersection.

ESTIMATION OF LIFE TABLE VALUES OTHER THAN $_nq_x$

The weights (k_0 and k_1, the "separation factors") that relate $_1L_0$ to
l_0 and l_1, and $_4L_1$ to l_1 and l_5, can be shown to equal the age at
death of those members of the life table population who die under
age 1 (for k_0), and the age at death minus 1 of those who die
between ages 1 to 4 (for k_1). The values of the factor k_0 were therefore determined by examining the average age at death under 1
year in the records of the populations whose mortality experience
was the basis for the four regional model life tables. No consistent
variation in average age of infant deaths was found at infant
mortality levels above 0.100, nor any consistent differences among
the "regions," except that k_0 for "East" infants was consistently less
than in the other "regions," and k_0 for males was generally slightly
less than for females. When $_1q_0$ is very low—in the range from
0.015 to 0.100—there is a clearly apparent tendency for k_0 to vary,
because at very low levels of infant mortality infant deaths are
much more concentrated immediately after birth. This tendency
was represented by allowing k_0 to rise linearly from a typical level
at the lowest observed infant mortalities until at $_1q_0 = 0.100$ it
reaches the plateau typical of higher infant mortality.

The value of k_1 in the expression $_4L_1 = k_1l_1 + (4 - k_1)l_5$ was
determined from the estimates of l_2, l_3, and l_4 given in Table XV at
the end of Chapter 3. In fact $_4L_1$ was calculated on the assumption
that l_x can be considered linear in each single-year age interval
from 1 to 5; thus $k_1 = 0.5 + \alpha_2 + \alpha_3 + \alpha_4$, when $l_2 = \alpha_2l_1 +
(1 - \alpha_2)l_5$, etc. Each α_i was based on the relation of l_i to l_1 and l_5
observed in the life tables upon which each family of model tables
was based. As with k_0, it was noted that when $_1q_0 > 0.100$, the
values of each α_i showed no tendency to vary as a function of the
level of mortality, although at levels of mortality below $_1q_0 = 0.100$
there is a tendency for the deaths from age 1 to age 5 to be more
evenly distributed, and for l_x in this range to be more nearly linear.
Hence at low levels of mortality, each α_i was allowed to fall as
$_1q_0$ increases in a manner analogous to the rise in k_0. The values
of α_2, α_3, and α_4 are somewhat different in each region. It is reassur-

299

ing to note that when $_1q_0 > 0.100$, the α_i's observed in the life tables underlying each regional set of model tables clustered about an average typical of the set, and differing from the average value of α_i for the other "regions." In other words, each region has a distinctive pattern of mortality from 1 to 5, as well as at other ages.

The formula for expectation of life at age 80 is based on a line fitted to observations in 70 life tables (35 for males and 35 for females) from the Netherlands, Norway, Sweden, Switzerland, and Japan. These tables were selected because age reporting in the census and mortality records in these countries for persons over 80 appeared especially trustworthy.

33. Estimates of Fertility and Mortality Based on Reported Age Distributions and Reported Child Survival

UNITED NATIONS (1967)

From *United Nations Manual IV: Methods of Estimating Basic Demographic Measures from Incomplete Data*, pp. 37—39. New York: United Nations.

$C(x)$ in this extract is the proportion under age x in the observed or model stable population. The Brass method of estimating child mortality that is referred to will be found in Brass and Coale (1968, 105—120).

301

A. Estimation of Birth and Death Rates from Childhood Survival Rates and a Single Enumeration by Reverse Projection

An accurate census that records the number of persons in each five-year age interval provides the basis for reconstructing recent birth and death rates, if migration either is known accurately or is negligible in magnitude, and if survival rates by age are known for recent periods. The method of estimation is simply to reverse the customary procedures of population projection—to reconstruct by reverse survival the births that brought into being the children recorded as under age five or ten, and to reconstruct the population among which the births occurred by reverse survival of persons over five or ten in the census.

The specific steps employed in reverse projection are to divide the population under five by $_5L_0/5 \times l_0$ from a life table representing the mortality of the preceding five years to obtain an estimate of births, and to divide each five-year age group by the appropriate survival factor from the same life table to reconstruct age group by age group the population five years before. The sum of all such estimated age groups is the estimated total population five years earlier. The estimated average annual number of births can then be divided by the average of the enumerated population at the end and the estimated population at the beginning of the period to give an average birth rate during the preceding five years. The average annual rate of increase $(1/5 \log_e P_t/P_{t-5})$ can be subtracted from the birth rate to estimate the death rate. Similarly, the population 5—9 can be projected back to provide the estimated birth rate in the next earlier five years, etc. The estimated total population becomes subject to increasing uncertainty as the reverse projection procedes, however, even if mortality is somehow accurately known: the oldest age group in the past population has no current survivors, and this segment of the past population must be estimated by some assumption about the nature of the past age distribution. For earlier and earlier dates, the portion of the population estimated in this way is larger and larger.

Given an accurate census, the crucial additional element for reverse projection is an appropriate life table. The Brass method of estimating child survival provides approximate values of l_2 (during the preceding four or five years) and l_3 (during the preceding six or eight). A model life can be selected with the given l_2, and survival factors from this table employed for the reverse projection. The value of $_5L_0$ in the model table is very close to the correct one, if the data from which l_2 was estimated are accurate. Differences in age patterns of mortality do not much affect the ratio of l_2 to $_5L_0$. On the other hand survival rates above age five are estimated by assuming that the age pattern of mortality conforms to the "West" family of model life tables, and if one judges by differences among life tables based on accurate data, the actual survival rates above five may diverge from the "West" family. However, differences in mortality above age five in life tables with a given l_2 would rarely produce estimates of over-all population 2.5 years earlier differing by more than one per cent, so that the error in the estimated birth rate from this source would rarely exceed half a point (e.g., an estimate of 50.5 per thousand instead of 50).

If the questions on children ever born and surviving have been asked and recorded separately for males and females, separate estimates of child mortality can be made for each sex and (if the internal consistency of the data is acceptable) the model life table selected for each sex can be based on this separate evidence. However, the questions on children ever born are typically asked (or tabulated) only for both sexes combined. One is tempted to derive, from values of l_2 and l_3 for the two sexes together, estimates for each sex on the assumption that the relation of female to male child mortality in the given population is the same as in the population whose experience underlies the "West" model tables. The typical relation between male and female mortality in these populations is that male and female life tables tend to be at about the same level. ...

However, the evidence available on sex differences in mortality in the less developed countries does not warrant the assumption that these differences always conform to the relations found in the experience— primarily from Europe, North America, and Oceania—underlying the model life tables. It is possible to find many examples of female mortality higher than male mortality and this resort to male and female model tables at the same level may introduce a mortality differential opposite to the actual one. ...

B. Estimation of Birth and Death Rates from Child Survival Rates and a Single Enumeration by Model Stable Populations

The use of model stable populations to estimate characteristics of a population requires the identification of a stable population among the tabulated age distributions that shares some of the observed or inferred characteristics of the recorded population. In chapter I, the identifying features used to locate a model population were the intercensal rate of increase (assumed equal to the rate of growth of the stable population), and the cumulative age distribution or ogive up to some age that depends on the apparent pattern of age-misreporting. ...

Two features of estimation by model stable populations selected by l_2 and $C(x)$ are worth noting: first, if an estimate of the birth rate is based on $C(5)$, the result is essentially identical with the results obtained by reverse projection, whether or not the population is stable; and second, estimates of the birth rate obtained from l_2 and $C(x)$ are insensitive to differences in age pattern of mortality, at least the differences found in the four families of model tables.

The birth rate in the model stable population with the same l_2 and $C(5)$ as the observed population is identical to that which would be obtained by applying reverse projection to the children under five in the stable population to obtain births, and reverse projection to the whole stable population to obtain an average number of persons (the denominator of the birth rate). The number of births estimated for the actual and the stable populations is identical, so that the only source of difference between the birth rate estimated by reverse projection and that found in the model stable population is in the denominator, which in each

303

case is a number obtained by applying reverse projection, with the same life table, to populations with the same number of persons, and the same number over and under five, but possibly differing in the internal age structure above age five. This point loses relevance as the choice of a model stable population is made dependent on C(10), C(15) and ogives to higher ages, because with possibly different internal age composition in the population (e.g., under fifteen) that is implicity or explicitly projected back to birth, and with differences in the size of the reverse projected denominator becoming more pronounced as the time period of reverse projection is extended, the virtual identity of the two estimates is lost. This feature does imply, however, that interpolation in stable populations is a convenient way of determining the birth rate implied by reverse projection of children under one, under five, one to five, five to ten or under ten.

The insensitivity of birth rate estimates from l_2 and $C(x)$ to differences in age patterns of mortality is an important advantage of such estimates. The advantage lies in the fact that the true age pattern of mortality is usually not known, and if estimates based on alternative plausible age patterns are widely different, the range of uncertainty is great. This point is discussed in greater detail in chapter IV.

C. Estimation of Birth and Death Rates from Child Survival Rates and Age Distribution in a Population Enumerated Several Times ...

1. Estimation of Birth and Death Rates in a Non-Stable Population

Suppose a population is enumerated in censuses at the beginning and end of a decade, and that the second census includes data permitting the calculation of l_2 and l_3. The best estimate of the over-all life table is obtained by accepting $\mathring{e}_5$ (and the m_x and q_x values for age five and over) from the model life table chosen as best fitting census survival rates, and by taking l_5 (and $_5m_0$) from the model life table with the value of l_2 or l_3 obtained by the Brass methods. The expectation of life at birth in this hybrid model table is easily calculated. The average death rate for the decate is then calculated, by applying the m_x values in this hybrid life table to a rough mid-decade age distribution obtained by averaging the distributions at the beginning and end. The birth rate can then be estimated as equal to the death rate plus the average annual rate of increase.

2. Estimation of Birth and Death Rates in a Stable Population Enumerated More than Once

If the population enumerated in two or more censuses has closely similar age distributions in each census, the methods just described can be supplemented by the following procedures: (a) estimate the birth rate by selecting a model stable population from l_2 and $C(x)$; and (b) estimate the death rate as the birth rate less the intercensal rate of increase. A slightly more elaborate procedure may be applied when an estimated life table, an adjusted age distribution, and other detailed parameters are sought. This procedure recognizes the superiority of the Brass estimation procedures for determining childhood mortality, and of stable population techniques using $C(x)$ and r to estimate mortality above age five. The procedure entails: (a) selecting a model stable population from l_2 and $C(x)$, and accepting the proportion under five and the child death rate in this population, and (b), selecting a model stable population from $C(x)$ and r, and accepting the age specific death rates above five, and the age distribution within the span above five in this population. The over-all death rate is then estimated as the sum of the death rate under five in (a) times the proportion under five in stable population (a) plus the death rate over five in (b) times one minus the proportion under 5 in stable population (a).

34. Methods of Analysis and Estimation

WILLIAM BRASS and ANSLEY J. COALE (1968)

From William Brass, et. al.: *The Demography of Tropical Africa*, pp. 127—132. Princeton: Princeton University Press.

In this extract Brass fits logit transformations of observed and model life table survival rates to a power function that preserves the extreme values $p(0)=1$, $p(\omega)=0$, where $p(x)$ is the probability of surviving from birth to age x. The work may be clearer if his equations (9) and (10) are rewritten slightly as

$$q(a)/p(a) = A\left[q_s(a)/p_s(a)\right]^{\beta} u(a), \tag{10a}$$

$$\ln\left[q(a)/p(a)\right] = \ln A + \beta \ln\left[q_s(a)/p_s(a)\right] + \ln u(a) \tag{9a}$$

with the multiplicative disturbance term $u(a)$, whose expected value in (9a) becomes $E(\ln u) = 0$. From (9a) the estimators $\hat{A}, \hat{\beta}$ are found by linear regression. The equation to generate the new survival estimates $\hat{p}(a)$ from the fitted regression is

$$\hat{p}(a) = \left\{1 + \hat{A}\left[q_s(a)/p_s(a)\right]^{\hat{\beta}}\right\}^{-1}.$$

Loosely, A is associated with life expectancy and β with the distribution of mortality between younger and older ages, relative to the standard table.

The Brass Model Life Tables and Stable Populations

The Brass method of constructing model life tables is to subject the survivor function in a life table chosen as a "standard" to the so-called *logit* transformation, and then to consider the life tables generated by assuming that their logits are linearly related to the logit of the standard table. In this way, a two-parameter set of model life tables can be constructed.

The logit function is as follows:

$$\text{logit}\,(x) = \frac{1}{2} \log_e \frac{1-x}{x} \tag{8}$$

For x in equation (8), Brass substitutes $p(a)$ (or l_a/l_0 in life-table notation). He then selects a standard table, $p_s(a)$, and constructs his model life tables by assigning different values to a and β in

$$\text{logit}\,p(a) = \alpha + \beta\,\text{logit}\,p_s(a) \tag{9}$$

This transformation implies that:

$$\frac{q(a)}{p(a)} = A \left(\frac{q_s(a)}{p_s(a)}\right)^{\beta} \tag{10}$$

where $A = e^{2\alpha}$.

The Nature of the Linear Logit Transformation of a Standard Life Table

The logit transformation can generate a survivor function $p(a)$ that passes through arbitrarily preassigned values at any two ages a_1 and a_2, by the selection of a and β.[19] The life table thus generated has the preassigned values of $p(a)$ at the selected ages and shares the shape, in a generalized sense, of the "standard" life table.

Figure 3.4 illustrates the effect of different values of a and β on the standard life table that Brass has constructed for Africa. The survival curves include: (a) the standard ($a = 0$, $\beta = 1.00$); (b) a curve that differs from the standard by having a β of 1.20 instead of 1.00; (c) a

---

[19] Given any preassigned $p(a_1)$ and $p(a_2)$, we obtain two equations:

$$a + \beta\,\text{logit}\,p_s(a_1) = \text{logit}\,p(a_1)$$
$$a + \beta\,\text{logit}\,p_s(a_2) = \text{logit}\,p(a_2)$$

The known values of $p_s(a_1)$ and $p_s(a_2)$ together with the preassigned values of $p(a_1)$ and $p(a_2)$ provide a unique determination of a and β, and hence a complete life table.

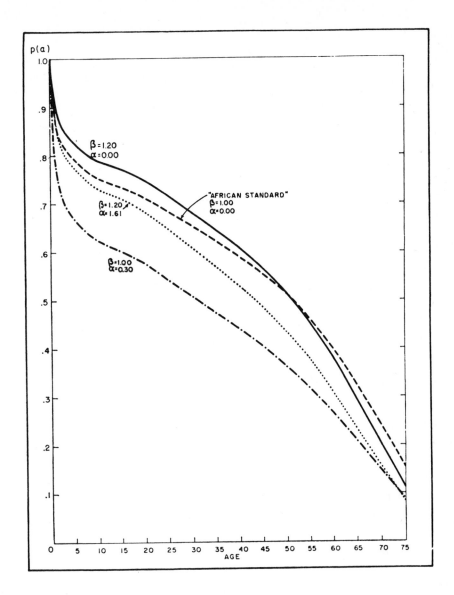

3.4. Proportion surviving to age a, $p(a)$, obtained from the
transformation $p(a) = \alpha + \beta$ logit $p_s(a)$ for various values
of α and β.

curve that has the same value of $p(2)$ as the standard, but a β of 1.20, and (d) a curve that differs from the standard by having a value of α of 0.30 instead of zero. Note that the curve with $\alpha = 0$ and $\beta = 1.20$ intersects the standard at an age where $p_s(a) = 0.50$. This is a general relation: all curves with the same α and different β's intersect at $\bar{a}$ determined by $p_s(\bar{a}) = 0.500$, because logit $0.5 = 0$. Hence variation of β produces a sub-family of survival curves, fixed at age 0 and ω, and with a node at $\bar{a}$ (about age 51 in the standard table). At ages below this node or intersection, the life table based on the larger β has higher values of $p(a)$, and at higher ages the opposite relation holds. As a result the area under these two curves (which in each instance is the expectation of life at birth) is very nearly the same.[20] However, this near constancy of $e°_0$ for different values of β is *not* a general relationship, but is the result of the selection of a standard life table with $p(51) = 0.50$. If a higher-mortality standard life table had been chosen, tables with the same a would intersect at a younger age (say 25) and higher values of β would lower the expectation of life at birth.

Note that the logit transformation permits the acceptance of a trustworthy estimate of childhood survival—e.g., an estimate of $p(2)$—and then the selection of some overall estimate of adult mortality, in the form of a choice of $p(30)$ or $p(50)$ given the accepted value of $p(2)$. The transformation of the standard table (with a and β chosen in this manner) then provides a whole model life table.

The Importance of the Standard Life Table in the Linear Logit Transformation

The logit transformation will generate life tables having any preassigned level of childhood mortality—expressed, for example, by a value of $q(2)$—and of adult mortality—expressed, for example, by $_{48}q_2$, or $p(50)/p(2)$. How well do such life tables conform to the observed age pattern of mortality in empirical populations? The Coale-Demeny families of model tables provide an interesting test. Each family of these tables expresses the particular age pattern of mortality found in the well-recorded experience of a group of populations. The groups of life tables underlying each family were assembled because of noticeable similarities in age patterns. Figure 3.5 shows the typical result of trying to express a model table within one of the families (a) as the linear logit transformation of another table in the same family, and (b) as the logit transformation of a table from a different family. Note that logit $p(a)$

---

[20] For the standard life table, $e°_0 = 43.6$, and for $\beta = 1.2$, $e°_0 = 44.1$. There is a kind of symmetry among curves with different values of β, and $a = 0$. The difference in p for each young age such that $p_s(a') > 0.5$ is matched by an equal (and opposite) difference in p for the older age a'' such that $p_s(a'') = 1 - p_s(a')$. Thus when $p_s(a) = 0.50$ occurs near the middle of the range zero to ω, life tables with $a = 0$ and different values of β have nearly the same $e°_0$.

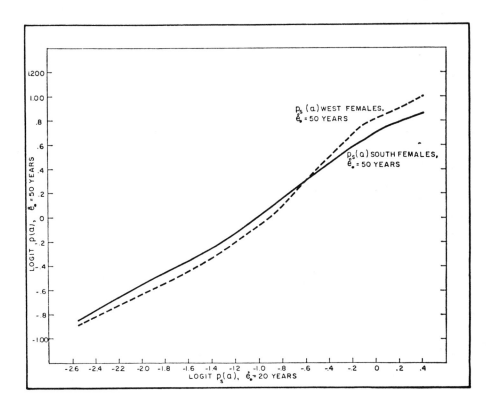

3.5. Logit $p(a)$ plotted against logit $p_s(a)$, when $p(a)$ is South model life table, females, $e°_0 = 20$ years, and $p_s(a)$ is alternatively West model and South model table, $e°_0 = 50$ years.

for the South model life table (females) with $e°_0 = 20$ has a very nearly linear relation to logit $p(a)$ for the South model table where $e°_0 = 50$ years, but that the relation to logit $p(a)$ for the *West* model table with $e°_0 = 50$ is by no means linear. Figure 3.6 shows the result of attempting to approximate the South model table with $e°_0 = 30$ years by a logit transformation of the South model table with $e°_0 = 50$. and the West model table with $e°_0 = 50$. Calculations of the same sort with the other Coale-Demeny families of model tables led to a similar result: linear logit transformations reproduce the model life tables *within* a family very closely, but between families not nearly so well. Table 3.10 shows the values of a and β needed to transform the female table with $e°_0 = 50$ into the tables with $e°_0$ of 20, 30, 40, 60, and 70 in each family. Note that in all families except South (where β is quite large at high mortality levels) the parameters needed to transform from one mortality level to another are almost identical.

311

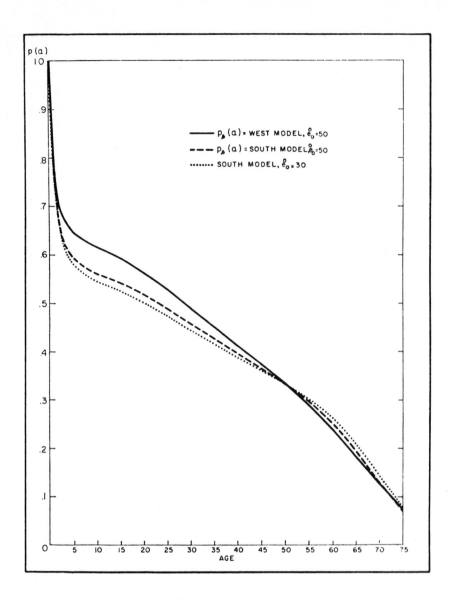

3.6. South model life table, females, $e°_\bullet = 30$ years compared with life tables obtained by linear logit transformation
of South and West model tables, $e°_\bullet = 50$ years.

Table 3.10. Values of α and β in logit $p(a) = \alpha + \beta$ logit $p_s(a)$ for North, South, East, and West model life tables (female), various levels of e^o_0, life table with $e^o_0 = 50$ serving as $p_s(a)$

		$e^o_0 = 20$	$e^c_0 = 30$	$e^o_0 = 40$	$e^o_0 = 50$	$e^o_0 = 60$	$e^o_0 = 70$
North	α	1.008	0.612	0.295	0.000	-0.316	-0.734
	β	1.296	1.154	1.064	1.000	0.950	0.900
West	α	1.036	0.632	0.307	0.000	-0.339	-0.819
	β	1.306	1.161	1.068	1.000	0.944	0.893
South	α	1.057	0.650	0.318	0.000	-0.356	-0.803
	β	1.533	1.318	1.155	1.000	0.810	0.599
East	α	1.000	0.618	0.302	0.000	0.338	0.813
	β	1.299	1.165	1.073	1.000	0.929	0.852

These examples show that the selection of a standard life table affects the detailed form of the survival curve in a life table fitted to the estimated mortality experience of a population by a linear logit transformation. The standard life table to be used in Africa, in other words, should ideally incorporate typical features (if such exist) of African age patterns of mortality.

35. Nuptiality, Fertility, and Reproductivity

S. D. WICKSELL (1931)

From Skandinavisk Aktuarietidskrift, pp. 149—157.

We omit Wicksell's discussion of Swedish nuptiality and fertility, and his distinction between net reproduction rates based on legitimate and illegitimate births. For his equation (31), a Pearson Type I distribution, the fitted constants should be

Case 1: $b_1 = 1.857$, $b_2 = 2.518$

Case 2: $b_1 = 1.689$, $b_2 = 2.593$.

Formulas for fitting the Type I distribution are given in Elderton and Johnson (1969, pp. 51—52). The fitting requires the mean and variance (μ, μ_2) and two next central moments (μ_3, μ_4) of the net maternity function, given by

$$\mu_0 = \int_{a_1}^{a_2} p(x)m(x)dx = R_0$$

$$\mu = \int_{a_1}^{a_2} xp(x)m(x)dx/\mu_0$$

$$\mu_i = \int_{a_1}^{a_2} (x-\mu)^i p(x)m(x)dx/\mu_0 \qquad i = 2,3,4\ldots$$

where $p(x)$ is the probability of survival from birth to age x and $m(x)$ the probability of giving birth in the interval $(x, x+dx)$. The lower and upper limits of the fertile age distribution are a_1 and a_2, respectively.

In Wicksell $l(x)f(x)$, or $r(x)$, replaces $p(x)m(x)$, and we have for the frequency function

$$r(x) = R_0 \frac{\Gamma(b_1+b_2)}{\Gamma(b_1)\Gamma(b_2)(a_2-a_1)^{b_1+b_2-1}} (x-a_1)^{b_1-1}(a_2-x)^{b_2-1},$$

with Γ the familiar gamma function. The equation is most easily solved from the moments by proceeding in steps, through the intermediate values c_1, c_2, c_3:

$$c_1 = -(\mu_2^3 + \mu_3^2 - \mu_2\mu_4) \left/ \left(\mu_2^3 + \frac{1}{2}\mu_3^2 - \frac{1}{3}\mu_2\mu_4 \right) \right.$$

$$c_2 = \frac{\mu_3(c_1+2)}{2\mu_2}$$

$$c_3 = |[c_2^2+4\mu_2(c_1+1)]^{1/2}|$$

$$a_1 = \mu - \frac{1}{2}(c_3-c_2)$$

$$a_2 = a_1+c_3$$

$$b_1 = \frac{1}{2}c_1[1-(c_2/c_3)]$$

$$b_2 = c_1-b_1$$

$$\text{Mode}=\mu-c_2/(c_1-2).$$

If age boundaries are preset, b_1 and b_2 can be found using only the first two moments of the distribution. Here we would have, on rearranging terms in the equations above,

$$b_1 = \frac{(\mu-a_1)^2(a_2-\mu)}{\mu_2(a_2-a_1)} - \frac{\mu-a_1}{a_2-a_1}$$

$$b_2 = b_1\frac{a_2-\mu}{\mu-a_1}$$

$$\text{Mode} = \mu + \frac{2\mu-(a_1+a_2)}{b_1+b_2-2}.$$

A fitting of the distribution using preset age limits has been made by Mitra (1967) for a large number of populations.

Wicksell's equation (21) uses the more easily fitted Type III distribution, for which $a_1=0$, $a_2=\infty$, and

$$r(x) = R_0\frac{\gamma^\beta}{\Gamma(\beta)}x^{\beta-1}e^{-\gamma x}$$

$$\gamma = \mu/\mu_2$$

$$\beta = \mu^2/\mu_2.$$

10. Putting

$$r(x) = f(x)l(x),$$

the function $r(x)$—being proportional to the distribution of the ages at which a generation of girls bear their daughters—plays a rather important part in the mathematical theory of population.

In Table 7 the function $r(x)$ is given, integrated over quinquennial age-groups (in the columns headed "total").

Table 7. Number of daughters that will be born to 1000 new-born girls if they have the maternity of Sweden in 1926, the mortality of Sweden in 1921—1925 and the nuptiality referred to as Case 1 and Case 2.

Age of mother	Case 1			Case 2		
	Number of daughters			Number of daughters		
	Legiti-mate	Illegiti-mate	Total	Legiti-mate	Illegiti-mate	Total
15—20	13	27	40	27	27	54
20—25	130	54	184	230	44	274
25—30	209	27	236	315	15	330
30—35	195	13	208	263	5	268
35—40	151	8	159	192	3	195
40—45	76	4	80	95	2	97
45—50	8	—	8	10	—	10
Sum	782	133	915	1132	96	1228

Both these series can be described with sufficiently good approximation by means of frequency functions of Pearson's type III, starting at $x=0$. If we put

(21) $$r(x) = R \cdot \frac{\gamma^{\beta}}{\Gamma(\beta)} x^{\beta-1} e^{-\gamma x}$$

and define the moments in the usual way, viz.

$$v'_0 = \int_0^\infty dx\, r(x),$$

$$v'_1 = \int_0^\infty dx\, x\, r(x),$$

$$v_2 = \int_0^\infty dx (x - v'_1)^2 r(x),$$

317

we get

$$R = v'_0; \qquad \gamma = \frac{v'_1}{v_2}; \qquad \beta = \frac{v'^2_1}{v_2}.$$

Thus, by means of the method of moments, we find

In Case 1: $R=0.915,$ $\quad \beta=19.6,$ $\quad \gamma=0.65.$

In Case 2: $R=1.228,$ $\quad \beta=19.5,$ $\quad \gamma=0.65.$

The curves are thus very nearly of the same form—differing only by the factor R —and we adopt for both curves the values of the parameters: $\beta=19.5$ and $\gamma=0.65$.
The expression (21) for $r(x)$ is a very convenient one. In the mathematical theory of population, as developed chiefly by Lotka, the integral

$$(22) \qquad R(t) = \int_0^\omega dx\, r(x) e^{-tx},$$

plays a very important part. In particular, the equation

$$(23) \qquad R(t) = 1$$

is of fundamental interest, and several rather complicated and laborious methods have been devised for solving it. But when $r(x)$ is given in the form (21)—and it seems that $r(x)$ can generally be expressed in this way—(22) and (23) assume very simple forms. We actually have (putting $\omega = \infty$)

$$(23^*) \qquad R(t) = \frac{R}{\left(1 + \dfrac{t}{\gamma}\right)^\beta},$$

when the real part of $t > -\gamma$ and the solution of (23) will simply be

$$(24) \qquad t = \gamma(R^{1/\beta} - 1).$$

We get

Case 1: $t = -0.0029$

Case 2: $t = +0.0070.$

The values of the t here calculated simply give the rates of increase per annum, corresponding to the net reproductivity rates given in Section 9. They are, what Lotka terms the "true rates of natural increase". The following table, calculated on the assumption that we have $\beta=19.5$ and $\gamma=0.65$, gives the value of the true rate of natural increase, t, for different values of the net reproductivity rate, R. The corresponding time for the population being doubled or halved is added.

318

This time is, as well known, given by the formula

$$(25) \qquad y = \left| \frac{0.693}{t} \right| = \left| \frac{1.066}{R^{1/\beta} - 1} \right|.$$

In Fig. 5 a graph of t as a function of R has been given ($\beta = 19.5$ and $\gamma = 0.65$). To save space also the negative values of t have been plotted to the positive side of the axis.

Table 8.

Net reproductivity rate R	Population growth in promille per annum $1000\,t$	Period of doubling or halving the population in years y	Net reproductivity rate R	Population growth in promille per annum $1000\,t$	Period of doubling or halving the population in years y
2.50	31.2	22	1.00	0.0	∞
2.00	23.5	29	0.95	− 1.7	410
1.80	19.9	35	0.90	− 3.5	197
1.60	15.9	44	0.85	− 5.4	128
1.40	11.3	61	0.80	− 7.3	94
1.30	8.8	79	0.70	−11.8	59
1.20	6.1	113	0.60	−16.8	41
1.15	4.7	148	0.40	−29.5	23
1.10	3.2	218	0.20	−51.5	13
1.05	1.6	426	0.10	−72.4	10

11. In a recent paper (Metron. Vol. VIII, 1930) Dublin and Lotka give tables of the function $r(x)$ for the U.S. (22 states) in the year 1925 and following years. The central ordinate is given in quinquennial age groups. Multiplying by 5 we get the following series for 1925:

Table 9. Number of daughters born at different ages to a generation of 1000 girls. (Mortality according to the U.S. life table 1925, maternity of U.S. (22 states) in 1925 and actual civil status distribution in the same registration area, 1925.)

Age	10–15	15–20	20–25	25–30	30–35	35–40	40–45	45–50	50–55	Sum
Number of daughters	0.6	100.6	309.0	291.0	218.9	149.3	53.0	5.6	0.0	1128

The moments are: $R = 1.128$; $v_1' = 28.33$; $v_2 = 45.88$, and we get $\beta = 17.49$; $\gamma = 0.6175$, and equation (24) gives

$$t = +0.00427.$$

319

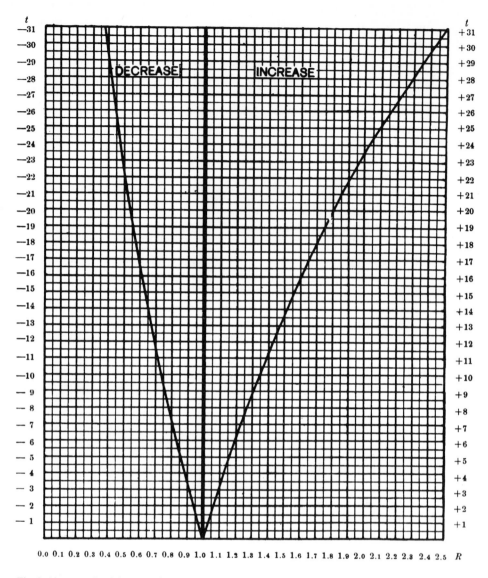

Fig. 5. Net reproductivity, R, and true rate of natural increase, t.

Dublin and Lotka give the same value of t, which value has, however, been computed by a quite different method.

In fact, Dublin and Lotka compute the value of t from the equation

(26) $$\frac{v_2}{2} t^2 - v_1' t + \log_e R = 0,$$

which is obtained by assuming $r(x)$ to be described by the normal, Laplace-Gauss'ian frequency function.

320

As the solution of equation (26) can, in our notation be written in the form

$$(27) \qquad t = \gamma \left(1 (+) \middle| \sqrt{1 - \frac{2}{\beta} \log_e R} \right),$$

it is easily seen that when $R-1$ is a small quantity (24) and (26) give the same value of the root. In fact (24) is the first term in an expansion of (27) in powers of $(R^{1/\beta} - 1)$.

The reason why our equation (24) gives the same values of t to three decimal places as obtained by Dublin and Lotka[1] is thus only that the values of R used, happen to deviate very little from unity.

Dublin and Lotka's equation for t, (26), is, as already mentioned, the consequence of assuming $r(x)$ to be given by the ordinary, normal frequency function. A glance at the asymmetry of table 9 shows, however, that the normal curve here gives rather a poor fit. Thus, when R deviates substantially from unity, equation (26) will not give a good value of t. To meet this difficulty Lotka has, in another paper, extended equation (26), so as to take account also of the higher moments. Equations of the third or fourth degree are the results. This more general method is obtained by neglecting the higher terms in the series

$$(28) \qquad R(t) = \log_e R + \sum_{k=1}^{\infty} \frac{\lambda_k}{k!} (-t)^k,$$

which series follows from the semi-invariant theory of Thiele (the λ_k being the so called semi-invariants).

Notwithstanding the fact that the rate of convergence of this series is, generally spoken, open to question, the method is at times certainly rather unwieldy from the numerical point of view. It therefore seems to me to be of practical importance that the formula (24), which is a skew curve, gives a better fit than the normal curve and that consequently formula (24) has a wider range of applicability than formula (26).

12. In concluding, the following additional remarks may be made. The curve (21) for $r(x)$ is a Pearson Type III curve with the starting point in $x=0$. This choice of the starting point causes the computations to depend only on the moments of the first and second order. It is clear, however, that a still better fit may at times be obtained by using the third order moment to determine the starting point. Putting

$$(29) \qquad r(x) = R \frac{\gamma^\beta}{\Gamma(\beta)} (x-a)^{\beta-1} e^{-\gamma(x-a)}$$

we get

$$(30) \qquad R(t) = \frac{R e^{-at}}{\left(1 + \dfrac{t}{\gamma} \right)^\beta}.$$

---

[1] This takes place not only in the example just cited but also in all the other examples given in the paper referred to.

The parameters are now obtained from the equations

$$\beta = \frac{(v_1' - a)^2}{v_2}; \qquad \gamma = \frac{v_1' - a}{v_2},$$

and

$$a = v_1' - 2\frac{v_2^2}{v_3}$$

The factor e^{-at} here makes the solution of the equation $R(t) = 1$ more difficult, and this is the reason why we have preferred to put $a = 0$, as a fairly good fit was nevertheless obtained. Cases may occur, however, where a substantial improvement is obtained by leaving the starting point a free, and using the third order moment to fix it. The equation $R(t) = 1$ may then be solved numerically by a graphical method.

A still better fit may be obtained by using Pearson's type I, i.e. by putting

$$(31) \qquad r(x) = R \cdot \frac{\Gamma(b_1 + b_2)}{\Gamma(b_1)\Gamma(b_2)(a_2 - a_1)^{b_1 + b_2 - 1}} (x - a_1)^{b_1 - 1} (a_2 - x)^{b_2 - 1}.$$

Using the ordinary moment-method of determining the parameters we find for the Swedish material (Table 7):

Case 1 $b_1 = 0.857$ $b_2 = 1.518$ $a_1 = 17.02$ $a_2 = 48.50$

Case 2 $b_1 = 0.689$ $b_2 = 1.593$ $a_1 = 17.65$ $a_2 = 48.83.$

The curves thus obtained give a nearly ideal fit in the cases in question, but of course here the solution of the equation $R(t) = 1$ is still more complicated.

The advantage of using the formula (21) of $r(x)$ with the corresponding formula (23*) of $R(t)$, instead of the expression (28) used by Lotka, is more clearly realized when we consider that in the mathematical theory of population growth not only the real root of the equation $R(t) = 1$ is required, but also a number of the complex roots.

36. Model Fertility Tables: Variations in the Age Structure of Childbearing in Human Populations

Ansley J. Coale and T. James Trussell (1974)

From *Population Index* 40. Excerpts are from pages 186—192.

The model tables and the authors' discussions of their uses and methods of fitting are omitted.

The Basis for the Model Schedules of Fertility

The basic assumption upon which the model schedules are calculated is that fertility conforms to the structure by age created by multiplying together two model subschedules: a sequence of model proportions ever married at each age and a model schedule of marital fertility. Thus, if the proportion ever married at age a in the model schedule of nuptiality is $G(a)$, and the proportion of married women at age a experiencing a live birth in the model schedule of marital fertility is $r(a)$, age-specific fertility is $f(a) = G(a) \cdot r(a)$. This construction applies exactly to a hypothetical population in which there is no fertility outside marriage, and no dissolution of marriage before the end of the childbearing span of ages. But it also duplicates quite adequately the age structure of fertility in actual populations through the selection of a $G(a)$ that differs slightly from the proportion ever married in the actual population, and of an $r(a)$ that differs slightly from the actual marital fertility schedules.

The representation $f(a) = G(a) \cdot r(a)$ makes possible the calculation of model fertility schedules from three specified parameters—two parameters required to specify a model schedule of proportions ever married, and one parameter required to specify a model schedule of marital fertility.

Age Structure of the Proportion Ever Married, G(a), Specified by Two Parameters

First-marriage frequencies, defined as the number of first marriages in a short age interval divided by the number of persons in that interval, have been shown to conform to a curve of the same shape in different populations (or more precisely in different cohorts). What differs from population to population is the age at which first marriage begins, the duration of the age span within which the majority of the marriages occur, and the proportion of the survivors in the cohort who, at advanced ages, have been married at some time. The similarity in structure of the age distribution of first marriages in different cohorts is analogous to the common shape characterizing different normal (Gaussian) distributions, which are alike only when the mean (location), standard deviation (horizontal scale), and vertical scale (number of cases, or size of population) are specified.

If the effect of differential mortality by marital status on the proportion ever married is neglected, the existence of a standard distribution of first marriage frequencies implies a standard curve describing the proportion ever married in different cohorts. The *form* of the curve is standard, but there are differences, of course, in the starting age of a tangible proportion ever married, in the pace at which the curve rises and in the ultimate proportion experiencing marriage—the proportion ever married by the age at which first marriage rates have fallen essentially to zero. If the standard proportion ever married x years after first marriages begin is $G_s(x)$, in any cohort $G(a) = C \cdot G_s((a - a_0)/k)$, where C is a factor determined by the ultimate proportion ever married, a_0 is the age at which first marriages begin, and k is the scale factor expressing the number of years of nuptiality in the given population equivalent to one year in the standard population. If k is 1.0, first marriages occur at the same pace as in the nineteenth-century Swedish population that served as the basis of the standard; if k is 0.5, or one-half, first marriages occur at twice the pace of the standard. Specifically,

324

according to the standard schedule half of the population that will ever marry has experienced first marriage ten years after the earliest age at which a consequential number of first marriages occur; if k is equal to 0.5, one-half the cohort has experienced first marriage five years after a_0.

The standard proportions ever married were published in an earlier article (Coale 1971), but for computational convenience, we have calculated $G(a)$ from a closed-form analytical expression for first marriage frequencies developed by Donald R. McNeil (Coale and McNeil 1972). This expression is:

(1) $g(a) = (0.19465/k) \exp \{(-0.174/k)(a-a_0-6.06k) - \exp [(-0.2881/k)(a-a_0-6.06k)]\}$

No analytical expression for $G(a)$ has been found, but $G(a)$ can be calculated by numerical integration of $g(a)$, since $G(a) = \int_{a_0}^{a} g(x)dx$. This representation of $G(a)$, with appropriate estimates of a_0 and k, provides an approximation of the proportion ever married in a cohort, if multiplied by a scale factor to allow for the particular proportion ultimately experiencing marriage. However, since the standard schedules of fertility that we have constructed represent only the age pattern of fertility and not the level, the proportion ultimately marrying is omitted here. Only the age of initiation and the pace of first marriages affect the structure of fertility; the proportion remaining celibate influences the level but not the age pattern of fertility.

The Age Structure of Marital Fertility, r(a), Specified By a Single Parameter.

Louis Henry found that there is a characteristic pattern of marital fertility in populations in which there is little or no voluntary control of births. He defined voluntary control as behavior affecting fertility that is modified as parity increases, and the absence of control—natural fertility—as behavior, whether affecting fertility or not, that is the same no matter how many children have been born (Henry 1961). The regularity in marital fertility that makes possible a single-parameter set of schedules is this: marital fertility either follows natural fertility (if deliberate birth control is not practiced), or departs from natural fertility in a way that increases with age according to a typical pattern. In a population in which fertility is voluntarily controlled, the ratio of marital fertility at each age, $r(a)$, to a schedule of natural fertility, $n(a)$, is given by:

(2) $$r(a)/n(a) = M \exp (m \cdot v(a))$$

The factor M is a scale factor expressing the ratio $r(a)/n(a)$ at some arbitrarily chosen age. Since we are concerned only with the age pattern of fertility (not its level), the value of M (like the value of the factor C in the model schedule of proportion ever married) is of no significance for the construction of our fertility schedules. The function $v(a)$ expresses the tendency for older women in populations practicing contraception or abortion to effect particularly large reductions of fertility below the natural level.

325

Model schedules of $r(a)$ are required at single years of age over the full range at which there is found both 1) a non-zero proportion cohabiting, and 2) non-zero marital fertility. The two functions $n(a)$ and $v(a)$, assumed to be invariant, must therefore be estimated by single years of age; the requisite family of model schedules is then obtained by assigning values to m, from zero, in which case $r(a)$ equals $n(a)$, to a maximum expressing the greatest likely departure of fertility from the age pattern of natural fertility resulting from a very high degree of voluntary control of births.

The functions $n(a)$ and $v(a)$ were derived from empirical data. There were two steps in the derivation: first, the estimation of approximate values of $n(a)$ and $v(a)$ by five-year age intervals above age 20, and second, determination of single-year values by freehand interpolation above age 20 plus extension to ages below 20 on somewhat arbitrary common sense principles.

Seven values of $n(a)$ at ages 20-24 through 45-49 were derived by calculating the arithmetical average of schedules designated by Henry as natural (Henry 1961). Henry's schedules begin at 20 because premarital conceptions have a large and irregular effect on teenage marital fertility. Ten schedules of natural fertility were averaged after discarding schedules known to be based on surveys in which age misreporting was especially prevelant and might have distorted the pattern of fertility. The effect of this selection (compared to the acceptance of all schedules listed by Henry) is minor, since the age pattern of all of those listed is broadly similar.

Seven values of $v(a)$, at ages 20-24 through 45-49, were obtained by calculations employing the marital fertility schedules listed in the United Nations Demographic Yearbook for 1965 (United Nations 1966). Again, schedules known or suspected to be distorted by age misreporting or other forms of faulty data were discarded. Each of the forty-three schedules not eliminated on this basis were provisionally accepted as embodying, each in its own degree, the typical pattern of departure from natural fertility.

For the ith schedule an individual $v_i(a)$ can be calculated by setting $m = 1.0$ in equation (2). For the ith schedule we find

$$(3) \qquad\qquad v_i(a) = \log\left[r_i(a)/(M \cdot n(a))\right]$$

M is chosen so that $v_i(a)$ is zero for the age interval 20-24. The arithmetical average of the forty-three values of $v_i(a)$ in each of the seven age intervals was then defined as $v(a)$ for each interval. The values of $n(a)$ and $v(a)$ are as follows:

	20-24	25-29	30-34	35-39	40-44	45-49
$n(a)$	0.460	0.431	0.396	0.321	0.167	0.024
$v(a)$	0.000	−0.316	−0.814	−1.048	−1.424	−1.667

The function $v(a)$ calculated in this way can be validated by substituting the tabulated values in equation (2) and seeing how well the result fits each marital fertility schedule. A value of M is chosen that equates $M \cdot n(a)$ with $r(a)$ at ages 20-24. One way of getting a visual impression of how well $v(a)$ fits a given marital fertility schedule is to calculate a separate value of m for each age interval. If equation (1) were fully valid, and $v(a)$ appropriately estimated, the separately determined values of m for age intervals 25-29 through 45-49 would all be the same. The sequence of m's calculated for the forty-three empirical marital fertility schedules is not in every instance highly uniform. However, the set of m's for most marital fertility schedules falls on a reasonably level plateau, and the difference in level of m between different populations is quite evident (see Figure 1).

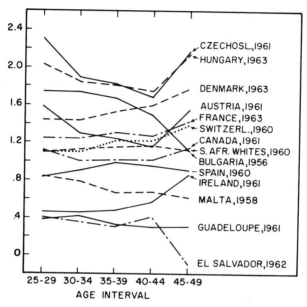

Fig. 1. Values of m, where $m = \log[r(a)/(M \cdot n(a))]/v(a)$, for selected marital fertility schedules

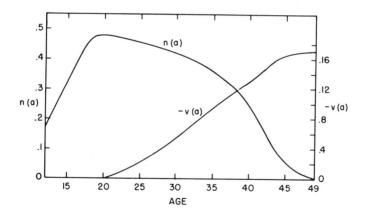

Fig. 2. Values of $n(a)$ (natural fertility), and $v(a)$ (logarithmic departure from $n(a)$)

327

Single-year values of $n(a)$ and $v(a)$ are shown in Figure 2, and tabulated as part of the FORTRAN program in Appendix A. The hand-fitted values of $n(a)$ above age 20 approximately match, in average value for each five-year interval, the values at five-year intervals listed earlier. The extension of $n(a)$ back to age 12 is based on general biomedical information that full reproductive capacity is reached a few years after menarche, and that the mean age at menarche varies from about 12 to 16 years in different populations. The particular choice of rates to represent $n(a)$ below age 20 is not of major importance because of the dominant role of $G(a)$ in determining the rise of age-specific fertility with age.

Values of $v(a)$ at single ages were chosen so that their sum over five-year age intervals matched (above age 25) the values at five-year intervals given earlier. To avoid a sharp change in the neighborhood of age 25, non-zero values were assumed to begin at age 20.

With single-year values of our three functions, we have the means of calculating a full range of fertility schedules for hypothetical populations in which there is no illegitimacy and no marital dissolution, and in which marriage begins at various initial ages and occurs over various age spans, and in which marital fertility ranges from the gradual decline with age characteristic of natural fertility to the much steeper decline characteristic of populations in which there is extensive control of fertility within marriage. The age pattern is given by equation (4):

$$(4) \qquad\qquad f(a) = G(a)n(a)e^{m \bullet v(a)}$$

where $f(a)$ is age-specific fertility, $G(a)$ is the proportion ever married (in a population where first marriage occurs according to a schedule characterized by selected values of the parameters a_0 and k), $n(a)$ is natural fertility, $v(a)$ is the characteristic pattern of departure from natural fertility, and m is the extent of that departure.

Model Schedules of Age-Specific Fertility, and Their Similarity to the Age Pattern of Fertility in Actual Populations

In actual populations, of course, births occur outside of marriage as well as within, and the proportion of the population currently married differs from the proportion ever married because of the presence of the widowed and divorced. However, the structure of fertility in an actual population may closely resemble that in a hypothetical population with no marital dissolution or extramarital fertility if the latter population has slightly different parameters of nuptiality and marital fertility from those found in the actual population. The effect of illegitimate births and of premarital conceptions on the age structure of fertility is equivalent to a schedule of first marriages that is slightly different from the observed one at early ages; the effect of illegitimate births at the older ages is equivalent to a slight increase in marital fertility at those ages. The proportion of the ever married population that is widowed and divorced rises monotonically with age, thus reducing fertility toward the end of childbearing in a way that is topographically similar to the effect of $v(a)$ on marital fertility. In other words, it is probable that the standard schedule of first-marriage frequencies, with a suitable choice of initial age and pace of occurrence of first marriages, can serve as a usable surrogate for the age of entry into sexual union (including unions that do not in fact involve marriage), and that modification of natural fertility by the proper choice of m by which to multiply $v(a)$ can serve to approximate the effect both of marital dissolution in reducing the fraction married at higher ages and of control of fertility on marital fertility. On the provisional assumption that such is the case, we have calculated a large array of model fertility schedules by single years of age; each schedule is composed of the product of an estimated proportion ever married and of marital fertility in each single-year age interval. The starting age of nuptiality was allowed to range from 12.5 to 18 years; the pace of marriage from 56 percent of the pace ($k = 1.7$) to five times the pace ($k = 0.2$) in the Swedish standard nuptiality schedule. The value of m was permitted to range from zero (natural fertility) to 3.9, on a scale in which 1.0 is the average value for forty-three schedules in the 1965 Demographic Yearbook. A total of 795 model schedules was tabulated. Each schedule has been normalized so that the sum of the fertility rates at all ages is 1.0; the schedules embody only an *age pattern* of fertility and carry no implication with respect to total fertility.

The tabulated schedules have been selected to produce mean ages at integral values from 24 to 34 years and values of standard deviation (achievable within the stipulated limits of the three underlying parameters) at intervals of half a year. The range of standard deviation is from 4.0 to 7.5, but some combinations (e.g. standard deviations of 7.0 or 7.5 with a mean age of 25) could not be attained within the limits of the three controlling parameters.

When a_0 was 15.0 or more, the single-year rates under age 20 were modified to conform to an observed feature of reliably recorded single-year schedules; non-zero fertility rates typically begin at about age 15 even when marriage begins relatively late. Positive fertility rates at ages 15 and 16 in such populations are probably the result primarily of extramarital conceptions that occur to a small number of adolescents. The requisite modification was achieved as follows: the value of fertility at exact age 20 and the cumulated value of fertility up to age 20 were accepted as initially calculated from equation (3). Values of n and R were found such that $f(a)$ equals $R(a - 15)^n$ matches the calculated value at age 20, and such that $R \int_{15}^{20}(a - 15)^n da$ matches cumulated fertility (as calculated) up to age 20.

A crucial question is whether this family of model fertility schedules provides a close fit to the fertility of actual populations. We have tried to determine how well the model schedules operate by finding a schedule (through interpolation among the printed values) that matches each of a number of recorded schedules in terms of the mean age and the standard deviation and the ratio of the average value of fertility in the interval from ages 15 to 20 to the average value from ages 20 to 25. Figure 3 shows the goodness of fit for three selected fertility schedules recorded by single years of age.

The schedules were chosen because they had the lowest and highest mean ages (Hungary, 1970, and Sweden, 1891-1900), and the lowest standard deviation (Japan, 1964) among the single-year fertility schedules that we examined; in spite of the fact that the schedules fitted are extreme, the fit in every case is quite close. In fact the absolute value of the area between the model schedule and the

329

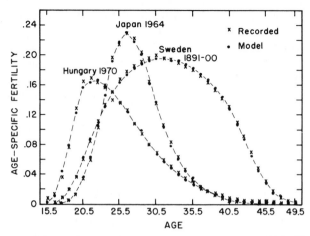

Fig. 3. Age-specific fertility rates of three populations fitted by model fertility schedules

recorded rates is in each instance less than 2.5 percent of the total area under either curve. We have fitted a number of other recorded fertility schedules with equal success.

Figure 4 shows the structure of fertility that results when entry into cohabitation is early and rapid or late and gradual, combined with natural fertility, and with fertility that is highly controlled. In interpreting Figure 4 the reader must keep in mind the normalization of each schedule so as to produce an arbitrary total fertility of 1.0. The figure illustrates the distribution of fertility by age, not differences in level of fertility associated with age patterns.

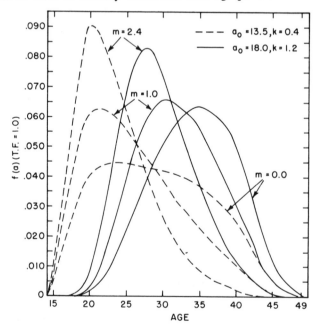

Fig. 4. Model fertility schedules, total fertility = 1.0. Combinations of early marriage with various degrees of fertility control and late marriage with various degrees of control

330

REFERENCES

Coale, Ansley J. 1971. Age pattern of marriage. *Population Studies* 25(2):193-214.

Coale, Ansley J., and McNeil, Donald R. 1972. The distribution by age of the frequency of first marriage in a female cohort. *Journal of the American Statistical Association* 67(340):743-749.

Henry, Louis. 1961. Some data on natural fertility. *Eugenics Quarterly* 8(2):81-91.

United Nations. Department of Economic and Social Affairs. 1966. *Demographic Yearbook 1965*. Sales No.: 66.XIII.1. New York.

37. A Note on the Law of Population Growth

Pierre-François Verhulst (1838)

Correspondence Mathématique et Physique Publiée par A. Quételet (Brussels) *10*, pp. 113—117. Translated by David Smith.

Of Verhulst's tables we have included only that for France, which best fits his formula and has an asymptotic population of approximately 40 million. Readers should note that errors in Verhulst's estimates of annual population change are serious and directional with respect to the official figures, and that the proportional error figures he gives are not the ones we most care about.

In a paper published some years later Verhulst (1845) gave the name *logistique* to his curve and discussed its properties in more detail. This article included his method of fitting the curve, which we reproduce in a note following his paper. He also considered more carefully than in the present work the quality of available population figures.

Malthusian

As is well known, the celebrated Malthus has established as a principle that the human population *tends* to increase in geometric progression so as to double after a certain period, for example every twenty-five years. This proposition is incontestable, if one neglects the ever increasing difficulty of procuring provisions after the population has acquired a certain degree of agglomeration; or [if he neglects] the resources that the population draws on in its growth, even while the society is still in its infancy—such as a greater division of labor, the existence of regular government and the means of defense which assure the public tranquility, etc.

In fact, *all other things equal*, if one thousand souls have become two thousand after twenty-five years, these two thousand will become four thousand after the same lapse of time.

In our old European societies, where suitable lands have long been cultivated, the work applied to improve a field already being farmed can add to its products only ever decreasing quantities. Admitting that in the first twenty-five years one could have doubled the product of the soil, in the second period one could barely succeed in perhaps making it produce a further third. The effective growth of the population thus finds a limit in the extent and fertility of the country, and the population, in consequence, tends more and more to become stationary.

It is not the same in certain cases, purely exceptional to be sure; for example, when a civilized people cultivates a fertile territory until then uninhabited, or when it exercises an ingenuity that gives large temporary benefits. A large family then becomes a source of wealth and the second generation finds it easier to establish itself than the first, because it need not like the former struggle against the obstacles that the untamed land offered to the first colonists.

To judge the speed with which the population grows in a given country, it is necessary to divide the increase of the population in each year by the population which furnished it. The relationship, being independent of the absolute size of the population, can be regarded as the measure of this speed. If it is constant the population increases in a geometric progression; if it is increasing the progression is greater than geometric, and less than geometric if it is decreasing.

One can make diverse assumptions about the resistance or the sum of the obstacles opposed to the indefinite expansion of the population. Mr. Quetelet supposes it proportional *to the square of the speed with which the population tends to increase* (Adolphe Quetelet: Essai de Physique Sociale, Vol. 1, p. 277. [Paris 1835]).

This is to liken the movement of the population to that of a body falling through a resistant medium. The results of this comparison accord, in a satisfying manner, with statistical data and with those that I obtained by my own formulas, when one supposes in the layers of the medium it passes through a density increasing indefinitely.

The growth of the population necessarily has a limit, if only in the extent of the soil indispensible for the lodging of this population. When a nation has consumed all the fruits of its fields, it can to be sure procure supplies from abroad through the exchange of its other products, and support in this way a new increase of population. But it is evident that these importations must have limits and stop a very long time before the entire of the country is converted to cities. All the formulas by which one attempts to represent the law of population must thus satisfy the condition of admitting a *maximum* which is attained only in an epoch infinitely distant. This *maximum* will be the number of the population when it becomes stationary.

I tried for a long time to determine by analysis the probable law of population; but I abandoned this type of inquiry because data from observations are too scarce for the formulas to be verified in such a way as not to leave doubt about their exactness. Still, as the steps that I followed appear to me to lead necessarily to knowledge of the true law when the data become sufficient, and as the results I have arrived at may be of interest, at least as an object of speculation, I thought I ought to accede to the invitation of Mr. Quetelet and bring them to the public.

Let p be the population, and let us represent by dp the infinitely small increase that it receives during an infinitely short time dt. If the population were growing in a geometric progression, we would have the equation $\dfrac{dp}{dt} = mp$. But as the rate of growth of the population is retarded by the augmentation of the number of inhabitants in itself, we must deduct from mp an unknown function of p; so that the formula to integrate becomes

$$\frac{dp}{dt} = mp - \phi(p).$$

The simplest hypothesis that one can make as to the form of the function ϕ is to suppose $\phi(p) = np^2$. One then finds for the integral of the equation above

$$t = \frac{1}{m}\left[\ln p - \ln(m - np)\right] + \text{constant},$$

and three observations will suffice for determining the two constant coefficients m and n and the arbitrary constant.

On solving the last equation for p, it becomes

$$p = \frac{mp' e^{mt}}{np' e^{mt} + m - np'}, \tag{1}$$

designating by p' the population that corresponds to $t=0$, and by e the base of Naperian logarithms. If one sets $t=\infty$, one sees that the corresponding value of p is $P = \dfrac{m}{n}$. That is thus *the upper limit of the population*.

Instead of supposing $\phi(p) = np^2$, one may take $\phi(p) = np^\alpha$, α being [any number] whatever, or $\phi(p) = n\ln p$. All these assumptions satisfy the observed facts equally well, but they give very different values for the upper limit of the population.

I supposed successively

$$\phi(p) = np^2, \quad \phi(p) = np^3, \quad \phi(p) = np^4, \quad \phi(p) = n\ln p;$$

335

and the differences between the calculated populations and the numbers observation furnished were appreciably the same.

When the population increases by a progression greater than geometric, the term $-\phi(p)$ becomes $+\phi(p)$; the differential equation then integrates as in the preceding cases, but one conceives that there can no longer be a *maximum* population.

I calculated the tables which follow by formula (1). The figures for France, Belgium, and the county of Essex were drawn from official documents. Those which relate to Russia are found in the work by Dr. Sadler, Law of Population [London 1830], and I cannot guarantee their authenticity, being ignorant by what manner they have been developed. I could have extended the tables for France and Belgium as far as 1837, by means of the *Annuaires* published in these

Table 1. Progress of the Population of France from 1817 to 1831, According to the *Annuaire* for 1834.

Year	According to the registrar	According to the formula	Proportional error	Logarithm of the calculated population
1817	29,981,336 195,902	29,981,336 208,281	0.0000	7.4768490
1818	30,177,238 161,948	30,189,500 204,500	+0.0004	7.4798565
1819	30,339,186 199,863	30,394,000 200,500	+0.0018	7.4827875
1820	30,539,049 188,227	30,594,500 197,300	+0.0018	7.4856461
1821	30,727,276 212,144	30,791,800 192,700	+0.0021	7.4884310
1822	30,939,420 198,634	30,984,500 189,500	+0.0014	7.4911453
1823	31,138,054 221,286	31,174,000 185,223	+0.0012	7.4937907
1824	31,359,340 220,546	31,359,340 182,777	0.0000	7.4963719
1825	31,579,886 175,974	31,542,000 178,000	−0.0012	7.4988859
1826	31,755,860 157,533	31,720,000 175,000	−0.0011	7.5013366
1827	31,913,393 189,071	31,895,000 172,000	−0.0005	7.5037257
1828	32,102,464 139,402	32,067,000 168,000	−0.0011	7.5060547
1829	32,241,866 161,074	32,235,000 164,500	−0.0002	7.5083251
1830	32,402,940 157,994	32,399,500 161,434	0.0000	7.5105385
1831	32,560,934*	32,560,934	0.0000	7.5126965

* Census figure (January 1)

two countries since 1833, and in this way verify my formula; but my pursuits have not left me the leisure. My work was terminated in 1833, and I have not touched it again since.

I will point out in passing that the table which relates to France appears to announce that the formula is especially exact, since the observations [for France] bear on the largest numbers and have been made with greater care. For the rest, the future alone can unveil for us the true mode of action of the retarding force which we have represented by $\phi(p)$.

Editors' Note: Fitting the Logistic

The form in which Verhulst presents the logistic makes its solution somewhat cumbersome. In a more standard form the equation becomes

$$r_t = b\left(1 - \frac{P_t}{k}\right)$$

$$P_t = \frac{k}{1 + \exp[a - bt]}.$$

Given the three population estimates $P_{t_0}, P_{t_0 + \delta}, P_{t_0 + 2\delta}$ (which we may index P_1, P_2, P_3), the constants a, b and asymptotic population k are found by the formulas (from Verhulst 1845, pp. 12—13; Croxton, Cowden and Klein 1967, pp. 274—275)

$$k = \frac{2 P_1 P_2 P_3 - P_2^2(P_1 + P_3)}{P_1 P_3 - P_2^2}$$

$$b = (1/\delta)\ln[P_2(k - P_1)/P_1(k - P_2)]$$

$$a = \ln[(k - P_1)/P_1] + bt_0^*$$

where t_0^* is a computational origin. The terms t_0^*, δ vanish under the linear transformation $t \to t^*$, defined by $t^* = (t - t_0^*)/\delta$.

To reconcile this form with Verhulst's equation (1) we write, using Verhulst's notation,

$$P_t = \frac{mp' e^{mt}}{np' e^{mt} + m - np'} = \frac{m/n}{1 + \left[\left(\dfrac{m}{n} - p'\right)/p'\right] e^{-mt}}$$

$$k = m/n$$

$$b = m$$

$$a = \ln\left[\left(\frac{m}{n} - p'\right)/p'\right].$$

Pearl and Reed, whose paper follows, also use a non-standard form for the logistic. Their equation and the conversion to standard form are

$$y_t = \frac{B}{\exp[-A(t - t_0^*)] + C}$$

$$= \frac{B/C}{1 + \exp[(A t_0^* - \ln C) - A t]}$$

$$k = B/C$$

$$b = A$$

$$a = A t_0^* - \ln C.$$

The inflection point, given by Pearl and Reed as (α, y_a) has as its coordinates $(a/b, k/2)$ in the standard form. An α not the same is given in their equation xvi.

Further treatment of the logistic, with its implied birth and death rates, will be found in Lotka (1939, pp. 48—62). Lotka centers the curve about its inflection point for convenience in developing its theoretical properties, by setting

$$t_0^* = a/b$$

$$P_{t^*} = P_{t-t_0^*} = \frac{k}{1+e^{-bt^*}} .$$

38. On the Rate of Growth of the Population of the United States since 1790 and its Mathematical Representation

RAYMOND PEARL and LOWELL J. REED (1920)

From *Proceedings of the National Academy of Science* 6. Excerpts are from pages 280—288.

Pearl and Reed begin their paper with a discussion of exponential and parabolic growth formulas for past U.S. population growth, which we have omitted.

It is quite clear on *a priori* grounds, as was first pointed out by Malthus in non-mathematical terms, that in any restricted area, such as the United States, a time must eventually come when population will press so closely upon subsistence that its rate of increase per unit of time must be reduced to the vanishing point. In other words, a population curve may start, with a convex face to the base, but presently it must develop a point of inflection, and from that point on present a concave face to the x axis, and finally become asymptotic, the asymptote representing the maximum number of people which can be supported on the given fixed area. ...

It would be the height of presumption to attempt to predict *accurately* the population a thousand years hence. But any real law of population growth ought to give some general and approximate indication of the number of people who would be living at that time within the present area of the United States, provided no cataclysmic alteration of circumstances has in the meantime intervened.

It has seemed worth while to attempt to develop such a law, first by formulating a hypothesis which rigorously meets the logical requirements, and then by seeing whether in fact the hypothesis fits the known facts. The general biological hypothesis which we shall here test embodies as an essential feature the idea that the rate of population increase in a limited area at any instant of time is proportional (*a*) to the magnitude of the population existing at that instant (amount of increase already attained) and (*b*) to the still unutilized potentialities of population support existing in the limited area.

The following conditions should be fulfilled by any equation which is to describe adequately the growth of population in an area of fixed limits.
1. Asymptotic to a line $y=k$ when $x=+\infty$.
2. Asymptotic to a line $y=0$ when $x=-\infty$.
3. A point of inflection at some point $x=\alpha$ and $y=\beta$.
4. Concave upwards to left of $x=\alpha$ and concave downward to right of $x=\alpha$.
5. No horizontal slope except at $x=\pm\infty$.
6. Values of y varying continuously from 0 to k as x varies from $-\infty$ to $+\infty$.
   In these expressions y denotes population, and x denotes time.
   An equation which fulfils these requirements is

$$y = \frac{be^{ax}}{1+ce^{ax}} \qquad\qquad\text{(ix)}$$

when a, b and c have positive values.
   In this equation the following relations hold:

$$x=+\infty \qquad y = \frac{b}{c} \qquad\qquad\text{(x)}$$

$$x=-\infty \qquad y = 0 \qquad\qquad\text{(xi)}$$

Relations (x) and (xi) define the asymptotes.

The point of inflection is given by $1 - ce^{ax} = 0$, or

$$x = -\frac{1}{a}\log c \qquad y = \frac{b}{2c} \tag{xii}$$

The slope at the point of inflection is $\frac{ab}{4c}$. ...

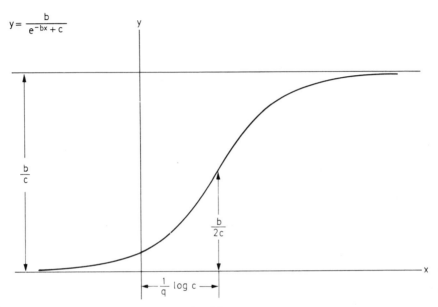

$$y = \frac{b}{e^{-bx} + c}$$

Fig. 2. General form of curve given by equation (ix).

The general form of the curve is shown in figure 2.

The question now is how well does (ix) represent the known historical facts as to the growth in population of the United States, and to what legitimate deductions as to the future course of population in this country does it lead?

It is obvious that equation (ix) as it stands cannot be fitted to observational data by the method of least squares. It is possible to write momental equations and fit by the method of moments, but at this time we do not care to develop that method because, as will presently appear, we do not regard equation (ix) as the final development of this type of equation for representing population, and we have no desire to encumber the literature with a mathematical discussion which we expect later to discard.

For present purposes it will be sufficient to fit (ix) to the observations by passing it through three points. Given three equally spaced ordinates, y_1, y_2 and y_3, the necessary equations are:

$$\frac{b}{c} = \frac{2y_1 y_2 y_3 - y_2^2(y_1 + y_3)}{y_1 y_3 - y_2^2} \tag{xiv}$$

343

$$a = \log_{10} \frac{y_2 \left(\dfrac{b}{c} - y_1 \right)}{y_1 \left(\dfrac{b}{c} - y_2 \right)} \div h \log_{10} e \tag{xv}$$

where h is the abscissal distance in years between y_1 and y_2, or y_2 and y_3.

$$c = \frac{1}{y_2 - y_1} \left(\frac{y_1}{e^{a\alpha}} - \frac{y_2}{e^{a(\alpha+h)}} \right) \tag{xvi}$$

where α is the abscissal distance in years from the origin to y_1.

Putting x_1 at 1790, x_2 at 1850, and x_3 at 1910, and taking origin at 1780 we have

$$y_1 = 3,929^9$$
$$a = 10$$
$$y_2 = 23,192$$
$$h = 60$$
$$y_3 = 91,972$$

and taking (ix) in the form

$$y = \frac{b}{e^{-ax} + c}, \tag{xvii}$$

we find these values for the constants:

$$y = \frac{2,930.3009}{e^{-0.0313395x} + 0.014854} \tag{xviii}$$

The closeness of fit of this curve is shown graphically in figure 3.

Though empirically arrived at this is a fairly good fit of theory to observations. ...

The significance of the result lies in this consideration. A curve which on *a priori* grounds meets the conditions which must be satisfied by a true law of population growth, actually describes with a substantial degree of accuracy what is now known of the population history of this country.

Let us examine some further consequences which flow from equation (xviii). The first question which interests one is this: when did or will the population curve of this country pass the point of inflection, and exhibit a progressively diminishing instead of increasing rate of growth? From (xii) it is easily determined that this point occurred about *April 1, 1914*, on the assumption that the numerical values of (xviii) reliably represent the law of population growth in this country. In other words, so far as we may rely upon present numerical values, the United

---

[9] Omitting 000 here and in the subsequent calculations till the end.

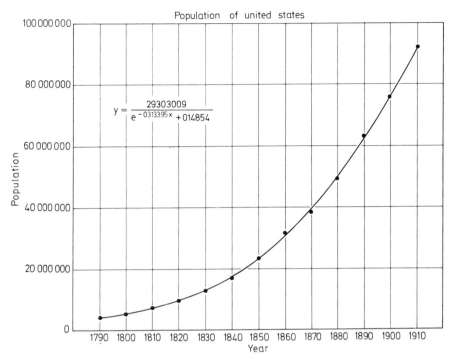

Fig. 3. Showing result of fitting equation (xviii) to population data.

States has already passed its period of most rapid population growth, unless there comes into play some factor not now known and which has never operated during the past history of the country to make the rate of growth more rapid. This latter contingency is improbable. While prophecy is a dangerous pastime, we believe, from the fragmentary results already announced, that the 1920 census will confirm the result indicated by our curve, that the period of most rapid population growth was passed somewhere in the last decade. The population at the point of inflection works out to have been 98,637,000, which was in fact about the population of the country in 1914.

The upper asymptote given by (xviii) has the value 197,274,000 roughly. This means that according to equation (xviii) the maximum population which continental United States, as now areally limited, will ever have will be roughly twice the present population. We fear that some will condemn at once the whole theory because the magnitude of this number is not sufficiently imposing. It is so easy, and most writers on population have been so prone, to extrapolate population by geometric series, or by a parabola or some such purely empirical curve, and arrive at stupendous figures, that calm consideration of real probabilities is most difficult to obtain. While, as will appear from the next section of this paper, we have no desire to defend the numerical results of this section, and indeed ourselves regard them only as a rough first approximation, it remains a fact that if anyone will soberly think of every city, every village, every town in this country having its present population multiplied by 2, and will further think

345

of twice as many persons on the land in agricultural pursuits, he will be bound, we think, to conclude that the country would be fairly densely populated. It would have about 66 persons per square mile of land area.

It will at once be pointed out that many European countries have a much greater density of population than 66 persons to the square mile, as for example Belgium with 673, Netherlands with 499, etc. But it must not be forgotten that these countries are far from self-supporting in respect of physical means of subsistence. They are economically self-supporting, which is a very different thing, because by their industrial development at home and in their colonies they produce money enough to buy physical means of subsistence from less densely populated portions of the world. We can, of course, do the same thing, provided that by the time our population gets so dense as to make it necessary there still remain portions of the globe where food, clothing material, and fuel are produced in excess of the needs of their home population. But in this, and in any other scientific discussion of population, it is necessary to limit sharply the *area* one is to talk about. This paper deals with population, and by direct implication the production of physical means of subsistence, within the present area of continental United States.

Now 197,000,000 people will require, on the basis of our present food habits,[10] about 260,000,000,000,000 calories per annum. The United States, during the seven years 1911—1918, produced as an annual average, in the form of human food, *both primary and secondary* (i.e., broadly vegetable and animal), only 137,163,606,000,000 calories per year.[11] So that unless our food habits radically change, and a man is able to do with less than 3000 to 3500 calories per day, or unless our agricultural production radically increases,[12] it will be necessary when our modest figure for the asymptotic population is reached, to import nearly or quite one-half of the calories necessary for that population. It seems improbable that the population will go on increasing at any very rapid rate after such a condition is reached. And is it at all reasonable to suppose that at such time, with all the competition for means of subsistence which the already densely populated countries of Europe will be putting up, there can be found any portion of the globe producing food in excess of its own needs to an extent to make it possible for us to find the calories we shall need to import?

Altogether, we believe it will be the part of wisdom for anyone disposed to criticise our asymptotic value of a hundred and ninety-seven and a quarter millions because it is thought too small, to look further into all the relevant facts.

III

With the above numerical results in hand it is desirable to discuss a little further the general theory of population growth set forth in the preceding section. At the outstart let it be said that we are convinced that equation (ix) represents

---

[10] Cf. Pearl, R., *The Nation's Food*, Philadelphia (W. B. Saunders Company), 1920 (247).

[11] Pearl, R., *loc. cit.*, p. 76.

[12] As a matter of fact East, in his able presidential address on "Population," before the American Society of Naturalists has shown that the United States has already entered upon the era of diminishing returns in agriculture in this country.

no more than a first approximation to a true law of population growth. There are several characteristics of this curve which are too rigid and inelastic to meet the requirements of such a law. In (ix) the point of inflection must of necessity lie exactly half-way between the two asymptotes. Furthermore the half of the curve lying to the right of the point of inflection is an exact reversal of the half lying to the left of that point. This implies that the forces which during the latter part of the population history of an area act to inhibit the rate of population growth are equal in magnitude, and exactly similarly distributed in time, to the forces which in the first half of the history operate to accelerate growth. We do not believe that such rigid and inelastic postulates as these are, in fact, realized in population growth. ...

We attach no particular significance to the *numerical* results of the preceding section. They obviously can give only the roughest approximation to probable future values of the population of the United States. Our only purpose in presenting them at all at this time is to demonstrate that the hypothesis here advanced as to the law of population growth, even when fitted by a rough and inadequate method, so closely describes the known facts regarding the past history of that growth, as to make it potentially profitable to continue the mathematical development and refinement of this hypothesis further. There is much that appeals to the reason in the hypothesis that growth of population is fundamentally a phenomenon like autocatalysis. In a new and thinly populated country the population already existing there, being impressed with the apparently boundless opportunities, tends to reproduce freely, to urge friends to come from older countries, and by the example of their well-being, actual or potential, to induce strangers to immigrate. As the population becomes more dense and passes into a phase where the still unutilized potentialties of subsistence, measured in terms of population, are measurably smaller than those which have already been utilized, all of these forces tending to the increase of population will become reduced.

39. The Measurement of Population Distribution

Otis Dudley Duncan (1958)

From *Population Studies* 11. Excerpts are from pages 27—37, 40—44.

We omit Duncan's discussion of residential classifications.

The analysis and explanation of patterns of population distribution are problems for the demographer, geographer, human ecologist, and location economist. None of these specialists, however, has taken responsibility for working out a comprehensive method for dealing with the subject. Demographers, in particular, have given it little systematic attention, although the literature is rich in elementary descriptive materials on population distribution in various regions. This paper attempts a summary of the major techniques of describing and measuring population distribution, indicating some unresolved problems of method that might well be the focus of further research.

The following is a tentative classification of measures of population distribution; it does not purport to be exhaustive and there is evident overlapping of some of the categories.

A. Spatial measures
      (1) Number and density of inhabitants by geographic sub-divisions
      (2) Measures of concentration
      (3) Measures of spacing
      (4) Centrographic measures
      (5) Population potential

B. Categorical measures
      (1) Rural-urban and metropolitan-non-metropolitan classification
      (2) Community size distribution
      (3) Concentration by proximity to centres or to designated sites

Within the compass of this paper it is possible to give only brief illustrations under these headings.

Numbers and Density by Geographic Sub-divisions

The basic information employed in most studies of population distribution is the census enumeration of population by geographic sub-divisions of a country or other territorial unit. It will appear below that summary measures of distribution may depend heavily on the areal units by which population enumerations are tabulated. Such units—geographic sub-divisions of the total territory over which distribution is described—are of three main types : (1) political units, (2) units consisting of combinations of political units, but not themselves political entities, and (3) units specially delineated for statistical purposes not necessarily conforming to political boundaries. International studies of distribution typically employ countries and combinations of countries as units, but there is interest as well in the more detailed analysis that combines intra-country with international comparisons.[1]

Geographers have developed several methods of portraying numbers cartographically such as the dot map or maps employing special symbols to designate places of specified sizes. Each of these, while appropriate for particular purposes, encounters difficulties when trying to depict with equal fidelity the distribution

---

[1] Glenn T. Trewartha, " A Case for Population Geography ", *Annals of the Association of American Geographers*, xliii (June, 1953), pp. 71-97 ; Abbott Payson Usher, " The History of Population and Settlement in Eurasia ", *Geographical Review*, xx (January, 1930), pp. 110-132, reprinted in *Demographic Analysis*, edited by Joseph J. Spengler and Otis Dudley Duncan (Glencoe, Ill. : The Free Press, 1956) pp. 3-25.

in regions of dispersed settlement and the large nucleations of population.[1] The basic difficulty with any cartographic method is, of course, that only relatively imprecise conclusions can be demonstrated with a map.

A step beyond the mere listing or cartographic portrayal of numbers by geographic sub-divisions is the computation of the ratio of population to area, i.e., population density. In view of the wide use and frequent calculation of density figures, it is surprising that the demographic literature contains little fundamental methodological discussion of the concept. A well-rounded treatment of the subject would require a separate paper. Here it is perhaps sufficient to indicate that there are serious unresolved problems that arise in the use of density statistics or maps to study population distribution. First, the results obtained will depend, to a significant degree, on the system of areal units for which densities are calculated, inasmuch as there is no way to assign a unique meaning to the notion of density in the vicinity of a given point. Second, in constructing dasymetric or isopleth density maps there is difficulty in deciding on a suitable set of density intervals. Each of the proposed solutions to this problem—whether it leans toward a " mathematical " or a " functional " criterion—is unsatisfactory in one way or another. Third, any attempt to refine density figures by basing them on " net " rather than " gross " area encounters a considerable indeterminacy in the notion of " net-ness ". Most presentations of " net " densities are neither clear nor convincing.

It would appear that there is need for cooperation between demographers and statistical geographers in developing techniques for studying density. Whereas the geographers have regarded the study of density as, primarily, a problem in cartographic presentation,[2] demographers should be in a position to clarify the analytical purposes and limitations of density measurements.

Measures of Concentration

Ordinarily, the major interest in studies of population density has to do with the variation of density over a territory rather than with just the overall or average density of the entire territory. Formally, this problem is identical with that of studying the " unevenness " or *concentration* of population. " Concentration ", it may be noted, is a term like so many in the scientific language having a similar Latin form, that has two specific meanings : (1) as referring to a state or degree of unevenness of population distribution at a given point in time, (2) as referring to the process of increase over time in the degree of unevenness (the reverse

[1] John W. Alexander, " An Isarithmic-Dot Population Map ", *Economic Geography*, XIX (October, 1943), pp. 431-432.
[2] Alexander, " An Isarithmic-Dot Population Map ", *op. cit.* : John W. Alexander and George A. Zahorchak, " Population-Density Maps of the United States : Techniques and Patterns ", *Geographical Review*, XXXIII (July, 1943), pp. 457-466 ; James A. Barnes and Arthur H. Robinson, " A New Method for the Representation of Dispersed Rural Population ", *Geographical Review*, XXX (January, 1940), pp. 134-137 ; Preston E. James, " The Geographic Study of Population", Chapter 4 in *American Geography : Inventory and Prospect*, edited by Preston E. James and Clarence F. Jones (Syracuse : Syracuse University Press, 1954) ; Eugene Mather, " A Linear-Distance Map of Farm Population in the United States ", *Annals of the Association of American Geographers*, XXXIV (September, 1944), pp. 173-180 ; John K. Wright, " A Method of Mapping Densities of Population, with Cape Cod as an Example ", *Geographical Review*, XXVI (January, 1936), pp. 103-110. See also, Philip M. Hauser, Otis Dudley Duncan, and Beverly Davis Duncan, *Methods of Urban Analysis : A Summary Report*, Research Report AFPTRC-TN-56-1 (San Antonio : Air Force Personnel & Training Research Centre, 1956), pp. 67-69.

change, decrease of unevenness, is called " deconcentration "). Abstracting from the spatial pattern of population distribution, any measure of concentration (in the first meaning of the term) seeks simply to make operational the notion of the " degree of unevenness ".

A device for graphic presentation and two index numbers associated therewith have been most widely used for measuring concentration (although many other measures are conceivable, some of which have been seriously proposed). These are (1) the Lorenz curve and Gini's " concentration ratio ", both originally suggested for measuring inequality of income or wealth[1] and adapted to the measurement of population concentration; and (2) what will be called here simply the " index of concentration ", or more generally the " index of dissimilarity ", $\triangle$. Figure 1A illustrates the principle on which the Lorenz curve is constructed. Area units are arrayed in order of decreasing density. Then,

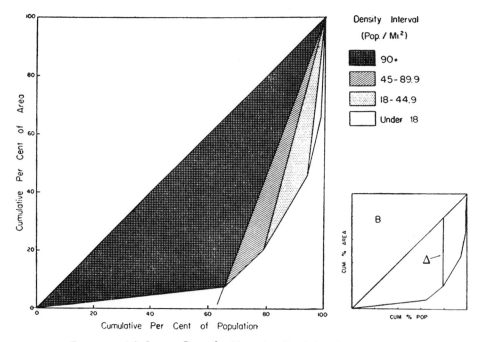

FIGURE 1. (*A*) Lorenz Curve for Measuring Population Concentration in the United States, in Relation to Density Intervals (County Basis): 1950. (*B*) Index of Concentration, $\triangle$, in Relation to Lorenz Curve.

treating these units separately or grouping them into intervals, one computes the cumulative percentage of population and of area with the addition of each unit (or interval of units), and plots the several cumulated percentages of area (Y_i) against the corresponding cumulated percentages of population (X_i). Such a

[1] Dwight B. Yntema, " Measures of the Inequality in the Personal Distribution of Wealth or Income ", *Journal of the American Statistical Association*, xxviii (December, 1933), pp. 423–433 ; Mary Jean Bowman, " A Graphical Analysis of Personal Income Distribution in the United States ", *American Economic Review*, xxxv (September, 1945), pp. 607–628.

curve would follow the diagonal throughout, in the case where all units had equal densities (even distribution), and would coincide with the X-axis if all population in the territory were concentrated at one mathematical point. Variation between these hypothetical extremes of complete evenness (no concentration) and complete concentration is indicated by the degree to which the curve departs from the diagonal.

The concentration ratio, CR, is given by the formula :

$$10,000 \ CR = \sum_{i=1}^{k} X_{i-1}Y_i - \sum_{i=1}^{k} X_i Y_{i-1},$$

where X and Y are the respective cumulative percentages, and k is the number of areal units (or intervals of units, if they have been grouped). Geometrically, this formula expresses the area on the graph contained between the curve and the diagonal as a proportion of the entire area below the diagonal. The Lorenz curve and concentration ratio have been studied, as measures of population distribution, most intensively by Wright, who, however, considers only the case in which all units are of equal area size, an unnecessary restriction and one which virtually precludes application of these measures to actual data.[1] Wright discusses also some index numbers, other than CR, that depend on the shape of the Lorenz curve as well as the degree to which it deviates from the diagonal.

The index of concentration, $\triangle$, algebraically is simply the maximum of the set of k values of $(X_i - Y_i)$. Geometrically, it is the maximum vertical distance from the diagonal to the curve, as illustrated in Figure 1B. ...

From the geometric relationships of CR and $\triangle$ to the Lorenz curve of population concentration, it can be deduced that

$$\triangle \le CR \le 2\triangle - \triangle^2.$$

Thus if $\triangle = 0 \cdot 5$ (50%) then CR is at least $0 \cdot 5$ and can be no greater than $0 \cdot 75$. Often CR approximates the average of these extreme values. For example, in Fig. 1, with $\triangle = 0 \cdot 587$, $2\triangle - \triangle^2 = 0 \cdot 830$, and the average is $0 \cdot 708$, as compared with the actual value, $CR = 0 \cdot 730$. For comparisons of situations of widely varying concentration, either index serves about as well as the other, though they would not, of course, rank closely similar situations exactly the same in all cases. Inasmuch as $\triangle$ can be computed without arraying areal units in order of density, it is usually the simpler measure to compute.

It can be shown that both CR and $\triangle$ can be expressed in terms of indexes of the dispersion of the densities of the areal units (the mean difference and the mean deviation, respectively). Hence a measure of concentration can be regarded as a measure of the dispersion of unit densities about the overall density. This being the case, the earlier remarks about the mathematical indeterminacy of the density concept apply as well as to measures of concentration. ...

---

[1] John K. Wright, " Some Measures of Distributions, *Annals of the Association of American Geographers* XXVII (December, 1937), 177–211.

An approach to the analysis of distribution that is closely related to density and concentration measures is the measurement of the spacing of population units. Theoretical contributions to this problem are due primarily to plant ecologists working on the spacing of members of species in the plant community[1] and to the geographers and economists who have developed hypothetical systems of economic areas on the assumption of even population distribution.[2] However, the most direct application of the idea of spacing to the representation of population distribution was developed by Barnes and Robinson[3] in their technique of the " linear distance map ". This type of map is recommended by the authors for displaying variations in density for relatively dispersed populations. It involves a transformation of density by the formula : Average distance $= 1 \cdot 11 /$ $\sqrt{\text{Density}}$. In their illustration of the technique, density is taken as number of farmhouses per square mile, and the average distance of a farmhouse to its six nearest neighbours is deduced on the assumption that farmhouses are evenly distributed. A correction of the formula is given by Mather,[1'] replacing the constant $1 \cdot 11$ by $1 \cdot 07$.[2'] In both these papers it is indicated that values of average distance computed by formula agree rather well with observed average distances separating farmhouses, even in areas where the distribution of farm houses appears somewhat irregular.

Evidently, the linear distance map commends itself primarily on impressionistic grounds. Since the quantity depicted is a simple function of density it can contribute nothing to the clarification of ambiguities inherent in the concept of density. But it is perhaps true that a linear measure is more " understandable " than a ratio of population to area, which is somewhat difficult to visualize.

Whereas the technique of linear distance maps employs the relationship between spacing and density mainly as a device for improving cartographic presentation, other studies have been concerned with measures and patterns of spacing in their own right. Clark and Evans state : " The pattern of distribution of a population of plants or of animals is a fundamental characteristic of that population, but it is a feature that is extremely difficult to describe in precise and meaningful terms ".[3'] Plant ecologists have devoted considerable effort to devising appropriate tests of the randomness of a distribution in space. Recent studies continue this interest, but endeavour as well to specify degree and pattern of departures from randomness. Some of the results of this work will be indicated briefly.

For a given universe of territory containing n units (individuals or defined groups of individuals) of population, let r_i be the linear distance of the ith unit to its nearest neighbouring unit, irrespective of direction. The r_i are measured

[1] Philip J. Clark and Francis C. Evans, " Distance to Nearest Neighbor as a Measure of Spatial Relationships in Populations ", *Ecology*, xxxv (October, 1954), pp. 445–453.
[2] August Lösch, *The Economics of Location* (New Haven : Yale University Press, 1954).
[3] " A New Method for the Representation of Dispersed Rural Population ", *op. cit.*

[1'] " A Linear-Distance Map of Farm Population in the United States ", *op. cit.*
[2'] A more precise value of this constant is $1 \cdot 0746$.
[3'] " Distance to Nearest Neighbor as a Measure of Spatial Relationships in Populations ", *op. cit.*

for the entire population of units, or a random sample thereof. Let ρ be the density of population units, area being measured on the same scale as is used in determining the r_i. If $\bar{r}$ is the mean of the r_i, it can be shown that $\bar{r}_E = 0 \cdot 5/\sqrt{\rho}$ is the expected mean in an infinitely large random distribution of density ρ. The observed mean, $\bar{r}_A$, varies below this, to a theoretical lower limit of zero, as distributions become more " clumped " or " aggregated ". In the limiting case each unit is contiguous to at least one other unit. The observed mean may vary above $\bar{r}_E$ to a limit of $1 \cdot 0746/\sqrt{\rho}$, which occurs for a perfectly uniform distribution in which each unit is equidistant from 6 other units. The ratio, R, of actual to expected distance, or $\bar{r}_A/\bar{r}_E$, may thus vary from zero to $2 \cdot 1491$, with a value of unity occurring for a random distribution. A significance test for the departure of R from unity is available.[4]

In studies of human populations individuals would, of course, be found in a pattern of aggregation to at least the household level. With the method referred to, it would be possible to make an exact test of the assumption of Barnes and Robinson and Mather that farm households in the United States tend to be spaced uniformly. No doubt there would be considerable regional variation in the value of R. The writer has experimented a little with measures of spacing of towns and cities in two regions where there was reason to believe that a tendency to uniform spacing would evidence itself. Values of R approximating $1 \cdot 4$, and significantly different from unity, were found for settlements in Iowa and in the northern two-thirds of Indiana, taking as units the urbanized areas and urban places outside urbanized areas. No doubt in other regions a tendency toward aggregation of communities, rather than uniformity of spacing, would be discovered by this technique. While this exercise produced only very gross results, owing to the use of rough measurements on a small scale map, it suggests that the tendency toward uniformity is not necessarily a function of the closeness of spacing, for $\bar{r}_A$ was around 13 miles in Indiana as compared with 18 miles in Iowa.

One can imagine patterns of distribution in which the technique described is of little value. For one thing, it fails to distinguish " clumped " patterns in which each unit has but one close neighbour from patterns in which the number of units per clump is larger and variable. Clark and Evans indicate some elaborations of the technique that would be helpful for some, but probably not all, such cases. It seems likely that the measurement of spacing, as applied to human population units, will prove more useful for detecting tendencies toward uniformity (which, as Lösch indicates, are of considerable theoretical importance) than for discriminating among patterns of aggregation.

p. 445.
[4] *Ibid.*

Owing to its popularization by the U.S. Bureau of the Census, one of the best known " centres of population " is the mean point, centroid, or centre of gravity, i.e., " the point upon which the United States would balance, if it were a rigid plane without weight and the population were distributed thereon with each individual being assumed to have equal weight and to exert an influence on a central point proportional to his distance from that point ".[1]   A measure of dispersion around the centre of gravity has been proposed[2] and criticized ;[3] but it seems to have been used very little, if at all, in empirical studies.

Two other measures of central tendency in areal distributions have received considerable attention. The " median point " is defined as the intersection of two orthogonal lines, each of which splits an area into two parts with equal numbers of inhabitants. The location of this point is to some degree indeterminate inasmuch as it varies with the rotation of the axes used in its calculation. Lines parallel to these axes may be calculated to divide the distribution into fourths in each direction. Their points of intersection are known as " quartilides ". The principle can be extended to " decilides ", " centilides ", etc., yielding a set of points that describes the degree and pattern of dispersion of the population over the territory.[1'] A somewhat different concept is that of the " median centre " or more descriptively, point of minimum aggregate travel. This is the point from which the sum of the linear radial deviations is a minimum. It has been discussed by Quinn as an optimum location in his " hypothesis of median location ."[2'] Methods are available for calculating the point of minimum aggregate travel where travel to the centre is along restricted routes rather than by shortest airline distance.[3']

It should be noted that the location of the centre of gravity is affected by a change in the position of any unit of the population, whereas there can be considerable movement without affecting the median point. Hart suggests that their mathematical properties render the centre of gravity most useful for studying the areal shifts of a distribution over time, the median point for comparing different distributions at the same time, and the median centre for investigating locational optima for centralized services.

---

[1] U.S. Bureau of the Census, 1950 *Census of Population*, Vol. II, Part I (Washington : Government Printing Office, 1953), p.9.

[2] D. Welty Lefever, " Measuring Geographic Concentration by Means of the Standard Deviational Ellipse ", *American Journal of Sociology*, XXXII (July, 1926), 88–94.

[3] P. H. Furfey, " Note on Lefever's ' Standard Deviational Ellipse' ", *American Journal of Sociology*, XXXIII (July, 1927), pp. 94–98.

[1'] John Fraser Hart, " Central Tendency in Areal Distributions ", *Economic Geography*, XXX (January, 1954), pp. 48–59 ; E. E. Sviatlovsky and Walter Crosby Eells, " The Centrographical Method and Regional Analysis ", *Geographical Review*, XXVII (April, 1937), pp. 240–254.

[2'] James A. Quinn, " The Hypothesis of Median Location ", *American Sociological Review*, VIII (April, 1943), pp. 148–156.

[3'] D. E. Scates and L. M. van Nortwich, " The Influence of Restrictive Routes upon the Center of

As was indicated, the centre of gravity is computed on the assumption that each individual exerts an influence proportional to his distance from the central point. The term "influence" here, of course, is construed in the sense of mathematical weight. But investigators concerned with social influence as affected by space have pointed out that influence is more probably an inverse than a direct function of distance. Professor Stewart's comment on the centre-of-gravity concept[4] expresses this point forcefully :

> The fact is that this population centre of gravity is principally bureaucratic hocus-pocus. . . . It is nonsense to compute a centre on a basis such that "the leverage of a dozen persons in California, at the long end of the teeter-totter, could conceivably counter-balance 100 persons in New York, the short end of the teeter-totter". . . . There is no evidence whatsoever that the impossible teeter-totter's fictitious fulcrum has any specific sociological meaning. There is on the contrary a wealth of evidence that people exert more influence close at hand than far away, and that a thousand people a hundred miles away exert about the same influence as five thousand people five hundred miles away.

If the "influence" of each individual at a point is considered to be inversely proportional to his distance from it, then the total potential of population at a point, L_o, is the sum of the reciprocals of the distances of all individuals in the population from the point. In practice, of course, the computation is made by assuming that all the individuals within a suitably small area are equidistant from L^o, whence,

$$\text{Potential at } L_o = \sum_{i=1}^{n} \frac{P_i}{D_i},$$

where the P_i are the populations of the n areas into which a territory is divided and the D_i are the respective distances of these areas from L_o (usually measured from the geographic centre or from the approximate centre of gravity of the population, in each area). After computing potential for a number of points such as L_o, it is possible to obtain values for other points by interpolation ; or one may construct isometric maps showing lines which are loci of points having equal potentials.[1]

It may be observed that the notion of potential at a point is in principle perfectly precise, whereas the concept of density at a point is meaningless and that of density in the vicinity of a point is ambiguous. While the density of any portion of a territory depends only on the number inhabiting that portion, potential at any point depends on the distribution of population over the entire territory. The two concepts are related, however, in that the configuration of equipotential lines is determined by the pattern of variation of density over a territory ; one could say the same, of course, with respect to the interrelations of the various other measures of distribution that have been described.

Minimum Aggregate Travel ", *Metron*, XIII (1937), pp. 78–81.
  [4] Taken from a letter to the *New York Times*, October 7, 1951.

  [1] For details see John Q. Stewart, "Empirical Mathematical Rules Concerning the Distribution and Equilibrium of Population ", *Geographical Review*, XXXVII (July, 1947), pp. 461–485 ; reprinted in *Demographic Analysis*, edited by Spengler and Duncan, *op. cit.*, pp. 344–371.

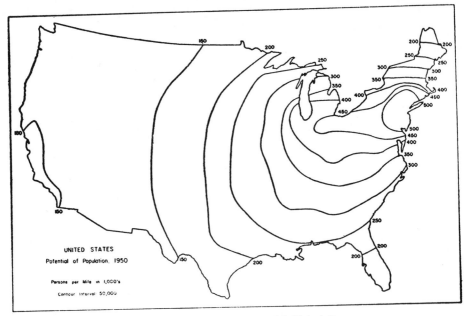

FIGURE 2. Isolines of Population Potential, United States : 1950.

Figure 2 is a map showing isolines of population potential for the United States. The principal determinant of the configuration of the equipotential lines is the massive concentration of metropolitan population in the Middle Atlantic States, but the contours are elongated to the West by the highly urbanized

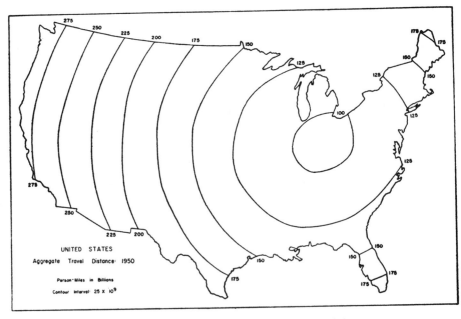

FIGURE 3. Isolines of Aggregate Travel Distance, United States : 1950.

zone stretching along the Great Lakes to Chicago. New York City, where the peak potential occurs, may be regarded as the " centre " of the country from the standpoint of demographic influence, granted the assumptions on which this map is constructed.

Figure 2 may be compared with Figure 3, which shows isolines of aggregate travel distance for the same underlying population distribution. This index reaches a minimum, i.e., the " median centre " occurs, at a point lying within the lowest contour shown. At that point aggregate travel distance is around 95 billion miles ; this is the total travel required to move every individual in the country to that point by shortest airline distance. The travel per person would be somewhat over 600 miles.

To the writer's knowledge it has not been observed previously that potential and aggregate travel distance are closely related concepts. The value of either at a point, L_o, can be expressed by the formula, $\Sigma\, P_i D_i^m$, where $m=-1$ for the potential computation and $m=1$ for calculating aggregate linear travel distance. The two should be regarded as alternative but not mutually exclusive concepts ; and it may well be that values of m other than ± 1 will be found suitable for certain problems.[1] ...

Several investigators have been intrigued with the observation that the size distribution of cities and towns exhibits a kind of regularity from place to place and over time. For any large territory in which urbanization is at all advanced it is almost uniformly found that small places outnumber large ones, and the size distribution is in general highly skewed. Auerbach[2] is credited with the first presentation of a formula to summarize the distribution of cities by size. He observed that when the cities of Germany were ranked in order of number of inhabitants the product of a city's rank and its size tended to be more or less constant. Moreover, he suggested that the average value of this product, expressed as a proportion of the national population, affords an index of population concentration suitable for comparing degrees of urbanization among countries. If the " rank-size rule " were to hold precisely, the product of rank and size would be equal to the size of the largest city. An average of the products may, therefore, be regarded as an " adjusted " or " estimated " size of the largest city. Hence, Auerbach's suggested coefficient comes down to taking the " adjusted " size of the largest city as a proportion of the national population.

---

[1] Theodore R. Anderson, " Potential Models and the Spatial Distribution of Population ", *Papers and Proceedings, the Regional Science Association*, II (1956), pp. 175–182. It is a curiosity that population potential is emphasized by an investigator (Stewart) working at Princeton University near the point of maximum potential, while the " hypothesis of median location " was set forth by an investigator (Quinn) at the University of Cincinnati, in close proximity to the point of minimum aggregate travel. It is left for students of *Wissenssoziologie* to speculate on the possible " demographic determination" of demographic concepts.

[2] Felix Auerbach, " Das Gesetz der Bevölkerungskonzentration ", *Petermanns Mitteilungen*, LIV (1913), pp. 74–76.

In a brief discussion of Auerbach's work, Lotka[1] indicated that for the 100 largest cities in the United States the " law of urban concentration " was more precisely (rank) $^{0.93}$ $\times$ size = constant.   Lotka stated, however : " It may be left an open question how much significance is to be attached to this empirical formula ".

Other early treatments of city-size distribution formulæ apparently were produced without knowledge of Auerbach's and Lotka's work.   Goodrich[2] noted that the number of cities above a given size limit tends to be equal to twice the number of cities above a limit of twice that size, a formula which is equivalent to the " rank-size rule ".   Saibante[3] experimented with fitting a Pareto curve to the city-size distribution and presented an analysis of concentration indexes analogous to Gini's indexes of income inequality or concentration.   Gibrat[4] suggested fitting the city-size distribution with a log-normal equation, and proposed an index of concentration related thereto.

Both the " rank-size " and the Pareto-curve approach have been further investigated in the last two decades.   Singer[5] drew attention to the parallel between the city-size distribution and Pareto's law of income distribution and presented evidence of goodness-of-fit of the Pareto curve to a number of city-size distributions.   Allen[6] recently reviewed and considerably extended Singer's work. The " rank-size rule " was much publicized by Zipf[7] who evidently began his work without reference to Auerbach's and Lotka's prior studies.   Mathematical as well as empirical contributions to the literature of the " rank-size rule " were made by Stewart[8] and Hammer.[9]   These investigators dealt with the generalization of the rule in which the rank of a city is allowed to have an exponent other than unity, and presented formulæ for estimating the number and aggregate population of cities within specified size limits from the parameters of the fitted curve. Another recent contribution is Madden's discussion of the temporal persistence of the " rank-size " relationship as evidence of stability in the growth patterns of cities.[1']   Aside from the empirical investigations incompletely listed here, a number of writers have commented on the significance of the " rank-size rule "

[1] Alfred J. Lotka, *Elements of Physical Biology* (Baltimore : Williams & Wilkins Co., 1925), pp. 306–307.

[2] Ernest P. Goodrich, " The Statistical Relationship between Population and the City Plan ", in *The Urban Community,* edited by Ernest W. Burgess (Chicago : University of Chicago Press, 1926).

[3] Mario Saibante, " La Concentrazione della Popolazione ", *Metron,* VII (1928), pp. 53–99.

[4] R. Gibrat, *Les Inégalités Économiques* (Paris : Librairie du Recueil Sirey, 1931).

[5] H. W. Singer, " The ' Courbe des Populations ' : A Parallel to Pareto's Law ", *Economic Journal,* XLVI (June, 1936), pp. 254–263.

[6] G. R. Allen, " The ' Courbe des Populations,' : A Further Analysis ", *Bulletin of the Oxford University Institute of Statistics,* XVI (May-June, 1954), pp. 179–189.

[7] George Kingsley Zipf, *National Unity and Disunity* (Bloomington, Ind. : The Principia Press, Inc., 1941) ; *Human Behavior and the Principle of Least Effort* (Cambridge, Mass. : Addison-Wesley Press, Inc., 1949).

[8] " Empirical Mathematical Rules Concerning the Distribution and Equilibrium of Population ", *op. cit.*

[9] Carl Hammer, " Rank Correlation of Cities and Refinement ", mimeographed report, Bureau of Applied Social Research, Columbia University, 1951.

[1'] Carl H. Madden, " On Some Indications of Stability in the Growth of Cities in the United States ", *Economic Development and Cultural Change,* IV (April, 1956), pp. 236–252.

from the standpoint of the general theory of population distribution and space-economy.[2]

It is curious that one finds in the literature little discussion of the relative merits of the " rank-size " and Pareto-curve approaches to summarizing the city-size distribution. The mathematical relationship between the two formulæ may be indicated as follows :

Let $x =$ size of city or town (number of inhabitants) and $y = y(x) =$ number of places of size x or larger (rank).[3] Then the formula for the Pareto curve is $\log y = \log A - \alpha \log x$, or $y = Ax^{-\alpha}$, or $yx^{\alpha} = A$, where A and α are constants determined from the data. The generalized " rank-size " formula is $\log x = \log B - \beta \log y$, or $x = By^{-\beta}$, or $xy^{\beta} = B$, where B and β are constants determined from the data. These equations imply that $\beta = 1/\alpha$ or $\alpha = 1/\beta$, and $A = B^{1/\beta}$ or $B = A^{1/\alpha}$. However, if A and α have been determined by minimizing the sum of the squares of the residuals in $\log y$ while B and β have been determined independently by minimizing the sum of the squares of the residuals in $\log x$, the foregoing relationships would not, in general, hold precisely.

The bulk of the work on fitting " rank-size " or Pareto curves has been carried out somewhat informally. In the majority of cases, constants have been determined graphically without any stated criterion of fit, or the data have simply been plotted on double-log paper and their approximation to a straight line has been noted by visual inspection. One finds in the literature little discussion of mathematical criteria and techniques for fitting curves to city-size distributions, and, correlatively, little formal attention to goodness of fit. Singer and Allen present calculations of " errors " and " average errors " in computing the number of cities in various size groups from the equation of the fitted curve. However, comparisons based on such calculations are invalidated to a degree, by variation from case to case in the number and size of intervals. In no case, to the writer's knowledge, has a formal test of goodness of fit, such as χ^2, been presented. One suspects that in the majority of cases where the fit has been judged acceptably good, a stringent mathematical test would detect statistically significant departures from the mathematical model. One consequence of the " rank-size " approach has been to focus attention unduly on the fit of the formula at the very upper end of the distribution Thus, in numerous verbal discussions the writer has encountered the statement that the " rank-size ". rule does not hold for such and such a country on the grounds that its largest city is much too large or too small. The net impression is that sceptics have often rejected the possibility of a simple formula adequately fitting city-size distributions on the basis of an examination of only a small part of the evidence whereas proponents of one or another formula may have been guilty of failing to give as much attention to negative as to positive evidence.

One might well be inclined to adopt a fairly liberal criterion of what constitutes an acceptable fit of a curve to the data if on such a criterion numerous independent

---

[2] For example, Lösch, *The Economics of Location, op. cit.* ; Rutledge Vining, " A Description of Certain Spatial Aspects of an Economic System ", *Economic Development and Cultural Change,* III (January, 1955), pp. 147–195 ; Walter Isard, *Location and Space-Economy* (New York : John Wiley & Sons, Inc., 1956).

[3] Singer and Allen let y (x) = number of places larger than x.

instances of acceptable fit were to be discovered and if instances of poor fit could be explained systematically rather than on *ad hoc* grounds differing for each instance. The generality of any proposed formula involves at least three considerations : the number or proportion of countries or regions where a good fit is obtained, the occurrence of a good fit in various time periods and the stability of the form of the size distribution over time, and the range of community sizes for which a good fit is obtained. On the last point, virtually all investigators recognize that the linear formula (in the logs) can be expected to hold only for sizes above a certain minimum. This minimum, however, may vary spatially and temporally ; moreover, it may sometimes fall within the range of available data and thus be approximately identifiable, or it may be lower than the size for which frequencies are tabulated. Both incautious statements and unjustified speculation on this matter are to be found in the literature ; however, Allen[1] presents careful empirical estimates of the minimum size for a number of countries. On the generality of " rank-size " and related formulas over space and time, it must be recognized that the fund of available data constitutes a quite biased sample of countries and periods. Moreover, the comparability of the available data is greatly impaired by variation in census methods of recording and tabulating community size. The largest collection of instances is presented by Allen who, however, worked mainly with data readily available in secondary sources. It is evident that cases are to be found where the Pareto or " rank-size " formula fits quite poorly, many, though not all, of which can be " explained away " by reference to deficiencies in the data. Intuitively, one feels that there are certain kinds of " countries " and, perhaps, certain " abnormal " periods in which such a formula would not be expected to fit well. However, until an adequate theoretical rationalization for the formula is available, it is difficult to organize these hunches.

A careful appraisal of the theoretical significance of the Pareto or the " rank-size " rule would probably assign it a position midway between two extremes : on the one hand, a merely empirical curiosity, and, on the other, a " law " rigorously deduced from an accepted theoretical scheme and verified under fully specified conditions. In other words, such a rule has " plausibility " as well as a modicum of empirical support. A " plausible " connection between this type of regularity and his doctrine of the hierarchical organization of economic regions was set forth by Lösch[2] and restated by Hoover.[3] Moreover, Vining has suggested the possibility of developing a relevant model based on the assumption " that some process of development leads in the limit to stable distributions of the sort that we have described ".[1'] As yet, however, none of these ideas has been pursued far enough to yield a theory of community-size distribution that predicts what conditions are both necessary and sufficient to

[1] " The ' Courbe des Populations ', A Further Analysis ", *op. cit.*

[2] *The Economics of Location, op. cit.*, pp. 431–438.

[2] Edgar M. Hoover, " The Concept of a System of Cities : A Comment on Rutledge Vining's Paper ", *Economic Development and Cultural Change*, III (January, 1955), pp. 196–198.

[1'] " A Description of Certain Spatial Aspects of an Economic System ", *op. cit.*, p. 185.

give rise to the Pareto or " rank-size " form. Under the circumstances, there is definitely indicated a need for concurrent efforts to discover the empirical limits of validity of these rules and attempts to formulate in theoretical terms the relationship of the city-size distribution to other aspects of population distribution and ecological organization. Pending these achievements, the Pareto curve may be accepted as a convenient means of summarizing community-size distributions under certain circumstances.

Probability Models of Conception and Birth

Pregnancy and birth distributions are mathematically unlike mortality in being repeated events, separated by intervals of non-risk and admitting as a matter of course a high degree of individual control. Corrado Gini (1924) was first to explore these distinctions and their implications, by his suggestion that birth intervals be treated as waiting time problems dependent on fecundability. This he defined as the probability that a woman at risk would conceive in a given monthly cycle.

Gini's article, paper 40, limits itself to first pregnancies and births under constant fecundability. The obvious possibilities for extending the work were not considered for about 30 years, when Louis Henry (1953) published the first of a series of investigations in the mathematical treatment of birth intervals. Henry's initial article, presented here in a translation by Mindel C. Sheps and Evelyne Lapierre-Adamcyk (1972), indicates how Gini's analysis is applied to birth orders beyond the first.

An equation of Henry's that figures in much later work is the proportion of women conceiving each month in a homogeneous population, given that at any point in time some women are already pregnant or in amenorrhea and hence not at risk. Henry shows that if the monthly probability of pregnancy to non pregnant women is p and the non susceptible period associated with pregnancy is s months, both constants, then the mean birth interval i approaches the stable value

$$i = \frac{1}{p} + s.$$

The inverse $1/i$ is the proportion conceiving each month. The result was extended by Tietze (1962) to show the effects of contraception; and by Potter (1963) to compare rates of abortion that would arise if it were used exclusively or used in conjunction with contraception, assuming in both cases that all pregnancies that occur are aborted. Contraception of effectiveness e changes the probability of pregnancy from p to $p(1-e)$, while abortion affects the sterile period s. The results show in general that if abortion is used alone as a means of controlling fertility it is required frequently. If $s' < s$ is the sterile period associated with abortion the intervals being compared are

$$\frac{1}{p(1-e)} + s'$$

365

and

$$\frac{1}{p} + s'.$$

With highly effective contraception, the first interval can be of great length, while the second is less than the normal interval between births.

In the paper that follows Henry's, D. Basu (1955) derives the monthly proportion conceiving independently, by means of difference equations. The solution is also attractively reached through matrix formulation, for which the reader may consult Keyfitz (1968, pp. 390—392). Henry's work is expanded in later papers to cover general fertility histories; several are available in English in Henry (1972).

Important contributions have been added by Robert G. Potter and M. P. Parker (1964), and Mindel C. Sheps (1964), whose related articles appeared together. Sheps, whose paper is included here, finds generating functions for the distribution of conception delays under Henry's assumption that fecundability differs among couples but is constant over time. Potter and Parker specify birth probabilities as a Pearson Type I distribution and find waiting times for this specific case.

The analysis of waiting times and conception probabilities has been extended very considerably beyond the level treated here, including allowance for pregnancy wasteage, for declining fecundability with age, and for contraceptive use. Much of Sheps' later work will be found in Sheps and Menken (1973). Sheps, Menken and Radick (1969) provide an introduction to other contributions.

40. First Investigations on the Fecundability of a Woman

CORRADO GINI (1924)

Proceedings of the International Mathematical Congress (Toronto) 2: 889—892. Translated by David Smith.

For the *Gini Index*, Gini's "approximate measure of the mean fecundability of primipares who had their child in the months of marriage from $x+9$ to $x+y+9$," we may write

$$F_G = \frac{n_{x+9} - n_{x+y+10}}{\sum\limits_{x+9} n_i} = \frac{n_1 - n_{y+1}}{\sum\limits_{1}^{y} n_i},$$

where n_i are births in the i'th, and n_i births in the $(i+9)$'th month of exposure. If fecundity differs among women but for each woman is constant over time, the Index has as its limiting value the arithmetic mean fecundability of the non-sterile population N^*. This is seen by noting that as y approaches infinity, $n_{y+1} \to 0$ and $\sum\limits_{1}^{y} n_i \to N^*$.

The *Pearl Index* (Pearl 1933), defined as the number of pregnancies experienced divided by person-months of exposure, is often more useful. It is

$$F_P = \frac{\sum\limits_{1}^{y} n_i}{\sum\limits_{1}^{y} in_i + y\left(N - \sum\limits_{1}^{y} n_i\right)},$$

where N is the population size. If analysis is again restricted to the non-sterile population this has as its limiting value the harmonic mean fecundability, whose inverse is the mean waiting time to pregnancy: $\left(\sum\limits_{1}^{\infty} in_i\right) / N^*$. The expression is formally a life expectancy, correct to the measurement error of the exposure time i.

I call *fecundability of a woman* the probability that a married woman becomes pregnant during the month, neglecting any Malthusian or neo-Malthusian practice intended to limit procreation.

It is unnecessary to stress the theoretical and practical importance of a measure for fecundability. We would have, in particular, the means to decide what fraction of the differences we observe between the coefficients of natality of different countries, different social classes, or different times arises from physiological causes and what fraction arises from voluntary causes.

It is also unnecessary to demonstrate that fecundability cannot be measured directly. Quite clearly, we are not in a position to observe the consequences of Malthusian and neo-Malthusian practices on natality. Nor do we know the frequency of ovulatory abortions that occur in the initial months of pregnancy.

We can, however, formulate an indirect measure of fecundability, based on the considerations that follow.

Let us suppose that married women able to bear children are n in number during the whole period being considered, and that all have the same fecundability p. On these assumptions the number of women who would conceive for the first time during their first month of marriage will be pn; the number who would conceive for the first time in the second month will be $(1-p)pn$; the number who would conceive for the first time in the third month will be $(1-p)^2 pn$, and so forth. The numbers obtained in this way form a geometric progression whose ratio is $1-p$.

If we assume that the percentage of abortions (or of abortions and stillbirths) and likewise the percentage of pregnancies longer or shorter than 9 months are the same for the products of conceptions that have taken place in the successive months of marriage, we can substitute for the ratio between first conceptions taking place in the month $x+1$ of marriage and first conceptions taking place in the preceding month x, the ratio between the firstborn (or first live births) coming into the world in the month $x+10$ of marriage, and the firstborn (or first live births) coming into the world in the month $x+9$.

It is evident that the value of p arrived at by this route is independent of the number, large or small, of women who are not in the childbearing population, either because of the sterility of the marriage or because of Malthusian or neo-Malthusian practices.

It is also evident that it is independent of the frequency of abortions and stillbirths.

The value of p thus determined measures, on the assumptions mentioned above, the fecundability of women able to conceive.

Among these assumptions there is one which clearly does not agree with reality however: it is that all fertile married women have the same fecundability p. This circumstance, which cannot be neglected, does not prevent our arriving at the measure of fecundability; though the route to follow is a bit longer.

Let $s \leq n$ be the number of groups into which the n married women can be classed according to their fecundability, $p_i|_{i=1,2,...,s}$ will be the fecundability of women in group i and n_{ix} the number of women of group i who, in the month x after marriage, have not yet conceived.

368

The number of women expected to conceive for the first time in the x'th month of marriage will be $\sum_{i=1}^{s} p_i n_{ix}$; and the number expected to conceive for the first time in the next month $x+1$ of marriage will be $\sum_{i=1}^{s} (1-p_i)p_i n_{ix}$. The expected value of the ratio between the second and first numbers (if we agree to give to each of the possible values of the ratio a weight proportional to the probability that this value is realized and to the number which forms its denominator) will be

$$\frac{\sum_{i=1}^{s} (1-p_i)p_i n_{ix}}{\sum_{i=1}^{s} p_i n_{ix}} = 1 - \frac{\sum_{i=1}^{s} p_i^2 n_{ix}}{\sum_{i=1}^{s} p_i n_{ix}}.$$

But $\sum_{i=1}^{s} p_i^2 n_{ix} \Big/ \sum_{i=1}^{s} p_i n_{ix}$ is precisely the expected value of the mean fecundability of women who conceive for the first time in the x'th month of marriage.

Thus we will be able to deduce the mean fecundability of women who conceive for the first time in the month x from the complement of the ratio between women who conceive for the first time in the month $x+1$ and women who conceive for the first time in the month x.

This calculation rests on the following assumptions, which can be accepted without difficulty:

(a) The number of women who were eligible for first conceptions in month x of marriage, and did not conceive in this month, will equal the number of women eligible for first conceptions in month $x+1$ of marriage. This is a way of saying that we can neglect the effects from one month to the next of mortality, migrations, and passages from the category of married women not able to conceive to the category of married women eligible for first conceptions (for example by cessation of contraceptive practices or by elimination of the cause of sterility), as well as passages from the second category to the first (for example, due to unexpected sterility or the adoption of contraception);

(b) in each of the s groups of women eligible for first conceptions after marriage, fecundability remains unchanged from month x to month $x+1$.

Supposing further (assumption c) that the probability of abortion (or abortion and stillbirth) and the probability of a gestation period greater or less than 9 months are the same for the products conceived in month x and for those conceived in month $x+1$, we will be able to substitute for the ratio between women who conceive for the first time in month $x+1$ and women who conceive for the first time in month x, the ratio between the firstborn (or first live births) of the marriage who came into the world in month $x+10$ and the firstborn (or first live births) of the marriage who came into the world in month $x+9$.

We are thus able to obtain a measure of the mean fecundability of primipares who, after marriage, conceived for the first time in month x. This fecundability can in turn be regarded as essentially equal to the fecundability of primipares who had their child in month $x+9$.

369

The complement of the ratio between the firstborn (or first live births) of the marriage coming into the world in month $x+10$ and the firstborn (or first live births) of the marriage coming into the world in month $x+9$ will give us, by assumptions (a), (b), and (c), an approximate measure of the mean fecundability of primipares who had their child in month $x+9$ of marriage.

If we calculate the mean of the ratios between the firstborn of the months of marriage $x+10$ and $x+9$, $x+11$ and $x+10$, ..., $x+y+10$ and $x+y+9$, giving each ratio a weight proportional to its denominator; and if we take the complement of this mean, we obtain by assumptions (a), (b), and (c) an approximate measure of the mean fecundability of primipares who had their child in the months of marriage from $x+9$ to $x+y+9$.

Here are the results obtained for several areas. They are based on data published by official sources, except for Italy, for which the data are the fruit of special inquiries, using the registers of vital events of 24 communes.

Area	Years	Mean fecundability of primipares who had their child during the months following marriage:			
		10—17	11—17	10—23	11—23
Berlin	1894—1895	26.1	24.4	23.0	20.5
Australian Confederation	1917—1921	19.4	20.6	18.5	19.2
Western Australia	1895—1915	23.9	22.7	21.9	20.2
New South Wales	1893—1905 and 1916—1921	20.6	22.2	18.9	19.7
Victoria	1898—1900	21.0	24.4	...	...
Tasmania	1905—1906				
Italy (24 Communes)	1900—1921	23.8	21.0	21.4	18.6

The uniformity of the results obtained is remarkable, above all when we exclude from the calculations the firstborn coming into the world during the tenth month of marriage, corresponding to conceptions of the first month of marriage. For the ratio between conceptions of the first and those of the second month of marriage, we cannot in fact accept the validity of assumption (a), fecundability being, in the first month, less than in the second due to the frequent virginity of the wife; nor that of assumption (c), ovulatory abortions being especially frequent for the products conceived in the first month because of honeymoon trips and perhaps other circumstances.

The obstacle to fertilization represented by virginity naturally increases in importance with the age of the wife, but beyond the first month it does not appear that the ability to conceive is weaker for older than for younger wives. That is a remarkable finding, and was brought to light by the data of the Australian Confederation for the period 1907—1914 and those of New South Wales for the period 1893—1898.

Another important finding is that the diminution of natality from 1901—02 to 1911—12 which has been verified in the Kingdom of Saxony and has reached

50%, does not appear to have been accompanied by a diminution of the ability of primipares to conceive.

The methods and results summarized in this note are set forth together with complementary details in two papers presented at the Instituto Veneto di Scienze, Lettere ed Arti on July 1, 1924, and in the article "Decline in the birthrate and 'fecundability' of women," in The Eugenics Review, January 1926. Further investigations are in process which can perhaps permit us to assemble other interesting results in the new area that the method proposed appears to open to statistical research.

41. Theoretical Basis of Measures of Natural Fertility

Louis Henry (1972 (1953))

From *On the Measurement of Human Fertility: Selected Writings of Louis Henry*, translated and edited by Mindel C. Sheps and Evelyne Lapierre-Adamcyk. New York: Elsevier. Excerpts are from pages 2, 4—9, 15—20.

In his remarks on Gini, Henry notes that the mean fecundability of women who conceive in the first month of marriage (in Gini's notation, paper 40 above: $\sum p_i^2 n_{i1} / \sum p_i n_{i1}$) may also be written $\bar{p}(1+c^2)$, with $\bar{p}$ the arithmetic mean fecundability of the non-sterile population and c the coefficient of variation. This permits calculation of the variance of fecundability V_p as

$$V_p = \bar{p}(1-\bar{p}) - n_2/N^*$$

where $\bar{p} = n_1/N^*$; n_1 and n_2 being the number of women conceiving in the first and second months of exposure respectively, and N^* the total non-sterile population (cf. Sheps and Menken 1973, pp. 129—130, 398—399).

We omit sections of the paper discussing sterility, natural fertility and fecundability, and the fitting of observed data to the models, and some remarks on the works of Gini and Pearl.

Introduction

One of the most interesting and difficult problems in demography is that of estimating natural fertility, i.e. the fertility of a human population that makes no deliberate effort to limit births. Natural fertility, which depends essentially on biological factors, is primarily a biological phenomenon, and a "natural" phenomenon has a particular attraction. It shares in the prestige of the natural sciences, overshadowing that of the social sciences which are still in their infancy. In addition, efforts to appraise the effectiveness of contraceptive practice lead to comparisons between actual fertility and the hypothetical fertility that a given population would have if it did not use any form of birth control.

The difficulties encountered in the study of natural fertility result from its very definition. Natural fertility is, for easily observable populations, hypothetical. No doubt one can find populations living under conditions of quasi-natural fertility; but, for the most part, little is known about them. For European populations, before the spread of contraceptive practice, only scanty data are available.

Thus, it is not surprising that indirect approaches are made to circumvent the difficulty. But in following indirect routes we risk losing sight of the goal: An investigation of the theoretical basis of measures of natural fertility is indispensable.

We will start with certain assumptions regarding the biological basis of fertility. These assumptions are not completely arbitrary; observed results are used to construct a mathematical model that is as close to reality as possible and yet simple enough to be easily manageable.

Fundamental Assumptions

It has long since been ascertained that in large families of a fixed size, the interval between successive births is more or less independent of birth order, except perhaps for the last few births. Now, a priori, the interval between births depends on the duration of pregnancy (independent of order), on the duration of the nonsusceptible period that follows confinement, and on the natural fecundity of the couple. If the mean interval between successive births does not vary, it is natural to think that these various factors also do not vary for a given couple as long as they are not sterile.

We were led, therefore, to characterize each fecund couple (couple able to produce living children, at the time or later):

1. by the duration, g, of the nonsusceptible period associated with pregnancy; g is the time that elapses between a conception and the first ovulation after delivery.

2. by fecundability or the probability of conception per unit time outside of the previously mentioned nonsusceptible periods.

Since conception is possible only at the time of ovulation, one ought, rigorously, to treat fecundability as discrete and define it as the probability of conceiving per menstrual cycle. In this case we designate it by p. But since it is often more convenient to treat time as continuous, we also introduce the probability ϕdx of conceiving during the infinitely small time interval dx. The term fecundability is also used for ϕ.

The quantity g is the sum of two terms: the duration of pregnancy, g_0, and the duration, $g - g_0$, of the period of nonsusceptibility following delivery. The latter varies appreciably between women, because, while it is very short in the absence of breastfeeding, it can be very long in the case of prolonged lactation. Moreover, fecundability certainly varies between women, whether because of the physiological characteristics of the couple or of the frequency of their sexual relations. To take these variations into account, we introduce two distributions: $h(g)$, the probability density function (p.d.f.) of the duration of nonsusceptibility, and $f(\phi)$, the p.d.f. of fecundability, or $f(p)$ in the discrete case.

Before going on, let us examine the foregoing. To characterize each couple by a pair (ϕ, g), is to assume that these quantities are invariant in time as long as the couple is not definitely sterile. It is certain that, in reality, ϕ and g vary over time. For g, this is obvious. Illness of the mother or death of the child may interrupt lactation; furthermore, if one considers conceptions, the duration of pregnancy also varies.[1]

It is more difficult to affirm, a priori, that ϕ varies. Doubtless our sense of continuity makes us think that women do not become completely sterile suddenly, without a preceding progressive decline in fecundability. But this feeling for continuity, though undeniable, does not constitute an argument. One could object to the assumption of a constant ϕ on the grounds of the existence of nonsusceptible periods outside of those associated with pregnancy, e.g. those due to separation or illness. But to the extent that these separations and illnesses are distributed uniformly in time, they intervene by modifying the probability of conception; they do not affect its assumed invariance.

On the other hand, we must admit that the distribution of these separations and illnesses is certainly not uniform. Long established marriages accept temporary separations more easily than do young married couples; also, the frequency of illness increases with age and therefore with duration of marriage. Thus, separations and illnesses tend to lower ϕ with increasing duration of marriage.

The relatively small variation in observed mean intervals between successive births shows, however, that variations in ϕ and g depend little on the duration of marriage, age or parity. With respect to g, this suggests that if this function varies for a given couple, the variation is independent of the duration of marriage. On the average, g will be the same at any point in a marriage. We shall see that under these conditions, the results are the same as if we assume a constant g for each couple.

---

[1] In the case of spontaneous abortion, pregnancy is shorter. For convenience, we often operate as if there were a fixed time between conception and birth and pass from one to the other at our choice. We are not unaware, however, that the existence of spontaneous abortions and stillbirths complicates matters. But these are secondary difficulties which it appears unnecessary to expound here.

Reproductive History of a Group of Couples

We begin by considering the reproductive history of a group of couples with the same characteristics (ϕ, g) in the absence of permanent sterility and of dissolution of the marriage by separation, divorce or death.

It is natural to take marriage as the time origin, under the assumption of no premarital sexual relations and, hence, no premarital conceptions. But, on this assumption, the period from marriage to the first live birth (if, as we are doing, one considers live births only) differs from subsequent intervals because g_0, the duration of pregnancy, rather than g, intervenes in this first interval. One could, obviously, retain this difference, but only at the cost of complicating the notation. It is simpler to change the origin, placing it at $g - g_0$ before marriage.

Let us designate the elapsed time from this translated origin by x. For $x < g$, all couples are childess; for $x > g$, the expected number of couples without births, taking the initial sample size as unity, is equal to $e^{-\phi(x-g)}$ and the number of first births in the interval $(x, x+dx)$ is $\phi e^{-\phi(x-g)} dx$.

For $x < 2g$, there are no second order births; it can easily be shown that, for $x \geq 2g$, the expected number in the interval $(x, x+dx)$ is equal to $\phi^2(x-2g)e^{-\phi(x-2g)} dx$.

It is then easy to show by recursion that, for $x \geq ng$, the expected number of births of order n in the interval $(x, x+dx)$ is equal to[2]

$$\phi^n \frac{(x-ng)^{n-1}}{(n-1)!} e^{-\phi(x-ng)} dx. \tag{1}$$

The expected number of births $B(x)dx$ in the interval $(x, x+dx)$, regardless of order, is equal to the sum of (1) for values of n from 1 to m such that $mg \leq x$. $B(x)$ is given by:

$$B(x) = \int_g^\infty B(x-t)\phi e^{-\phi(t-g)} dt \tag{2}$$

which is of the same form as equations studied by Lotka. Its solution is of the form $\sum_s Q_s e^{r_s x}$. The Q_s are coefficients and the r_s are roots of the equation

$$1 = \int_g^\infty \phi e^{-rt - \phi(t-g)} dt. \tag{3}$$

---

[2] If one had considered couples with fecundability ϕ and $g_0, \ldots g_{n-1}$ for the duration of nonsusceptability corresponding to each successive birth order, one would have obtained for the number of births of order n in the interval $(x, x+dx)$:

$$\phi^n \frac{(x-g_0-g_1-\ldots-g_{n-1})^{n-1}}{(n-1)!} e^{-\phi(x-g_0-g_1-\ldots-g_{n-1})} dx,$$

for $x \geq g_0 + g_1 + \ldots + g_{n-1}$, given that g_k is not dependent upon the duration of marriage. For a set of couples in which $g_0 + g_1 + \ldots + g_{n-1} = ng$, we again arrive at the earlier result. There is no need, therefore, to introduce variations of g independent of marital duration.

The real root is zero and the real component of the imaginary roots is negative. Consequently, $B(x)$ is the sum of a constant and of damped periodic functions. With increasing duration of marriage, this sum approaches the constant part of $B(x)$, which we designate by ϕ' (asymptotic fertility rate). This is equal to $1/i$, where i is the mean interval between births. Now,

$$i = \int_g^\infty \phi x e^{-\phi(x-g)} dx = g + \int_g^\infty \phi(x-g) e^{-\phi(x-g)} dx \tag{4}$$

whence $i = g + \dfrac{1}{\phi}$. Hence

$$\phi' = \frac{1}{g+1/\phi} = \frac{\phi}{1+g\phi}. \tag{5}$$

Often in demography, rates are only intermediate results in the attempt to derive the mean number of events per capita over a long period: here, the mean number of births per marriage. Hence, we are led to study cumulative fertility as a function of time, that is, the expected number of children born in x years of marriage. Let us designate it by $E(x)$. We have:

$$E(x) = \int_0^x B(\zeta) d\zeta = \phi' x + \sum_s \int_0^x Q_s e^{rs\zeta} d\zeta. \tag{6}$$

$E(x)$ is the sum of a linear function of x and of damped periodic functions. After a number of oscillations, that is, after a given duration of marriage, $E(x)$ approaches $\phi' x + E_0$, where E_0 depends on ϕ and g. Given ϕ and g, $E(x)$ can be calculated from appropriate tables, e.g. χ^2 tables [since Eq. (1) is equivalent to a χ^2 (chi square) distribution with $2n$ degrees of freedom]. From this, one deduces E_0. But it is not necessary to consider all pairs (ϕ, g). Let us assume two values, g and λg. The maximum value of n in time x given g is the same as the maximum value in time λx given λg. If ϕ is the fecundability associated with g, let us consider the pair $(\phi/\lambda, \lambda g)$. We then have:

$$E(x|\phi,g) = \sum_1^n \int_{ng}^x \frac{\phi^n(\xi-ng)^{n-1}}{(n-1)!} e^{-\phi(\xi-ng)} d\xi \tag{7}$$

$$E(\lambda x|\phi/\lambda, \lambda g) = \sum_1^n \int_{\lambda ng}^{\lambda x} \left(\frac{\phi}{\lambda}\right)^n \frac{(\zeta-\lambda ng)^{n-1}}{(n-1)!} e^{-(\phi/\lambda)(\zeta-\lambda ng)} d\zeta. \tag{8}$$

Let $\zeta = \lambda \xi$; it follows that $E(\lambda x \mid \phi/\lambda, \lambda g) = E(x \mid \phi, g)$. The two terms in x are equal to $\dfrac{\phi}{1+g\phi} x$ for the pair (ϕ, g), and to

$$\frac{\phi}{\lambda(1+g\phi)} \lambda x$$

377

for the pair $(\phi/\lambda, \lambda g)$. They are therefore equal and we have finally, $E_0(\phi, g) = E_0(\phi/\lambda, \lambda g)$. Hence, when $g\phi$ is a constant, E_0 is fixed. Values of E_0 for selected $g\phi$ were calculated, and a regression fitted as:

$$E_0 = -0.5 + \frac{0.115}{g\phi}. \tag{9}$$

Table 1 shows calculated and estimated values.

Table 1. E_0 According to $g\phi$

$g\phi$	0.5	1	2	3	4	5	6	∞
E_0 calculated	−0.275	−0.375	−0.445	−0.470	−0.480	−0.485	−0.490	−0.500
E_0 estimated from Eq. (9)	−0.270	[−0.385]	−0.443	−0.462	−0.471	−0.477	−0.481	−0.500

We now pass to the more realistic case of a heterogeneous group where the couples have all possible characteristics (ϕ, g). Assume that there is no correlation between ϕ and g: the group is then characterized by two probability densities, $f(\phi)$ and $h(g)$. This group, by virtue of the heterogeneity of ϕ and g, has diverse values for ϕ'; the distribution of ϕ' has a probability density $k(\phi')$.

In such a group, the expected number of births in the interval $(x, x+dx)$ is equal to the sum of the births of the homogeneous subgroups (ϕ, g) which constitute the entire group. If x is large, the expected number in the subgroup (ϕ, g) departs little from $\phi' dx$; the expected number of births for the total, therefore, is equal to $dx \int \phi' k(\phi') d\phi' = \bar{\phi}' dx$, and the cumulative fertility of the group is approximately equal to

$$x \int \phi' k(\phi') d\phi' + \int \int E_0 f(\phi) h(g) d\phi dg \tag{10}$$

or $x\bar{\phi}' + \bar{E}_0$, where $\bar{E}_0$ is the mean of the values of E_0 corresponding to the various combinations (ϕ, g).

If we substitute Eq. (9), the approximation for E_0 referring to a sufficiently long time, into Eq. (10), the result is:

$$\bar{E}_0 = -0.500 + 0.115 \int \frac{f(\phi) d\phi}{\phi} \int \frac{h(g) dg}{g} = -0.500 + \frac{0.115}{\bar{\phi}_h \bar{g}_h}, \tag{11}$$

where $\bar{\phi}_h$ and $\bar{g}_h$ are the harmonic means of ϕ and of g. ...

The Discrete Case: Gini's Method

To study the fecundability of newlyweds, it is preferable to abandon the continuous notation used until now. Let p be the probability of conception in the course of a

378

menstrual cycle, where the cycle is assumed to be of a fixed duration equal to one month. Let us consider, then, a homogeneous group with fecundability p in which the couples were married on the same date and had no premarital conceptions. If the original number is put equal to unity, there are p conceptions expected the first month, $p(1-p)$ the second, $p(1-p)^2$ the third, and so on.

In heterogeneous groups characterized by a density $f(p)$, the expected number of conceptions in the first month is $\bar{p}$ and in the second month it is $\bar{p}-\bar{p}^2(1+c^2)$, where c is the coefficient of variation of p [c is equal to the standard deviation divided by the mean]. The complement of the ratio of the second month's conceptions to those of the first is expected to be $1-\left[\dfrac{\bar{p}-\bar{p}^2(1+c^2)}{\bar{p}}\right]=\bar{p}(1+c^2)$.

This is the same as the mean fecundability of women conceiving the first month, which may be written as:

$$\int p\,\frac{p f(p)}{\bar{p}}\,dp=\bar{p}(1+c^2).\tag{22}$$

More generally, the complement of the ratio of the expected conceptions of the $(n+1)$th month to those of the nth month is equal to the mean fecundability of women conceiving in the nth month. The mean fecundability of women conceiving in the course of the first n months of marriage is equal to a weighted mean of the preceding quantities, the weights being the number of women conceiving the first, second, ..., nth month, i.e. the denominators of the preceding quantities. It is written:

$$\sum_{i=0}^{i=n-1}\frac{\int p\cdot p(1-p)^i f(p)\,dp}{\int p(1-p)^i f(p)\,dp}\times\frac{\int p(1-p)^i f(p)\,dp}{\sum\limits_{0}^{n-1}\int p(1-p)^i f(p)\,dp}.\tag{23}$$

As n increases indefinitely, this quantity approaches $\bar{p}$, which is thus the mean fecundability of women who conceive at least once.

The essentials of this theory are due to C. Gini. ...

Intervals Between Births

We have already seen that in the case of a homogeneous group (ϕ,g), the mean interval between births is $g+1/\phi$; by a change of origin we have equated marriage to a birth; if one returns to the real origin at marriage, the mean interval between marriage and the first birth is reduced by $g-g_0$ and becomes g_0+1/ϕ.

In discrete notation, the mean number of ovulations (or of months) between marriage and the first conception is equal to $p+2p(1-p)+\ldots+np(1-p)^{n-1}$, that is, on setting $1-p=q$,

$$p\,\frac{d}{dq}\left[\sum_{i=1}^{\infty}q^i\right]=p\,\frac{d}{dq}\left[\frac{q}{1-q}\right]=\frac{1}{p}.\tag{24}$$

379

If k is the mean interval, counted in menstrual cycles (or months), between marriage and the first ovulation, the mean interval between marriage and first conception is equal to $(k-1) + \dfrac{1}{p}$. If we assume k equal to $1/2$, the mean interval is $1/p - 1/2$. Division by 13 (or by 12) gives the mean interval in years; on adding g_0 we have the interval between marriage and the first live birth.

In a heterogeneous group the mean interval, $\bar{\imath}$, between births is given, in the continuous case, by the relation

$$\bar{\imath} = \iint (g + 1/\phi) f(\phi) h(g) d\phi\, dg = \bar{g} + 1/\bar{\phi}_h. \tag{25}$$

One can also write, since $i = 1/\phi'$,

$$\bar{\imath} = \int \frac{k(\phi') d\phi'}{\phi'} = \frac{1}{\bar{\phi}'_h} \tag{26}$$

Since $\bar{\phi}'_h < \bar{\phi}'$, it follows that $1/\bar{\imath} < \bar{\phi}'$, $\bar{\imath} > 1/\bar{\phi}'$. Analogous relations hold in the discrete case. In particular, the mean interval between marriage and the first conception is equal to $k - 1 + (1/\bar{p}_h)$. Evidently, $\bar{p}_h < \bar{p}$; but, the inverse of the mean interval to conception, equal to $\dfrac{\bar{p}_h}{1 - (1-k)\bar{p}_h}$ is also greater than $\bar{p}_h$. In practice the difference between $\bar{p}_h$ and $\bar{p}$ is generally expected to be large; the inverse of the mean interval must therefore be less than $\bar{p}$.

The foregoing holds in the absence of sterility; its presence may modify the mean intervals between births. Let $n(x) dx$ be the births of a given order that would occur in $(x, x + dx)$ after the preceding births in the absence of sterility, and let $F'(x)$ be the proportion of couples still fecund x years after the preceding birth, when all were, by definition, fecund. The mean interval between births becomes

$$\bar{\imath}' = \frac{\int x n(x) F'(x) dx}{\int n(x) F'(x) dx}. \tag{27}$$

This is a new weighted mean of x with larger weights for low values of x and smaller weights for high values; $\bar{\imath}'$ is therefore smaller than $\bar{\imath}$; its inverse is consequently larger than that of $\bar{\imath}$. One then wishes to determine whether $1/\bar{\imath}'$ can equal or surpass $\bar{\phi}'$. For a homogeneous group of medium or high fecundability, the reduction in the mean interval by sterility at younger ages, and even up to about 40 years, is almost negligible. The reduction at older ages, on the contrary, may be very great with low and very low values of fecundability.

From the little we know of the distribution of fecundability, its mean is sufficiently high so that sterility will not affect the interval between births until older ages. However, the dispersion of fecundability is apparently great. Hence, those with low fecundability, although probably small in number, may have a sufficient effect to reduce the mean interval appreciably.

From available information on the mean interval between marriage and the first birth by the woman's age, there does not appear to be a reduction with

380

increasing age. No doubt these data relate to populations that already practice contraception and should be verified. We think, however, we may assume that, except at older ages, the reduction in the mean interval to the first birth caused by the onset of sterility is not very important. It follows that $\overline{T}' \approx \overline{T}$, and therefore, the inverse of the mean interval between marriage and the first birth, $1/\overline{T}'_1$ will be less than $\overline{\phi}$, given that age at marriage is still low. If $\overline{T}$ is the mean interval between two births calculated for relatively low ages (the first and the second birth for example), $1/\overline{T}$ should, for the same reasons, be less than $\overline{\phi}'$.

Instead of the mean interval between two births of a given order, we could calculate the mean interval between all births. Let us examine what would happen in this case. Consider women who marry young; let x be the duration of marriage and dS the risk of sterility between x and $x+dx$. When age at marriage is low, $\dfrac{dS}{dx}$ is negligible when x is low; when $\dfrac{dS}{dx}$ is not negligible, x is large enough so that cumulative fertility is approximately equal to $\phi'x$ for a homogeneous group of fertility ϕ'.

Let us pass to the calculation of intervals. For women who will become sterile at x, let ζ be the duration of marriage at a given birth. In the interval $(\zeta, \zeta+d\zeta)$, there are $\phi'd\zeta$ births (except when ζ is low, which is unimportant for what follows). If ζ is between 0 and $x-g$, then $\phi'e^{-\phi(x-\zeta-g)}d\zeta$ of these children are last births. If ζ is between $x-g$ and x, all $\phi'd\zeta$ births are last births. The mean duration of marriage at the last childbirth is, therefore,

$$\left(x - \frac{g}{2}\right)\phi'g + \int_0^{x-g} \phi'\zeta e^{-\phi(x-\zeta-g)}d\zeta;\tag{28}$$

that is,

$$\phi'\left[g\left(x - \frac{g}{2}\right) + \frac{x-g}{\phi} - \frac{1}{\phi^2} + \frac{e^{-\phi(x-g)}}{\phi^2}\right],\tag{29}$$

or, ignoring the term in $e^{-\phi(x-g)}$,

$$\phi'\left[(x-g)\left(g + \frac{1}{\phi}\right) + \frac{g^2}{2} - \frac{1}{\phi^2}\right] = x-g+\phi'\left(\frac{g^2}{2} - \frac{1}{\phi^2}\right).\tag{30}$$

The duration of marriage at first birth is equal to

$$\int_{g_0}^x \zeta\phi e^{-\phi(\zeta-g_0)}d\zeta,$$

which is

$$(g_0 + 1/\phi)(1 - e^{-\phi(x-g_0)}) - (x-g_0)e^{-\phi(x-g_0)},\tag{31}$$

which, for sufficiently large x, reduces to $g_0 + 1/\phi$. Then, the time between first and last births reduces to $x-g-g_0-1/\phi+\phi'(g^2/2-1/\phi^2)$. The number of

381

births of order 2 and higher is, on the other hand, equal to

$$\phi'x - 1 + e^{-\phi(x-g_0)} \tag{32}$$

being, for all practical purposes, $\phi'x - 1$. For a heterogeneous group, the time between first and last births is equal to $x - k$, where

$$k = \iint \left[g + g_0 + \frac{1}{\phi} - \phi'\left(\frac{g^2}{2} - \frac{1}{\phi^2}\right) \right] f(\phi) h(g) d\phi dg. \tag{33}$$

The total number of births is $\bar{\phi}'x - 1$. For all values of x, one then has $\int (\phi'x - 1)dS = \phi' \int x \, dS - \int dS$ births and, for the durations: $\int x \, dS - k \int dS$. In practice, only large values of x enter the picture. Therefore, the ratio of births to durations is in practice reduced to $\bar{\phi}'$. We emphasize, however, that this result is valid only for the assumptions made; in particular that the population is non-contracepting and that the distribution of fecundability is such that the frequency of low values (those for which $e^{-\phi x}$ and $xe^{-\phi x}$ are not negligible) is negligible.

42. A Note on the Structure of a Stochastic Model Considered by V. M. Dandekar

D. Basu (1955)

Sankhya 15: 251—252.

Mr. Dandekar starts with the stochastic process $x_0, x_1, x_2, \ldots$, ad inf. which may be characterised by the following defining postulates :

Π_1 : Each x_i can take only the two values 0 ($=$ failure) and 1 ($=$ success) and the probability that $x_0 = 1$ is p.

Π_2 : For any m (or less) consecutive x_i's at most one can be 1.

Π_3 : If any $m-1$ (or more) consecutive x_i's are known to be zeros then the next x_i is 1 with probability p.

Π_4 : If $x_0 = 0$ then the conditional stochastic process $x_1, x_2, \ldots$ is the same as the original process $x_0, x_1, x_2, \ldots$.

Let $P_n = P(x_n = 1)$, $n = 0, 1, 2, \ldots$ ($P_0 = p$, $q = 1-p$) and let $\varphi(t)$ be the generating function $\sum_{0}^{\infty} P_n t^n$.

The recurrence relation

$$P_n = pP_{n-m} + qP_{n-1}, \quad n = 1, 2, \ldots$$

where $P_r = 0$ for negative r is easily verified.

Hence

$$\varphi(t) = p + \sum_{1}^{\infty} P_n t^n$$

$$= p + \sum (pP_{n-m} + qP_{n-1}) t^n$$

$$= p + pt^m \varphi(t) + qt\varphi(t),$$

or

$$\varphi(t) = p/(1 - qt - pt^m).$$

Therefore

$$\varphi(1/z) = pz^m/(z^m - qz^{m-1} - p)$$

$$= z\{pz^{m-1}/(z-\alpha_1) \ldots (z-\alpha_m)\}.$$

It is easily checked that the zeros $\alpha_1, \alpha_2, \ldots, \alpha_m$ of the polynomial $z^m - qz^{m-1} - p$ are all distinct. Hence we have

$$\varphi(1/z) = z \sum_{1}^{m} c_i (z-\alpha_i)^{-1} \quad \text{where} \quad c_i = p\alpha_i \{m\alpha_i - (m-1)q\}^{-1},$$

or

$$\varphi(t) = \sum_{1}^{m} c_i (1 - \alpha_i t)^{-1}.$$

Equating co-efficients of t^n we have

$$P_n = \sum_{1}^{m} c_i \alpha_i^n.$$

Now, it is easily seen that one of the α_i's is unity and that all the others lie within the unit circle (provided $q > 0$).

Therefore

$$\lim_{n \to \infty} P_n = \frac{p}{m - (m-1)q} = \frac{p}{1 + (m-1)p} \quad \text{(if } q > 0).$$

---

* Dandekar, V. M. (1955): Certain modified forms of binomial and Poisson distributions. *Sankhyā*, **15**, Part 3.

Mr. Dandekar considers the following problem:

If we make an abrupt start on the stochastic process $x_0, x_1, x_2,...$, i.e. if we take an x_n without knowing what n is and without knowing what happened to the previous x_i's then what is the probability that $x_n = 1$?

The question stated as above has no answer. If n were known then the answer is P_n. If we have a priori knowledge about n being a random variable then the answer is $\Sigma P_i q_i$ where $q_i = P(n = i)$. Mr. Dandekar arrives at the conclusion

$$P(x_n = 1) = p/\{1+(m-1)p\}$$

by an ingenious argument.

As we have noted before this is the limit of P_n as $n \to \infty$. When Mr. Dandekar makes an abrupt start on the stochastic process $(x_0, x_1, x_2, ...)$ he implicitly assumes that the process is in operation for an indefinitely long time. He then gets a new stochastic process $(y_0, y_1, y_2, ...)$ with the following characteristics:

Π_1' : The marginal distribution of each y_i is the same—each taking the two values 0 and 1 with probabilities $1-\pi$ and π respectively.

Π_2' : Same as Π_2 with x_i replaced by y_i.

Π_3' : Same as Π_3 with x_i replaced by y_i.

It is easy to verify that the above three properties may be taken as the defining postulates of the stochastic process $(y_0, y_1, y_2, ...)$. That $\pi = p/\{1+(m-1)p\}$ may then be proved as follows :

By Π_1', $P(y_{m-1} = 1) = \pi$. By Π_2', the event $y_{m-1} = 1$ can happen only if $y_0 = y_1 = ... = y_{m-2} = 0$ and by Π_3', $P(y_{m-1} = 1 | y_0 = y_1 = ... = y_{m-2} = 0) = p$. By Π_1' and Π_2', the probability that at least one of the first $(m-1)$ y_i's is 1 is $(m-1)\pi$.

Hence

$$\pi = \{1-(m-1)\pi\}p$$

or

$$\pi = p/\{1+(m-1)p\}.$$

If it is known that r $(r < m-1)$ consecutive y_i's are zeros then the probability that the next y_i is 1 is

$$1 - \frac{1-(r+1)\pi}{1-r\pi} = \pi/(1-r\pi).$$

(This follows from Π_1' and Π_2').

I wish to thank Professor C. R. Rao for drawing my attention to this problem.

43. On the Time Required for Conception

MINDEL C. SHEPS (1964)

From *Population Studies* 18. Excerpts are from pages 85—92.

We omit the appendix, and sections of the paper discussing correlations between two successive conception delays and numerical results. (The maximum likelihood estimates for the mean and variance of fecundability, developed in the appendix, are the $\bar{p}$ and V_p given in our introduction to Henry, paper 41 above.)

Introduction

The purpose of this paper is to investigate some characteristics of rates of conception in a mixed or heterogeneous population. Consider that one has data showing, for a sample of couples, the number of women conceiving in each consecutive month of observation. These data on the monthly incidence of conception may be used to characterize the fecundability (monthly chance of conception) of the couples or to estimate the effectiveness of a contraceptive method in use by them. The data may also be regarded as showing the distribution of the interval to conception. This interval constitutes one component of the interval between successive births and hence is a subject of interest in the study of fertility.[1]

If we accept the reasonable hypothesis that the couples in the sample vary among themselves in their fecundability, what can we expect our data to show? How will the monthly incidence of conception vary? What is the expected distribution of the intervals to conception? Given a set of data as described above, can we arrive at any conclusions about the underlying distribution of fecundability? The results to be given below, in the attempt to answer these questions, are general in the sense that they do not depend on any assumptions regarding the specific form of the underlying distribution of fecundability.

The discussion is related to part of the subject matter considered by Potter and Parker in the preceding paper.[2] As is well illustrated in that paper among others,[3] a reasonable model for the distribution of the intervals to conception assists both in interpreting observed data and in considering reasonable expectations for a group of women. A number of investigators have considered this problem and have developed theoretical formulations for it, as many of these formulations have either: (1) involved the admittedly unrealistic assumption that all the couples in a population have an equal probability of conception per month, (2) utilized numerical examples to illustrate what might result if the population consisted of some mixtures of couples having two or three different levels of fecundability, or (3) assumed that fecundability has a continuous frequency distribution of a specified form.

The most recent example of the last procedure is given in the preceding paper by Potter and Parker, who assume that the frequency distribution of fecundability

---

[1] The importance of birth intervals as an approach to fertility analysis and their components are discussed for example by: L. Henry, 'Intervals between confinements in the absence of birth control', *Eugenics Quarterly*, **5**, 1958, pp. 200—11; R.G. Potter, Jr., 'Birth intervals: Structure and change', *Population Studies*, **17**, 2, 1963, pp. 155—166; and M.C. Sheps and E.B. Perrin, in 'Changes in birth rates as a function of contraceptive effectiveness: some applications of a stochastic model', *American Journal of Public Health*, **53**, 1963, pp. 1031—46, and in 'The distribution of birth intervals under a class of stochastic fertility models', *Population Studies*, **17**, 3, 1964, 321—31.

[2] R.G. Potter, Jr., and M.P. Parker, 'Predicting the time required to conceive', *Population Studies* (this issue).

[3] See, for example: C. Tietze, 'Differential fecundity and effectiveness of contraception', *The Eugenics Review*, **50**, 1959, pp. 231—4; R.G. Potter, Jr., *loc cit.*, and 'Length of the observation period as a factor affecting the contraceptive failure rate', *Milbank Memorial Fund Quarterly*, **38**, 1960, 142—4 and other references cited by Potter and Parker.

is a unimodal Type I (Beta type) distribution. They derive a number of important and interesting results, including expressions for the mean and variance of the conception delay and the correlation between two successive delays in the same woman. Further, they estimate the parameters of Type I distributions from two sets of data and illustrate their results numerically by using the estimated values. Many of their inferences have general applications.

Gini and Henry investigated some properties of the expected results for more general distributions of fecundability.[4] The present paper will extend this investigation, considering some of the same problems as are discussed by Potter and Parker but the assumption of a specified frequency distribution of fecundability is relaxed to permit the distribution to assume practically any shape.[5] It can, for example, be a bimodal distribution in which an appreciable proportion of couples use highly effective contraceptives (with resultant low or zero fecundability) while the fecundability of the remainder depends on the interaction between inherent fecundity, patterns of sexual relations and somewhat careless contraceptive use. Accordingly, some of the conclusions constitute a generalization of earlier results; specifically they support and generalize some of the conclusions in the preceding paper. In addition, parameters that characterize a distribution that may underline a set of observations are suggested. Estimators of these parameters are derived for situations where only some of the women have conceived when the data were collected, as well as from complete observations.

Assumptions

The assumptions made here agree with those of Potter and Parker, with the exception already indicated. They are:
1. Conception is a chance (random) occurrence.
2. The fecundability of each couple in a population remains constant during the period of observation.
3. Fecundability varies between couples in the population in an unspecified way.

Conceptions, regardless of the outcome of the pregnancy, are considered here. In conformity with the preceding paper, let the fecundability of a particular couple be denoted as p. Postulate an arbitrary distribution of fecundability $\Phi(p)$, where $0 \leq p \leq 1$ and $\int_0^1 \Phi(p) dp = 1$.[6] A group of couples with identical values of p will be referred to as a homogeneous population, and a group with a variety of values, distributed as $\Phi(p)$, will be referred to as a heterogeneous population. A

---

[4] C. Gini, 'Premières recherches sur la fécondabilité de la femme', *Proc. of the International Mathematics Congress*, 1924, 889—92, and L. Henry, 'Fondements théoriques des mesures de la fécondité naturelle', *Rev. de l'Inst. Int. de Statistique*, **21**, 3, 1953, 135—51. Among others, the results in expressions (1), (2) and (5) below were given by Henry.

[5] With some restrictions, as defined below.

[6] Since p is bounded and non-negative all moments of p and of $1-p$ are finite and non-negative. See H. Cramér, *Mathematical Methods of Statistics*, Princeton, 1946, p. 175.

heterogeneous population can be considered as an aggregate of homogeneous sub-populations.

The Proportion Conceiving Each Month

In a homogeneous population, under the assumptions given, the expected proportion of women who conceive in the first month of exposure is p. In the second month it is $p(1-p)$ which is equivalent to a proportion p of those who failed to conceive the first month. Generally, the expected proportion conceiving monthly decreases geometrically at a ratio of $(1-p)$. At every month, the expected number conceiving is p times the number still at risk, i.e. the conditional probability of conceiving is constant.

Under the same assumptions, in a heterogeneous population the expected proportion conceiving during the first month of exposure is

$$C_1 = \int_0^1 p\Phi(p)dp = \bar{p} \tag{1}$$

which is equal to the arithmetic mean fecundability of the population. During the second month the expected proportion conceiving is:

$$C_2 = \int_0^1 p(1-p)\Phi(p)dp = E(p) - E(p^2) = \bar{p} - \bar{p}^2 - \sigma_p^2 \tag{2}$$

where $E(u)$ is the expected value of u and σ_p^2 is the variance of p in the population. Hence, a smaller proportion of the total group may be expected to conceive in the second month than would be the case for a homogeneous population with $p=\bar{p}$. The expected number conceiving in the second month, considered as a proportion of those who failed to conceive during the first month (conditional probability of conceiving in the second month), is:

$$\frac{C_2}{1-C_1} = \frac{\int_0^1 p(1-p)\Phi(p)dp}{1 - \int_0^1 p\Phi(p)dp} = \bar{p} - \frac{\sigma_p^2}{(1-\bar{p})}, \tag{3}$$

a smaller quantity than $\bar{p}$.

As shown in the appendix, similar results hold for all subsequent months, i.e. the proportion expected to conceive out of those still at risk (conditional probability) decreases monthly. Accordingly, we may conclude that in a heterogeneous population where the assumption of constant fecundability for any couple holds reasonably well, the proportion of conceptions—as in a 'contraceptive failure rate'—will tend to fall during each successive month of exposure.

This conclusion, which was previously reached by Tietze and Potter using a less general approach,[7] indicates that it may be misleading to estimate contraceptive effectiveness by calculating the number of conceptions per 100 'person years' of use. As just indicated, 100 'person years' may have very different expectations of conception rates, depending on whether they represent two years for each of 50 couples, a half year for each of 200 couples, or a mixture of long and short periods of observation. Instead, a method utilizing a life table approach, such as was recently suggested by Potter,[8] would provide a more appropriate description.

Furthermore, as is shown in the appendix and illustrated numerically, expressions (1) and (2) and their analogues lead to a method of describing the distribution of fecundability (or of contraceptive effectiveness) from knowledge of the proportion (of a sample) conceiving each month. Inferences about the distribution of fecundability, such as estimates of its mean, variance and third central moment, can be made from observations on the first three months only. They do not require information on the conception delays of all the women in the sample; nor is it necessary to exclude sterile couples for whom $p = 0$.

Distribution of Waiting Times

The probability that conception occurs in a given month after the initial exposure is simultaneously the probability that the time required for conception ('waiting time') will be that number of months. In a homogeneous population, the time required for conception is, on the defined assumptions, distributed geometrically. Thus the mean, the variance and other moments are known functions of p, and more specifically of the powers of $(1/p)$.

A heterogeneous population, as previously defined, may be considered to generate a compound geometric distribution of waiting times. In such a population the expected values of the moments of the waiting times can be obtained by calculating the expected values of the appropriate functions of p. This procedure will be demonstrated after a review of the moments of the simple, homogeneous geometric distribution.

(a) Distribution of Waiting Times of a Homogeneous Population

In the literature there is some variation in the definitions of the geometric distribution describing the process of conception in a homogeneous population, and therefore variation in the formulae derived for the mean conception delay. The apparent disagreement originates simply in two different ways of counting the months of exposure. Thus one may say that, given a constant value of p, the probability of conception during the xth month is $p(1 - p)^{x-1}$. One may also say that the probability that the delay to conception will be exactly x months (i.e. that conception will occur during month $(x + 1)$ is $p(1 - p)^x$. The first formulation

---

[7] See the papers cited in footnotes 2 and 3.
[8] R.G. Potter, 'Additional measures of use—effectiveness in contraception', *Milbank Memorial Fund Quarterly* **41**, 1963, pp. 400—18.

might be said to refer to the number of *trials* necessary (each cycle being defined as a single trial), whereas the second refers to the conception *delay*. The two alternatives may be shown as follows:

Cycle (i.e. trial) during which conception occurs	Delay (in cycles)	Probability
1st	0	p
2nd	1	$p(1-p)$
–	–	–
–	–	–
xth	$x-1$	$p(1-p)^{x-1}$

If one uses the values in the first column, counting the month when conception occurs, the probability of conceiving after zero trials is zero, and the mean number of trials needed is $\frac{1}{p}$. If one uses the values in the second column, not counting the month when conception occurs, the probability of a delay of zero months is p, and the mean delay is equal to $\frac{(1-p)}{p}$ months.

Depending on the choice made, the expression relating to month x is, as shown before, $p(1-p)^{x-1}$ or $p(1-p)^x$. Letting $q=1-p$, the moment generating functions[9] of these two distributions are:

(a) for the number of trials:[10]

$$C(s) = \sum_{x=1}^{x=\infty} pe^s(qe^s)^{x-1} = \frac{pe^s}{1-qe^s} \tag{4A}$$

and (b) for the conception delay:

$$D(s) = \sum_{x=0}^{x=\infty} p(qe^s)^x = \frac{p}{1-qe^s}. \tag{4B}$$

From (4A) or (4B), the kth moment about zero of the waiting time, i.e. $E(x^k)$, is obtained by differentiating the generating function k times with respect to s and then evaluating the derivative at $s=0$. This gives, for the two homogeneous distributions, the first three moments as shown in Table 1.[11]

---

[9] M.G. Kendall and A. Stuart, *The Advanced Theory of Statistics*, Vol. 1, London, 1958.

[10] Since the probability of a conception after zero trials is zero in this case, summation from $x=1$ is equivalent to summation from $x=0$.

[11] The third moments are derived to exemplify the general approach and because these moments may be needed for studies of the distribution of the number of births to a group of women in a specified interval following marriage. See E. B. Perrin and M. C. Sheps, 'Human Reproduction. A Stochastic Process'. *Biometrics*, **20**, 1, 1964, 28—45.

(b) Distribution of the Waiting Times of a Heterogeneous Population

For the discussion that follows, only fecund women may be included (i.e. $0 < p \le 1$) and it is assumed that conception times of the whole fertile population are known. The kth moment of x in a heterogeneous population is the expected value of the components of the kth moment of x in a homogeneous population.

Table 1. Moments of the homogeneous geometric distributions.

	Number of trials (4A)	Delay (4 B)
$E(x)$ (mean)	$\dfrac{1}{p}$	$\dfrac{q}{p} = \dfrac{1}{p} - 1$
$E(x^2)$	$\dfrac{1+q}{p^2} = \dfrac{2}{p^2} - \dfrac{1}{p}$	$\dfrac{q}{p} + \dfrac{2q^2}{p^2} = \dfrac{2}{p^2} - \dfrac{3}{p} + 1$
$E(x^3)$	$\dfrac{6}{p^3} - \dfrac{6}{p^2} + \dfrac{1}{p}$	$\dfrac{6}{p^3} - \dfrac{12}{p^2} + \dfrac{7}{p} - 1$
$E(x^2) - [E(x)]^2$ (variance)	$\dfrac{q}{p^2} = \dfrac{1}{p^2} - \dfrac{1}{p}$	
$E[x - E(x)]^3$ (third central moment)	$\dfrac{2}{p^3} - \dfrac{3}{p^2} + \dfrac{1}{p}$	

Thus, for example, the second moment of the delay (distribution 4 B) will be $E\left(\dfrac{2}{p^2} - \dfrac{3}{p} + 1\right) = E\left(\dfrac{2}{p^2}\right) - E\left(\dfrac{3}{p}\right) + 1$. We accordingly need to derive the expected values of powers of $\dfrac{1}{p}$ in a compound geometric distribution. From the usual definition of the harmonic mean as $\left[\displaystyle\int_0^1 \dfrac{f(u)\,du}{u}\right]^{-1}$, it is clear that $E\left(\dfrac{1}{p}\right)$ is the reciprocal of the harmonic mean. Let the harmonic mean be designated p' and its reciprocal as R. Then:

$$E\left(\frac{1}{p}\right) = \int_0^1 \frac{1}{p}\,\Phi(p)\,dp = \frac{1}{p'} = R. \tag{5}$$

Further, let the mean square of the deviations of $\dfrac{1}{p}$ about this reciprocal be denoted Q. Q is thus a measure of dispersion, a kind of 'variance', of the reciprocals of p about the harmonic mean. In order that the results to be given be meaningful it is necessary that the distribution $\Phi(p)$ be such that Q is finite.[12]

---

[12] An analogous point is made by Potter and Parker with reference to the Beta distribution.

Now

$$Q = \int_0^1 \left(\frac{1}{p} - \frac{1}{p'}\right)^2 \Phi(p)\,dp = E\left(\frac{1}{p^2}\right) - \left[E\left(\frac{1}{p'}\right)\right]^2, \tag{6}$$

whence

$$E\left(\frac{1}{p^2}\right) = Q + R^2. \tag{7}$$

In terms of the Type I distribution (Beta distribution) discussed by Potter and Parker, $R = \dfrac{a+b-1}{a-1}$ and $Q = \dfrac{b(a+b-1)}{(a-1)^2(a-2)}$. Therefore, in this type of distribution, given a fixed value of $p', Q = \dfrac{bR}{(a-1)(a-2)}$. It increases with increasing b, and decreases as a increases.

Similarly, we may define the third moment of $\dfrac{1}{p}$ about its mean (a measure of absolute skewness) as:

$$\Delta = \int_0^1 \left(\frac{1}{p} - \frac{1}{p'}\right)^3 \Phi(p)\,dp = E\left(\frac{1}{p^3}\right) - 3\left(\frac{1}{p'}\right)E\left(\frac{1}{p^2}\right) + 2\left(\frac{1}{p'}\right)^3,$$

where a finite value of Δ is assumed. Then

$$E\left(\frac{1}{p^3}\right) = \Delta + 3QR + R^3. \tag{8}$$

We may now apply the result of expressions (5)—(8) to the values in Table 1 to obtain the moments of the heterogeneous distribution. For example, as already shown, the second moment ($M^{(2)}$) of the conception delay (distribution 4 B) is $E\left(\dfrac{2}{p^2}\right) - E\left(\dfrac{3}{p}\right) + 1$. From (5) and (7) this is equal to $2(Q+R^2) - 3R + 1$. The variance of the delay is then $M^{(2)} - M^2$ which is equal to $2Q + R(R-1)$. More complete results are displayed in Table 2.

In the case of a homogeneous distribution $Q = 0, \Delta = 0$, and the results in Table 2 reduce to those in Table 1. The mean delay of a heterogeneous population is the same as for a homogeneous population with $p = p'$. Since the harmonic mean of any distribution is smaller than its arithmetic mean, the mean delay in a heterogeneous population is accordingly longer than might be expected from the arithmetic mean of the distribution or the proportion conceiving in the first month of exposure. In other words, $R = 1/p'$ is always greater than $1/\bar{p}$, where $\bar{p}$ is the (arithmetic) average fecundability in the population.

Further, the variance of the waiting times in a heterogeneous population exceeds that of a homogeneous population with the same mean delay by twice the

Table 2. Moments of the heterogenous (compound) geometric distributions

	Number of trials (4A)	Delay (4B)
Mean number of months: (M)	$E\left(\dfrac{1}{p}\right) = R = 1/p'$	$E\left(\dfrac{1}{p}\right) - 1 = R - 1 = (1 - p')/p'$
Second moment: $M^{(2)}$	$2Q + 2R^2 - R$	$2Q + 2R^2 - 3R + 1$
Third moment: $M^{(3)}$	$6\Delta + 18QR - 6Q$ $+ 6R^3 - 6R^2 + R$	$6\Delta + 18QR - 12Q$ $+ 6R^3 - 12R^2 + 7R - 1$
Variance: $M^{(2)} - M^2$	$2Q + R(R-1)$	
Third central moment: $M^{(3)} - 3(M)M^{(2)} + 2M^3$	$6\Delta + 12QR - 6Q + R(R-1)(2R-1)$	

$$\text{where } Q = \int_0^1 \left(\frac{1}{p} - \frac{1}{p'}\right)^2 \Phi(p)\,dp$$

$$\text{and } \Delta = \int_0^1 \left(\frac{1}{p} - \frac{1}{p'}\right)^3 \Phi(p)\,dp.$$

quantity Q. This difference is always positive, since Q, the mean of the squared deviations $\left(\dfrac{1}{p} - \dfrac{1}{p'}\right)^2$, is always greater than zero in a mixed or compound distribution.

Branching Theory and Other Stochastic Processes

Branching theory has its origin in a question concerning the extinction of family names posed by A. DeCandolle (1873) and Francis Galton (Galton and Watson 1874): did the fact that many of the great men of the past had no living descendents imply that the genetic constitution of the race was deteriorating? Galton, responding to DeCandolle's suggestion that all names disappear in time, put the problem in its correct context: "[If] by the ordinary law of chances, a large proportion of families are continually dying out, it evidently follows that, until we know what that proportion is, we cannot estimate whether any dimunition of surnames among families whose history we can trace is or is not a sign of their diminished fertility" (Galton and Watson 1874, p. 138).

H. W. Watson provided Galton with a solution for the probability x of *eventual* extinction of surnames in the form

$$x = p_0 + p_1 x + p_2 x^2 + \cdots,$$

where the terms in $p_n x^n$ are the probabilities that a man has n sons (p_n) all of whom give rise to lines that eventually vanish. In other words, the probability that a man's line is extinguished will be equal to the probability that the lines of all of his sons are extinguished. Galton and Watson were not correct in their conclusion, which had been DeCandolle's, that if all names have a finite probability of extinction virtually all become extinct; their error having been to neglect roots to the equation other than 1. Their article is included here as paper 44 and is followed by a modern presentation of the correct solution, from Feller (1968). The probability of extinction was first found by J. F. Steffensen (1930, 1933); it is the lesser root x in the interval $p_0 \leq x \leq 1$.

The application of stochastic theory to demography has its more general development in works by G. Udny Yule (1924) and W. H. Furry (1937) exploring pure birth processes; later expanded by Feller (1939) and others to incorporate both birth and death. For the simplest case, which assumes constant birth and death probabilities independent of age, Feller showed that a population with birth rate b and death rate d would have the expected size and variance at time t:

$$E\left[\frac{N(t)}{N_0}\right] = e^{(b-d)t} = e^{rt}$$

$$\text{Var}\left[\frac{N(t)}{N_0}\right] = \frac{b+d}{b-d}e^{rt}(e^{rt}-1) \qquad b \neq d$$

$$= 2bt \qquad\qquad b = d.$$

The mean is the same for both deterministic and stochastic models, but the variance term of the stochastic case establishes a finite probability of extinction, becoming certainty as $t \to \infty$ if $b \leq d$. In deterministic models a stationary population does not become extinct, in stochastic models it does. Feller's and related works are reviewed by David Kendall (1949), paper 46. Kendall also discusses stochastic treatment of the two-sex problem: this section of his article is presented in chapter 7 as paper 52.

A stochastic treatment of population where birth and death rates are age specific is due to W.A. O'N. Waugh (1955), whose paper follows Kendall. Waugh develops the generating function for population size and finds extinction probabilities for the case in which individual lives end by death or fission, and shows that this may be generalized for other birth processes. The difficulty of manipulating generating functions where fertility is distributed among several age groups and is not coincident with death establishes severe practical limitations to their use.

J. H. Pollard (1966), paper 48, has shown that under very general assumptions an alternative of great flexibility is provided through matrix analysis. Applications to population projections are given by Z. M. Sykes (1969), who considers the treatment of variances and covariances in different projection models, and Tore Schweder (1971).

44. On the Probability of the Extinction of Families

Francis Galton and H. W. Watson (1874)

Journal of the Royal Anthropological Institute 4: 138—144.

THE decay of the families of men who occupied conspicuous positions in past times has been a subject of frequent remark, and has given rise to various conjectures. It is not only the families of men of genius or those of the aristocracy who tend to perish, but it is those of all with whom history deals, in any way, even of such men as the burgesses of towns, concerning whom Mr. Doubleday has inquired and written. The instances are very numerous in which surnames that were once common have since become scarce or have wholly disappeared. The tendency is universal, and, in explanation of it, the conclusion has been hastily drawn that a rise in physical comfort and intellectual capacity is necessarily accompanied by diminution in "fertility" —using that phrase in its widest sense and reckoning abstinence from marriage as sterility. If that conclusion be true, our population is chiefly maintained though the "proletariat," and thus a large element of degradation is inseparably connected with those other elements which tend to ameliorate the race. On the other hand, M. Alphonse De Candolle has directed attention to the fact that, by the ordinary law of chances, a large proportion of families are continually dying out, and it evidently follows that, until we know what that proportion is, we cannot estimate whether any observed diminution of surnames among the families whose history we can trace, is or is not a sign of their diminished "fertility." I give extracts from M. De Candolle's work in a foot-note,* and may add that, although I have not hitherto published anything on the matter, I took considerable pains some years ago to obtain numerical results in respect to this very problem. I made certain very simple, but not very inaccurate, suppositions, concerning average fertility, and I worked to the nearest integer, starting with 10,000 persons, but the computation became intolerably tedious after a few steps, and I had to abandon it. More recently, having first privately applied in vain to some mathematicians, I put the problem into a shape suited to mathematical treatment, and proposed it in the pages of a well-known mathematical periodical of a high class, the "Educational Times."

* "Au milieu des renseignements précis et des opinions très-sensées de MM. Benoiston de Châteauneuf, Galton, et autres statisticiens, je n'ai pas rencontré la réflexion bien importante qu'ils auraient dû faire de l'extinction *inévitable* des noms defamille. Évidemment tous les noms doivent s'éteindre ....Un mathématicien pourrait calculer comment la réduction des noms ou titres aurait lieu, d'après la probabilité des naissances toutes féminines ou toutes masculines ou mélangées et la probabilité d'absence de naissances dans un couple quelconque," etc.—Alphonse de Candolle, "Histoire des Sciences et des Savants," 1873.

It met with poor success at first, because the answer it received was from a correspondent who wholly failed to perceive its intricacy, and his results were totally erroneous. My friend the Rev. H. W. Watson then kindly, at my request, took the problem in hand, and published his first results in the above-mentioned periodical. These have since been considerably extended, and form the subject of the following paper. They do not give what can properly be called a general solution, but they do give certain general results. They show (1) how to compute, though with great labour, any special case; (2) a remarkably easy way of computing those special cases in which the law of fertility approximates to a certain specified form; and (3), how all surnames tend to disappear. I therefore feel sure that Mr. Watson's memoir will be of interest to the Anthropological Institute, and I beg to submit it to their notice, both for its intrinsic value and in hopes that other mathematicians may pursue the inquiry and attain still nearer to a complete solution of this very important problem.

The form in which I originally stated the problem is as follows. I purposely limited it in the hope that its solution might be more practicable if unnecessary generalities were excluded :—

A large nation, of whom we will only concern ourselves with the adult males, N in number, and who each bear separate surnames, colonise a district. Their law of population is such that, in each generation, a_0 per cent. of the adult males have no male children who reach adult life; a_1 have one such male child; a_2 have two; and so on up to a_5 who have five. Find (1) what proportion of the surnames will have become extinct after r generations; and (2) how many instances there will be of the same surname being held by m persons.

Discussion of the problem by the Rev. H. W. Watson.

Suppose that at any instant all the adult males of a large nation have different surnames, it is required to find how many of these surnames will have disappeared in a given number of generations upon any hypothesis, to be determined by statistical investigations, of the law of male population.

Let, therefore, a_0 be the percentage of males in any generation who have no sons reaching adult life, let a_1 be the percentage that have one such son, a_2 the percentage that have two, and so on up to a_q, the percentage that have q such sons, q being so large that it is not worth while to consider the chance of any man having more than q

adult sons—our first hypothesis will be that the numbers a_0, a_1, a_2, etc., remain the same in each succeeding generation. We shall also, in what follows, neglect the overlapping of generations—that is to say, we shall treat the problem as if all the sons born to any man in any generation came into being at one birth, and as if every man's sons were born and died at the same time. Of course it cannot be asserted that these assumptions are correct. Very probably accurate statistics would discover variations in the values of a_0, a_1, etc., as the nation progressed or retrograded ; but it is not at all likely that this variation is so rapid as seriously to vitiate any general conclusions arrived at on the assumption of the values remaining the same through many successive generations. It is obvious also that the generations must overlap, and the neglect to take account of this fact is equivalent to saying, that at any given time we leave out of consideration those male descendants of any original ancestor who are more than a certain average number of generations removed from him, and compensate for this by giving credit for such male descendants, not yet come into being, as are not more than that same average number of generations removed from the original ancestors.

Let then $\dfrac{a_0}{100,}$ $\dfrac{a_1}{100,}$ $\dfrac{a_2}{100,}$ etc., up to $\dfrac{a_q}{100,}$ be denoted by the symbols t_0, t_1, t_2, etc., up to t_q, in other words, let t_0, t_1, etc., be the chances in the first and each succeeding generation of any individual man, in any generation, having no son, one son, two sons, and so on, who reach adult life. Let N be the original number of distinct surnames, and let $_rm_s$ be the fraction of N which indicates the number of such surnames with s representatives in the rth generation.

Now, if any surname have p representatives in any generation, it follows from the ordinary theory of chances that the chance of that same surname having s representatives in the next succeeding generation is the coefficient of x^s in the expansion of the multinomial

$$(t_0 + t_1 x + t_2 x^2 + , \text{ etc. } + t_q x^q)^p$$

Let then the expression $t_0 + t_1 x + t_2 x^2 +$ etc. $+ t_q x^q$ be represented by the symbol T.

Then since, by the assumption already made, the number of surnames with no representative in the $r-1$th generation is $_{r-1}m_0$ N, the number with one representative $_{r-1}m_1 .$ N, the number with two $_{r-1}m_2 .$ N and so on, it follows, from what we last stated, that the number of surnames with s representatives in the rth generation must be the coefficient of x^s in the expression

$$\left\{ {_{r-1}m_0} + {_{r-1}m_1}\mathrm{T} + {_{r-1}m_2}\mathrm{T}^2 + \text{ etc. } + {_{r-1}m_{q^{r-1}}}\mathrm{T}^{q^{r-1}} \right\} \mathrm{N}$$

If, therefore, the coefficient of N in this expression be denoted by $f_r(x)$ it follows that $_{r-1}m_1$, $_{r-1}m_2$ and so on, are the coefficients of x, x^2 and so on, in the expression $f_{r-1}(x)$.

If, therefore, a series of functions be found such that

$$f_1(x) = t_0 + t_1 x + \text{ etc. } + t_q x^q \text{ and } f_r(x) = f_{r-1}(t_0 + t_1 x \text{ etc. } + t_q x^q)$$

then the proportional number of groups of surnames with s representatives in the rth generation will be the coefficient of x^s in $f_r(x)$ and the actual number of such surnames will be found by multiplying this coefficient by N. The number of surnames unrepresented or become extinct in the rth generation will be found by multiplying the term independent of x in $f_r(x)$ by the number N.

The determination, therefore, of the rapidity of extinction of surnames, when the statistical data, t_0, t_1, etc., are given, is reduced to the mechanical, but generally laborious process of successive substitution of $t_0 + t_1 x + t_2 x^2 +$ etc., for x in successively determined values of $f_r(x)$, and no further progress can be made with the problem until these statistical data are fixed; the following illustrations of the application of our formula are, however, not without interest.

(1) The very simplest case by which the formula can be illustrated is when $q=2$ and t_0, t_1, t_2 are each equal to $\frac{1}{3}$.

Here $f_1(x) = \dfrac{1+x+x^2}{3}$ $f_2(x) = \dfrac{1}{3}\left\{ 1 + \dfrac{1}{3}(1+x+x^2) + \dfrac{1}{9}(1+x+x^2)^2 \right\}^2$

and so on.

Making the successive substitutions, we obtain

$$f_2(x) = \frac{1}{3}\left\{ \frac{13}{9} + \frac{5x}{9} + \frac{6x^2}{9} + \frac{2x^3}{9} + \frac{x^4}{9} \right\}$$

$$f_3(x) = \frac{1249}{2187} + \frac{265x}{2187} + \frac{343x^2}{2187} + \frac{166x^3}{2187} + \frac{109x^4}{2187} + \frac{34x^5}{2187} + \frac{16x^6}{2187} + \frac{4x^7}{2187} + \frac{x^8}{2187}$$

$f_4(x) = {\cdot}63183 + {\cdot}08306x + {\cdot}10635x^2 + {\cdot}07804x^3 + {\cdot}06489x^4 + {\cdot}05443x^5 + {\cdot}01437x^6$
$\qquad + {\cdot}01692x^7 + {\cdot}01144x^8 + {\cdot}00367x^9 + {\cdot}00104x^{10} + {\cdot}00015x^{11} + {\cdot}00005x^{12}$
$\qquad\qquad + {\cdot}00001x^{13} + 00000x^{14} + {\cdot}00000x^{15} + {\cdot}00000x^{16}$

and the constant term in $f_5(x)$ or $_5m_0$ is therefore

$${\cdot}63183 + \frac{{\cdot}08306}{3} + \frac{{\cdot}10635}{9} + \frac{{\cdot}07804}{27} + \frac{{\cdot}06489}{81} + \frac{{\cdot}05443}{243} + \frac{{\cdot}01437}{729} + \frac{{\cdot}01692}{2187} + \frac{{\cdot}01144}{6561}$$

$$ + \frac{{\cdot}00367}{19683} + \frac{{\cdot}00104}{59049} + \frac{{\cdot}00015}{177147} + $$

The value of which to five places of decimals is ${\cdot}67528$.

The constant terms, therefore in f_1, f_2 up to f_5 when reduced to decimals, are in this case ${\cdot}33333$, ${\cdot}48148$, ${\cdot}57110$, ${\cdot}64113$, and ${\cdot}65628$ respectively. That is to say, out of a million surnames at starting, there have disappeared in the course of one, two, etc., up to five generations, 333333, 481480, 571100, 641130, and 675280 respectively.

The disappearances are much more rapid in the earlier than in the later generations. Three hundred thousand disappear in the first generation, one hundred and fifty thousand more in the second, and so on, while in passing from the fourth to the fifth, not more than thirty thousand surnames disappear.

All this time the male population remains constant. For it is evident that the male population of any generation is to be found by multiplying that of the preceding generation, by $t_1 + 2t_2$, and this quantity is in the present case equal to one.

If axes Ox and Oy be drawn, and equal distances along Ox represent generations from starting, while two distances are marked along every ordinate, the one representing the total male population in any generation, and the other the number of remaining surnames in that generation, of the two curves passing through the extremities of these ordinates, the *population* curve will, in this case, be a straight line parallel to Ox, while the *surname* curve will intersect the population curve on the axis of y, will proceed always convex to the axis of x, and will have the positive part of that axis for an asymptote.

The case just discussed illustrates the use to be made of the general formula, as well as the labour of successive substitutions, when the expressions $f_1(x)$ does not follow some assigned law. The calculation may be infinitely simplified when such a law can be found ; especially if that law be the expansion of a binomial, and only the extinctions are required.

For example, suppose that the terms of the expression $t_0 + t_1 x +$ &c. $+ t_q x_q$ are proportional to the terms of the expanded binomial

$(a + bx)^q$ *i. e.,* suppose that $t_0 = \dfrac{a^q}{(a+b)^q}$ $t_1 = q\dfrac{a^{q-1}b}{(a+b)^q}$ and so on.

$$\text{Here } f_1(x) = \frac{(a+bx)^q}{(a+b)^q} \text{ and } {}_1 m_0 = \frac{a^q}{(a+b)^q}$$

$$f_2(x) = \frac{1}{(a+b)^q}\left\{ a + b\frac{(a+bx)^q}{(a+b)^q} \right\}^q$$

$$_2 m_0 = \frac{1}{(a+b)^q}\left\{ a + b_1 m_0 \right\}^q$$

$$\text{Generally } _r m_0 = \frac{1}{(a+b)^q}\left\{ a + b_{r-1} m_0 \right\}^q = \frac{b_q}{(a+b)^q}\left\{ \frac{a}{b} + {}_{r-1} m_0 \right\}^q$$

If, therefore, we wish to find the number of extinctions in any generation, we have only to take the number in the preceding generation, add it to the constant fraction $\dfrac{a}{b}$, raise the sum to the power of q, and multiply by $\dfrac{b_q}{(a+b)^q}$

With the aid of a table of logarithms, all this may be effected for a great number of generations in a very few minutes. It is by no means unlikely that when the true statistical data t_0, t_1, etc., t_q are ascertained, values of a, b, and q may be found, which shall render the terms of the expansion $(a + bx)^q$ approximately proportionate to the terms of $f_1(x)$. If this can be done, we may *approximate* to the determination of the rapidity of extinction with very great ease, for any number of generations, however great.

For example, it does not seem very unlikely that the value of q might be 5, while t_0, $t_1 \ldots t_q$ might be ·237, ·396, ·264, ·088, ·014, ·001, or nearly, $\frac{1}{4}$, $\frac{1}{3}$, $\frac{7}{24}$, $\frac{1}{23}$, $\frac{1}{138}$, and $\frac{1}{1000}$.

Should that be the case, we have $f_1(x) = \dfrac{(3 + x)^5}{4^5}$, $_1m_0 = \dfrac{3^5}{4^5}$

and generally $_r m_0 = \dfrac{1}{4^5} \left\{ 3 + _{r-1}m_0 \right\}^5$

Thus we easily get for the number of extinctions in the first ten generations respectively

$$·237, ·346, ·410, ·450, ·477, ·496, ·510, ·520, ·527, ·533$$

We observe the same law noticed above in the case of $\dfrac{1 + x + x^2}{3}$

viz., that while 237 names out of a thousand disappear in the first step, and an additional 109 names in the second step, there are only 27 disappearances in the fifth step, and only 6 disappearances in the tenth step.

If the curves of surnames and of population were drawn from this case, the former would resemble the corresponding curve in the case last mentioned, while the latter would be a curve whose distance from the axis of x increased indefinitely, inasmuch as the expression

$$t_1 + 2t_2 + 3t_3 + 4t_4 + 5t_5$$

is greater than one.

Whenever $f_1(x)$ can be represented by a binomial, as above suggested, we get the equation

$$_r m_0 = \frac{1}{(a + b)^q} \left\{ a + b_{r-1}m \right\}^q$$

whence it follows that as r increases indefinitely the value of $_r m_0$ approaches indefinitely to the value y where

$$y = \frac{1}{(a + b)} \left\{ a + by \right\}^q$$

that is where $y = 1$.

405

All the surnames, therefore, tend to extinction in an indefinite time, and this result might have been anticipated generally, for a surname once lost can never be recovered, and there is an additional chance of loss in every successive generation. This result must not be confounded with that of the extinction of the male population; for in every binomial case where q is greater than 2, we have $t_1 + 2t_2 +$ &c. $+ qt_q > 1$, and, therefore an indefinite increase of male population.

The true interpretation is that each of the quantities, $_rm_1$, $_rm_2$, &c., tends to become zero, as r is indefinitely increased, but that it does not follow that the product of each by the infinitely large number N is also zero.

As, therefore, time proceeds indefinitely, the number of surnames extinguished becomes a number of the *same order of magnitude* as the total number at first starting in N, while the number of surnames represented by one, two, three, etc., representatives is some infinitely smaller but finite number. When the finite numbers are multiplied by the corresponding number of representatives, sometimes infinite in number, and the products added together, the sum will generally exceed the original number N. In point of fact, just as in the cases calculated above to five generations, we had a continual, and indeed at first, a rapid extinction of surnames, combined in the one case with a stationary, and in the other case an increasing population, so is it when the number of generations is increased indefinitely. We have a continual extinction of surnames going on, combined with constancy, or increase of population, as the case may be, until at length the number of surnames remaining is absolutely insensible, as compared with the number at starting; but the total number of representatives of those remaining surnames is infinitely greater than the original number.

We are not in a position to assert from *actual calculation* that a corresponding result is true for every form of $f_1(x)$, but the reasonable inference is that such is the case, seeing that it holds whenever

$f_1(x)$ may be compared with $\dfrac{(a + bx)^q}{(a + b)^q}$ whatever a, b, or q may be.

45. Extinction Probabilities in Branching Processes

WILLIAM FELLER (1968 (1950))

From *An Introduction to Probability Theory and its Applications*, Volume 1, third edition (revised printing), pages 295—298. New York: John Wiley & Sons.

Denote by $\mathbf{Z}_n$ the size of the nth generation, and by P_n the generating function of its probability distribution. By assumption $\mathbf{Z}_0 = 1$ and

$$(4.1) \qquad P_1(s) = P(s) = \sum_{k=0}^{\infty} p_k s^k.$$

The nth generation can be divided into $\mathbf{Z}_1$ clans according to the ancestor in the first generation. This means that $\mathbf{Z}_n$ is the sum of $\mathbf{Z}_1$ random variables $\mathbf{Z}_n^{(k)}$, each representing the size of the offspring of one member of the first generation. By assumption each $\mathbf{Z}_n^{(k)}$ has the same probability distribution as $\mathbf{Z}_{n-1}$ and (for fixed n) the variables $\mathbf{Z}_n^{(k)}$ are mutually independent. The generating function P_n is therefore given by the compound function

$$(4.2) \qquad P_n(s) = P(P_{n-1}(s)).$$

This result enables us to calculate recursively all the generating functions. In view of (4.2) we have $P_2(s) = P(P(s))$, then $P_3(s) = P(P_2(s))$, etc. The calculations are straightforward, though explicit expressions for P_n are usually hard to come by. We shall see presently that it is nevertheless possible to draw important conclusions from (4.2)....

The first question concerning our branching process is whether it will continue forever or whether the progeny will die out after finitely many generations. Put

$$(4.5) \qquad x_n = \mathbf{P}\{\mathbf{Z}_n = 0\} = P_n(0).$$

This is the probability that the process terminates at or before the nth generation. By definition $x_1 = p_0$ and from (4.2) it is clear that

$$(4.6) \qquad x_n = P(x_{n-1}).$$

The extreme cases $p_0 = 0$ and $p_0 = 1$ being trivial, we now suppose that $0 < p_0 < 1$. From the monotone character of P we conclude then that $x_2 = P(p_0) > P(0) = x_1$, and hence by induction that $x_1 < x_2 < x_3 < \cdots$. It follows that there exists a limit $x \leq 1$, and from (4.6) it is clear that

$$(4.7) \qquad x = P(x).$$

For $0 \leq s \leq 1$ the graph of $P(s)$ is a *convex* curve starting at the point $(0, p_0)$ above the bisector and ending at the point $(1, 1)$ on the bisector. Accordingly only two situations are possible:

Case (i). The graph is entirely above the bisector. In this case $x = 1$ is the unique root of the equation (4.7), and so $x_n \to 1$. Furthermore, in this case $1 - P(s) \leq 1 - s$ for all s, and letting $s \to 1$ we see that the derivative $P'(1)$ satisfies the inequality $P'(1) \leq 1$.

Case (ii). The graph of P intersects the bisector at some point $\sigma < 1$. Since a convex curve intersects a straight line in at most two points, in this case $P(s) > s$ for $s < \sigma$ but $P(s) < s$ for $\sigma < s < 1$. Then $x_1 = P(0) < P(\sigma) = \sigma$, and by induction $x_n = P(x_{n-1}) < P(\sigma) = \sigma$. It follows that $x_n \to \sigma$ and so $x = \sigma$. On the other hand, by the mean value theorem there exists a point between σ and 1 at which the derivative P' equals one. This derivative being monotone, it follows that $P'(1) > 1$.

We see thus that the two cases are characterized by $P'(1) \leq 1$ and $P'(1) > 1$, respectively. But

$$(4.8) \qquad \mu = P'(1) = \sum_{k=0}^{\infty} kp_k \leq \infty$$

.is *the expected number of direct descendants*, and we have proved the interesting

Theorem. *If* $\mu \leq 1$ *the process dies out with probability one. If, however,* $\mu > 1$ *the probability* x_n *that the process terminates at or before the nth generation tends to the unique root* $x < 1$ *of the equation* (4.7).

In practice the convergence $x_n \to x$ is usually rapid and so with a great probability the process either stops rather soon, or else it continues forever. The expected size of the nth generation is given by $\mathbf{E}(Z_n) = P'_n(1)$. From (4.2) we get by the chain rule $P'_n(1) = P'(1)P'_{n-1}(1) = \mu\mathbf{E}(X_{n-1})$, and hence[9]

$$(4.9) \qquad \mathbf{E}(X_n) = \mu^n.$$

It is not surprising that the process is bound for extinction when $\mu < 1$, but it was not clear a priori that a stable situation is impossible even when $\mu = 1$. When $\mu > 1$ one should expect a geometric growth in accordance with (4.9). This is true in some average sense, but no matter how large μ there is a finite probability of extinction. It is easily seen that $P_n(s) \to x$ for all $s < 1$ and this means that the coefficients of s, s^2, s^3, etc., all tend to 0. After *a sufficient number of generations it is therefore likely that there are either no descendants or else a great many descendants* (the corresponding probabilities being x and $1 - x$).

---

[9] For further details see the comprehensive treatise by T. E. Harris, *The theory of branching processes*, Berlin (Springer), 1963.

46. Stochastic Processes and Population Growth

DAVID G. KENDALL (1949)

From *Journal of the Royal Statistical Society*, B, 11. Excerpts are from pages 235—237.

Kendall's paper reviews a number of contributions to the stochastic theory of population. Besides the sections reproduced here and in chapter 7, they include birth, death and immigration processes; processes that are time-dependent or periodic; maximum likelihood estimation of the birth rate in pure birth processes; comments on the stochastic treatment of logistic growth processes; and stochastic fluctuations in the age distribution.

In the section reproduced here, the differential equation (20) for the birth and death process defines the probability generating function

$$\phi(z,t) = \left(\frac{\mu(e^{(\lambda-\mu)t}-1)-z(\mu e^{(\lambda-\mu)t}-\lambda)}{\lambda e^{(\lambda-\mu)t}-\mu}\right)\left(1-z\frac{\lambda(e^{(\lambda-\mu)t}-1)}{\lambda e^{(\lambda-\mu)t}-\mu}\right)^{-1}.$$

On setting $\mu=0$ this becomes the generating function for the simple birth process, given by equation (16). For $\lambda=0$ a simple death process results.

The probability of extinction by time t is given by

$$\phi(0,t) = \frac{\mu[e^{(\lambda-\mu)t}-1]}{\lambda e^{(\lambda-\mu)t}-\mu}.$$

To find the waiting time to extinction where $\lambda\leq\mu$ note that $\phi(0,t)$ is equivalent to the cumulative proportion dying in the life table, and that the time to extinction is formally a life expectancy. We therefore have by application of standard formulas:

$$\phi(0,t) = [1-l(t)]/l_0$$

$$\mathring{e}_0 = -\int_0^\infty t l'(t)\,dt/l_0 = \int_0^\infty t\phi'(0,t)\,dt$$

$$= \mu(\lambda-\mu)^2 \int_0^\infty \frac{t e^{(\lambda-\mu)t}}{[\lambda e^{(\lambda-\mu)t}-\mu]^2}\,dt$$

$$= \begin{cases} -\dfrac{1}{\lambda}\ln\left(1-\dfrac{\lambda}{\mu}\right) & \lambda<\mu \\[2ex] \infty & \lambda=\mu. \end{cases}$$

411

For $\lambda > \mu$ the probability of extinction will be

$$\lim_{t \to \infty} \phi(0, t) = \mu/\lambda$$

and we have the life expectancy

$$\mathring{e}_0 = \frac{\mu}{\lambda} {}_\infty a_0 + \left(1 - \frac{\mu}{\lambda}\right) \infty = \infty.$$

The conditional life expectancy ${}_\infty a_0$ for those who die is given by

$$\begin{aligned}
{}_\infty a_0 &= - \int_0^\infty t l'(t)\,dt \Big/ \int_0^\infty l'(t)\,dt \\
&= -\frac{1}{\mu} \ln\left(1 - \frac{\mu}{\lambda}\right).
\end{aligned}$$

Further treatment of birth and death processes will be found in Bailey (1964, pp. 84—105) and Keyfitz (1968, pp. 362—368: the waiting time to extinction given in this source contains a superfluous factor μ, corrected in the second edition now in press).

2. Stochastic Fluctuations in the Total Population Size

(i) *The simple birth-and-death process.*—The first example of a population process of the present type was apparently that introduced by W. H. Furry in 1937. The physical application which he had in mind will be mentioned in section 2 (iv); for the moment his model will be considered with its biological interpretation. From this point of view the system under discussion is a population of organisms multiplying in accordance with the following rules:

    (*a*) the sub-populations generated by two co-existing individuals develop in complete independence of one another;

    (*b*) an individual existing at time *t* has a chance

$$\lambda \, dt + o(dt)$$

of multiplying by binary fission during the following time-interval of length dt;

    (*c*) the "birth rate" λ is the same for all individuals in the population at all times *t*.

The rule (*b*) will usually be interpreted in the sense that at each birth just one new member is added to the population, but of course mathematically (and because the age-structure of the population is being ignored) it is not possible to distinguish between this and an alternative interpretation in which a parent always dies at birth and is replaced by two new members.

Let N_0 be the number of individuals at the initial time $t = 0$, and suppose for the moment that $N_0 = 1$; let $p_n(t)$ be the probability that the population size $N(t)$ has the value n at time t. Then

$$\frac{d}{dt} p_n(t) = (n - 1) \lambda p_{n-1}(t) - n\lambda p_n(t) \qquad (n > 1),$$

and

$$\frac{d}{dt} p_1(t) = \qquad\qquad - \lambda p_1(t), \qquad . \qquad . \qquad . \qquad . \qquad . \quad (13)$$

and so the probability-generating function

$$\varphi(z, t) = \sum_{(n)} z^n p_n(t) \qquad . \qquad . \qquad . \qquad . \qquad . \qquad . \quad (14)$$

must satisfy the partial differential equation

$$\frac{\partial \varphi}{\partial t} = \lambda z(z - 1) \frac{\partial \varphi}{\partial z}, \qquad . \qquad . \qquad . \qquad . \qquad . \quad (15)$$

the most general solution to which is of the form

$$\varphi(z, t) = \Phi\{(1 - 1/z) \, e^{\lambda t}\}.$$

But

$$\varphi(z, 0) = \Phi\{1 - 1/z\} = z,$$

and so

$$\Phi(Z) = 1/(1 - Z),$$

and

$$\varphi(z, t) = z e^{-\lambda t} [1 - z(1 - e^{-\lambda t})]^{-1} \qquad . \qquad . \qquad . \qquad . \quad (16)$$

Expansion in powers of z now gives the probability distribution for the population size at any time t; it is

$$p_n(t) = e^{-\lambda t} (1 - e^{-\lambda t})^{n-1} \qquad (n \geqslant 1), \qquad . \qquad . \qquad . \quad (17)$$

a distribution of geometric form, the common ratio of which approaches unity as t tends to infinity.

From this simple example one can begin to appreciate a number of points which prove to be of more general importance. In the first place, because of the assumption (*a*), the population size at time t, when $N_0 > 1$, is the sum of N_0 independent variables for each of which the probability distribution is given by the law (17), and so the distribution of $N(t)$ in general can readily be obtained by raising the function given by (16) to the power N_0, and then expanding in powers of z as before. The resulting distribution is obviously of negative-binomial type.

Again, when $N_0 = 1$, the mean $\overline{N}$ of $N(t)$ can readily be calculated to be

$$\overline{N}(t) = \frac{\partial}{\partial z} \varphi(1, t) = e^{\lambda t} \, . \qquad . \qquad . \qquad . \qquad . \quad (18)$$

and similarly for the variance of $N(t)$ one finds

$$\text{Var } \{N(t)\} = \frac{\partial^2}{\partial z^2}\, \varphi(1, t) - \overline{N}(\overline{N} - 1) = e^{\lambda t}(e^{\lambda t} - 1) \qquad . \qquad . \qquad . \quad (19)$$

When $N_0 > 1$, the right-hand side of each equation is to be multiplied by N_0.

It will be noticed that the mean growth of the process, given by (18), follows the same exponential law as that which appeared in (2) as the solution to the simplest deterministic equation (with λ now taking the place of ν). All the other stochastic models to be examined, with one exception, similarly mimic in their mean behaviour the corresponding deterministic model; this is the justification of the assertion sometimes made that the deterministic theory is simply an account of the expectation behaviour of the random variables which occur in the stochastic formulation. That this is *not* generally true was first pointed out by Feller (1939)....

In the deterministic theory it made no difference whether the intrinsic rate of growth ν was purely reproductive in origin, or was really a balance, $\lambda - \mu$, between a birth rate λ and a death rate μ. In the stochastic theory this is no longer true, and the birth-and-death process is quite distinct from the pure birth process just described.

The birth-and-death process was introduced by Feller (1939), who obtained the formulae (23) and (24) below. It was reconsidered in connexion with a physical application by N. Arley (1943), and the equations giving its probability-structure were first solved by C. Palm in an unpublished letter to Arley. (See Arley and Borchsenius (1945), for an account of Palm's work.) The difference from the Furry process is that there is now a mortality effect acting in the following way:

(d) an individual existing at time t has a chance

$$\mu dt + o(dt)$$

of dying in the following time-interval of length dt;

(e) the "death rate" μ is the same for all individuals at all times t.

The equation to be satisfied by the generating function is now

$$\frac{\partial \varphi}{\partial t} = (\lambda z - \mu)(z - 1)\frac{\partial \varphi}{\partial z} \qquad . \qquad . \qquad . \qquad . \qquad . \quad (20)$$

and in accordance with a remark made in one of the preceding paragraphs it will be enough to state the solution when $N_0 = 1$, since the solution for a general number of ancestors can be obtained from this. Palm's formulae can be written

$$p_0(t) = \xi_t \text{ and } p_n(t) = [1 - p_0(t)]\,(1 - \eta_t)\,\eta_t^{n-1} \qquad (n > 1) \qquad . \qquad . \quad (21)$$

where

$$\frac{\xi_t}{\mu} = \frac{\eta_t}{\lambda} = \frac{e^{(\lambda - \mu)t} - 1}{\lambda e^{(\lambda - \mu)t} - \mu}\, . \qquad . \qquad . \qquad . \qquad . \quad (22)$$

so that the distribution (when $N_0 = 1$) is now a geometric series with a modified zero term, the mean population size being

$$\overline{N}(t) = e^{(\lambda - \mu)t}. \qquad . \qquad . \qquad . \qquad . \qquad . \quad (23)$$

while

$$\text{Var } \{N(t)\} = \frac{\lambda + \mu}{\lambda - \mu}\, e^{(\lambda - \mu)t}\, \{e^{(\lambda - \mu)t} - 1\} \qquad . \qquad . \qquad . \quad (24)$$

When $\lambda = \mu$, so that the net expected rate of growth is zero and the mean population size is stationary, the solution to (20) assumes a somewhat different form which can most readily be obtained by letting $\lambda - \mu$ approach zero in the preceding formulae. It then appears that

$$p_0(t) = \xi_t = \eta_t = \lambda t/(1 + \lambda t) \qquad . \qquad . \qquad . \qquad . \quad (25)$$

while

$$\overline{N} = 1 \quad \text{and} \quad \text{Var } \{N(t)\} = 2\lambda t \qquad . \qquad . \qquad . \qquad . \quad (26)$$

(when $N_0 = 1$), and of course these results can also be obtained by a direct solution of the original equation.

A most important observation is that for *all* N_0,

$$\lim_{t \to \infty} p_0(t) = 1 \qquad \qquad \text{when } \lambda \leqslant \mu$$

and

$$\lim_{t \to \infty} p_0(t) = (\mu/\lambda)^{N_0} \qquad \text{when } \lambda > \mu \qquad . \qquad . \qquad . \quad (27)$$

414

This limit can be interpreted as the probability of the extinction of the population in a finite time, so that one can say that there will be "almost certain" extinction whenever $\lambda \leqslant \mu$, and in particular when $\lambda = \mu$ and the mean size of the population is constant. The above result, which is true whatever the initial number of individuals may be, brings out very clearly the inadequacy of the deterministic description of the growth of a population, and it throws fresh light on the questions first discussed by de Candolle, Galton and Watson in 1873. It was first established in this form by M. S. Bartlett in 1946, although a similar result for his "life-energy" process was given by Feller in 1939.

References

Arley, N. (1943), *On the Theory of Stochastic Processes, and their Application to the Theory of Cosmic Radiation*. Copenhagen.
——, and Borchsenius, V. (1945), "On the theory of infinite systems of differential equations and their application to the theory of stochastic processes and the perturbation theory of quantum mechanics," *Acta Math.*, **76**, 261–322 (especially pp. 298–9).

Bartlett, M. S. (1947), *Stochastic Processes*. (Notes of a course given at the University of North Carolina in the Fall Quarter, 1946.) (There is a copy in the Society's library.)

de Candolle, A. (1873), *Histoire des Sciences et des Savants depuis deux siècles*. Geneva. (See especially pp. 388–9.)

Feller, W. (1939), "Die Grundlagen der Volterraschen Theorie des Kampfes ums Dasein in wahrscheinlichkeitstheoretischer Behandlung," *Acta Biotheoretica*, **5**, 11–40. (Abstract in *Math. Reviews*, **1**, 22).

Furry, W. H. (1937), "On fluctuation phenomena in the passage of high energy electrons through lead," *Phys. Rev.*, **52**, 569–81.

Watson, H. W., and Galton, F. (1873), Problem 4001, in *Mathematical Questions, with their Solutions, from the 'Educational Times,'* **19**, 103–5.

47. An Age-Dependent Birth and Death Process

W. A. O'N. Waugh (1955)

From *Biometrika* 42. Excerpts are from pages 291—294.

The later sections of Waugh's paper trace the implications of birth and death processes where the birth and death rates have Pearson Type III (Gamma) distributions. They are omitted, along with Waugh's discussion of extinction probabilities for the more general case, which is expressed in terms of the fission problem.

1. Introduction

Many examples of the stochastic processes known as branching or population processes have been developed. The assumptions leading to the various models have been decided partly by mathematical convenience and partly by the attempt to reproduce features controlling the development of biological or physical populations. The model we shall develop involves a continuous time parameter and an enumerable set of states, and we can compare it with several other processes of this type, e.g. the Yule–Furry birth process, Feller's birth and death process (see Feller, 1939, 1950), and the non-Markovian birth process due to D. G. Kendall (1948).

There are two motives for the development of the present model. Mathematically, it furnishes a conveniently soluble example of a non-Markovian process, in which transitions from a given state may be to one of several other states, and in this it is a generalization of Kendall's process in which transitions are always from the given state to its successor, i.e. from state number n to state number $n+1$. For purposes of application it is based on assumptions which are not proposed as ideal but which bring in features which are perhaps a step in the direction of realism. Thus it may help in the development of yet more realistic models when they are required, and in indicating both the scope and the limitations of the models constructed under an assumption which has been widely adopted in population problems; viz. that the individuals reproduce independently of one another.

Bellman & Harris (1948, 1952) have described a stochastic process similar to the one we shall examine, and we shall adopt their notation. They consider populations of particles which reproduce by fission independently of one another. They suppose that a 'generation time' is defined for these populations which is a random variable having a general distribution function $G(t)$ (which does not depend on the size of the population, in accordance with the assumption of independence just mentioned), so that $dG(t)$ expresses the probability that a particle born at time $t = 0$ ends its life in the interval $(t, t+dt)$.

They assume that with given probabilities $q_0, q_1, q_2, \ldots$ any particle whose life has just ended is replaced by $0, 1, 2, \ldots$ new particles.

Their detailed work is carried out for the pure birth process involving binary fission only:

$$q_2 = 1, \quad q_0 = q_1 = q_3 = \ldots = 0.$$

They state that their results can easily be generalized to other modes of fission. Note that when $q_0 \neq 0$ the process is a birth *and death* process. Further developments in this direction have also been obtained by Ramakrishnan (1951) and by Reid (1953).

It may be convenient to note here that a focal point of the present investigation will be the coefficient of variation of the population size, considered as a function of the time. The importance of this function in the study of populations has been emphasized previously, e.g. by Kendall (1952a).

2. Age-dependent process involving two possible fates: death or fission

For clarity and because it leads to a model which may be useful in applications, we shall not adopt, in the main part of our investigation, the most general assumptions that are possible. We shall indicate at various points throughout the paper certain further generalizations that merely involve difficulties of technique and alterations in detail which might be worked out if applications seemed to demand it.

We shall suppose that at the end of its life an individual either disappears or is replaced by two new individuals (i.e. death without issue or reproduction by binary fission). These two new individuals then develop and reproduce independently of one another in the same way as the parent individual, i.e. if the parent individual split at time t to form them, then the probability that either of the two new individuals ends its life during $(t+\tau, t+\tau+d\tau)$ is $dG(\tau)$, the distribution $G(\tau)$ being fixed, and the same for all individuals of the population. In the notation of Bellman & Harris this would be equivalent to putting

$$q_0 \neq 0, \quad q_2 \neq 0, \quad q_1 = q_3 = \ldots = 0,$$

but we shall make the more general assumption that these probabilities depend on the age at which the parent individual's life ends. A further generalization to admit the possibility of an individual's facing more than two alternative fates is quite simple. We shall suppose that at time $t = 0$ the population consists of just one *newly born* individual. Let us name the alternatives which face an individual of the population; risk (b) is the event that its life ends with the birth of two new individuals, risk (d) is the event that it dies without issue. We shall derive the functions $q_j(t)$ and $G(t)$ as follows. Suppose that there are two functions $\lambda(t)$ and $\nu(t)$ such that, given that an individual born at time 0 is alive at time t,

$$\text{Pr}\{\text{individual succumbs to risk } (b) \text{ in } (t, t+dt)\} = \lambda(t)\,dt + o(dt),$$

$$\text{Pr}\{\text{individual succumbs to risk } (d) \text{ in } (t, t+dt)\} = \nu(t)\,dt + o(dt),$$

thus $\qquad \text{Pr}\{\text{individual's life ends in } (t, t+dt)\} = [\lambda(t) + \nu(t)]\,dt + o(dt).$

The function $[\lambda(t) + \nu(t)]$ might be called the 'force of mortality'. In the argument which follows, and which leads to the fundamental integral equation, it is convenient if $\lambda(t)$ and $\nu(t)$ are both continuous and do not vanish together at any point in $0 < t < \infty$ and are such that $\int_0^t [\lambda(u) + \nu(u)]\,du \to \infty$ as $t \to \infty$. In the latter part of the paper we assume a certain form for $\lambda(t)$ and $\nu(t)$, and it happens that these requirements of continuity, etc., are fulfilled. However, it is possible to extend the argument to cover cases in which certain discontinuities are allowed, in particular the case where an instantaneous chain reaction can occur at the moment of fission (a possibility which is mentioned by Bellman & Harris (1952)). In this case our assumption of age-dependence permits the probabilities governing the numbers of progeny at instantaneous fission to be different to those corresponding to fission after a finite life length.

From $\lambda(t)$ and $\nu(t)$ we obtain the following conditional probabilities, given that the individual's life ends in $(t, t+dt)$:

$$\text{Pr}\{\text{life ends by risk } (b)\} \equiv q_2(t) \equiv \frac{\lambda(t)}{\lambda(t) + \nu(t)},$$

$$\text{Pr}\{\text{life ends by risk } (d)\} \equiv q_0(t) \equiv \frac{\nu(t)}{\lambda(t) + \nu(t)}.$$

Let
$$G(t) \equiv 1 - \exp\left\{-\int_0^t [\lambda(u) + \nu(u)]\,du\right\}$$

$$= \text{Pr}\{\text{individual born at time 0 lives for a time less than or equal to } t\}.$$

We can imagine an idealized population in which either the risk (d) is suspended and the population develops under risk (b) alone $(\nu(t) \equiv 0)$ or vice versa. Then

$$B(t) \equiv 1 - \exp\left\{-\int_0^t \lambda(u)\,du\right\}$$

$$= \text{Pr}\{\text{individual born at time 0 lives a shorter time than } t, \text{ when risk } (d) \text{ is suspended}\},$$

$$D(t) \equiv 1 - \exp\left\{-\int_0^t \nu(u)\,du\right\}$$

$$= \text{Pr}\{\text{individual born at time 0 lives a shorter time than } t, \text{ when risk } (b) \text{ is suspended}\}.$$

If
$$g(t) = \frac{dG(t)}{dt}, \quad d(t) = \frac{dD(t)}{dt}, \quad b(t) = \frac{dB(t)}{dt},$$

then
$$q_0(t) = \frac{d(t)\,[1 - B(t)]}{g(t)}$$

and
$$q_2(t) = \frac{b(t)\,[1 - D(t)]}{g(t)},$$

and also
$$1 - G(t) = [1 - B(t)]\,[1 - D(t)].$$

3. THE INTEGRAL EQUATION FOR THE GENERATING FUNCTION OF THE POPULATION SIZE

In order to obtain the mean and variance of the population size as functions of the time we start from an integral equation for the generating function of the distribution of the population size, which is very similar to the one from which Bellman & Harris (1952) derive their results. It is possible to give a definition of a suitable sample space Ω and a function $r_t(\omega)$ over Ω which may be called the population size at time t, and by considerations of measurability to show that $\{r_t : t \geqslant 0\}$ is a well-defined stochastic process, and furthermore to derive the fundamental integral equation from this definition. However, we will content ourselves with this reference to the more rigid approach and will merely give the following intuitive argument which leads to the integral equation.

Let $p_r(t)$ be the probability that the population size is r at time t, given that the population consisted of a single newly born individual at time 0.

Let
$$\phi(z, t) \equiv \sum_{r=0}^{\infty} p_r(t)\, z^r \equiv E z^{r_t},$$

and let
$$h(z, t) \equiv \sum_{n=0}^{\infty} q_n(t)\, z^n.$$

Then the situation at time t may have come about in either of two ways.

(i) The initial individual may still be alive.

(ii) The initial individual's life may have ended in the interval $(u, u + du)$ where $0 \leqslant u < t$. In this case, if its life ends in fate (b), then because of the independence of the members of the population and the fact that all individuals are assumed to develop under the same probabilities as regards life length and number of progeny, we have effectively two populations developing each for a period $t - u$, from initial ancestors newly born at time u. On the other hand, if its life ends in fate (d) then the population size at time t is zero.

Hence we have the integral equation
$$\phi(z, t) = \int_0^t [q_0(u) + q_2(u)\, \phi^2(z, t - u)]\, g(u)\, du + z[1 - G(t)],$$
$$= \int_0^t h[\phi(z, t - u), u]\, g(u)\, du + z[1 - G(t)],$$

the latter form of which can easily be seen to hold when modes of fission other than binary are considered.

In terms of the functions $B(t)$, $D(t)$ and their derivatives this can be written
$$\phi(z, t) = \int_0^t [\phi(z, t - u)]^2\, b(u)\, [1 - D(u)]\, du + \int_0^t d(u)\, [1 - B(u)]\, du + z[1 - B(t)]\,[1 - D(t)].$$

If we differentiate this equation with respect to z and then put $z = 1$ we get the following equation for the mean population size at time t,

420

$$\mu_1(t) = 2 \int_0^t \mu_1(t-u)\, b(u)\, [1 - D(u)]\, du + [1 - B(t)]\, [1 - D(t)],$$

while if we differentiate twice with respect to z and then put $z = 1$ we get the equation

$$\mu_2(t) = 2 \int_0^t [\mu_1(t-u)]^2\, b(u)\, [1 - D(u)]\, du + 2 \int_0^t \mu_2(t-u)\, b(u)\, [1 - D(u)]\, du,$$

for the second factorial moment of the population size.

REFERENCES

BELLMAN, R. & HARRIS, T. E. (1948). On the theory of age-dependent stochastic branching processes. *Proc. Nat. Acad. Sci., Wash.*, **34**, 601–4.

BELLMAN, R. & HARRIS, T. E. (1952). On age-dependent binary branching processes. *Ann. Math.* **55**, 280–95.

FELLER, W. (1939). Die Grundlagen der Volterraschen Theorie des Kampfes ums Dasein in wahrscheinlichkeitstheoretischer Behandlung. *Acta biotheor., Leiden*, **5**, 11–40.

FELLER, W. (1950). *An Introduction to Probability Theory and its Applications.* New York: John Wiley and Sons, Inc.

KENDALL, D. G. (1948). On the role of variable generation time in the development of a stochastic birth process. *Biometrika*, **35**, 316–30.

KENDALL, D. G. (1952a). On the choice of a mathematical model to represent normal bacterial growth. *J. R. Statist. Soc.* B, **14**, 41–4.

RAMAKRISHNAN, ALLADI (1951). Some simple stochastic processes. *J. R. Statist. Soc.* B, **13**, 131–40.

REID, A. T. (1953). An age-dependent stochastic model of population growth. *Bull. Math. Biophys.* **15**, 361–5.

48. On the Use of the Direct Matrix Product in Analyzing Certain Stochastic Population Models

J. H. POLLARD (1966)

From *Biometrika* 53. Excerpts are from pages 397—398, 401—405.

The sections of Pollard's article that are omitted include an introduction to the Leslie matrix; stochastic treatment of the case of multiple births; and an extension of the methods outlined here to higher order moments.

1. INTRODUCTION

Deterministic models of population growth exist in two forms: those using a continuous time-variable and a continuous age-scale (following Sharpe & Lotka, 1911), and those using a discrete time-variable and a discrete age-scale (Bernadelli, 1941; Leslie, 1945, 1948). Both types have their advantages, but the discrete formulation is the closer to actuarial practice and is preferable when the age-specific birth and death rates are to be given on the basis of empirical data, rather than as analytical formulae. In the discrete formulation, however, difficulties arise due to the effects of grouping: we need to decide the method of calculating the group survivorship probabilities $\{P_x\}$ and the group fertility measures $\{F_x\}$. Clearly these depend upon the age distribution *within* the age group. Usually it is assumed that this age distribution follows the stable age distribution, which depends only upon the mortality function l_x, and indeed for many human populations this assumption is approximately true.

Small age groups seem preferable, but in practice the decision about grouping is determined to some extent by available data. A time unit of one year for human populations has much to recommend it, but it can lead to a large amount of calculation. It is surprising how little results differ (at least with deterministic theories) when different groupings are used. Keyfitz (1964), for example, obtains 1·796 % per annum rate of natural increase for Australian human females using 1960 data and a time unit of 5 years. In the present paper we find a rate of natural increase of about 1·785% per annum using 1960 data and a time unit of one year. This latter result is obtained using actual unsmoothed birth rates in 1960 and the 1954 Life Table. Using the integral equation theory Keyfitz finds a rate of natural increase of 1·7922 % per annum.

The main point to be borne in mind is that, in general, the effect on a Markov model of lumping states together is to cause the model to become non-Markovian. It is usually assumed that the underlying continuous-time population model is Markovian, and the discrete-time model, *under this assumption*, cannot be Markovian. We are not considering the continuous-time process, and will assume that the discrete-time process is Markovian. This implies that another discrete-time model with a different time unit will be non-Markovian. However, for a sufficiently small time unit, the errors due to grouping are likely to be small compared with those due to the simplified nature of the unisexual model and therefore we shall ignore them.

A stochastic version of the continuous model was constructed by Kendall (1949); this leads to integral equations for the expectation behaviour which are identical with those of the deterministic model, but his integral equations for the quadratic moments are difficult to deal with, save in some unrealistic special cases.

Kendall's continuous stochastic model was inspired by an earlier investigation by Bartlett (1946), in which discrete time and discrete age groups were used, but, as Bartlett regarded this as a first step towards the construction of a continuous model, his analysis was designed to yield asymptotic forms for the linear and quadratic moments when the age groups and time-intervals were both made small. From this point of view, we may say that Bartlett's work lies half-way between a discrete and a continuous formulation. The exact stochastic analogue of Leslie's theory, using fixed (*not necessarily small*) age-steps and time-steps appears never to have been worked out as such, although it can be regarded as a special case of the general theory of the multi-type Galton–Watson process. The object of the present paper is to fill this gap, and therefore, in dealing with the problem, methods have been employed which are different from those usually associated with Branching Processes (Harris, 1963). ...

424

4. Some preliminary results

Let M and N be discrete correlated random variables taking non-negative integral values. Let their means be μ and ν respectively, their variances σ^2 and τ^2 respectively, and their covariance cov (M, N). Let M_1' be a random variable conditional on M, and having the conditional binomial distribution $B(M, p_1)$. Similarly, let M_2' and N' be random variables having the conditional binomial distributions $B(M, p_2)$ and $B(N, p)$ respectively. Then

$$E(M_1') = p_1 \mu, \tag{11}$$

$$\mathrm{var}\,(M_1') = p_1^2 \sigma^2 + p_1 q_1 \mu, \tag{12}$$

$$\mathrm{cov}\,(M_1', M_2') = p_1 p_2 \sigma^2, \tag{13}$$

$$\mathrm{cov}\,(M_1', N') = p_1 p\, \mathrm{cov}\,(M, N), \tag{14}$$

where $q_1 = 1 - p_1$. Furthermore, if M is a binomial random variable, then M_1' is also a binomial random variable.

We omit the elementary proof.

We must also define the *direct product of two matrices* $\mathbf{X}$ and $\mathbf{Y}$. Let $\mathbf{X} = (X_{ij})$ and and $\mathbf{Y} = (Y_{ij})$ be matrices of dimension $l \times m$ and $r \times s$ respectively. Then the direct product of $\mathbf{X}$ and $\mathbf{Y}$ is denoted by $\mathbf{X} \times \mathbf{Y}$ and is defined by

$$\mathbf{X} \times \mathbf{Y} = \begin{bmatrix} X_{11}\mathbf{Y} & X_{12}\mathbf{Y} & \dots & X_{1m}\mathbf{Y} \\ X_{21}\mathbf{Y} & X_{22}\mathbf{Y} & \dots & X_{2m}\mathbf{Y} \\ & & \cdot & \\ & & \cdot & \\ X_{l1}\mathbf{Y} & X_{l2}\mathbf{Y} & \dots & X_{lm}\mathbf{Y} \end{bmatrix}, \tag{15}$$

which is a matrix of dimension $lr \times ms$.

Using the theory of classical canonical matrices (Turnbull & Aitken, 1951, pp. 58–63), it may be shown that $\mathbf{A} \times \mathbf{A}$ is positive regular if $\mathbf{A}$ is so. The dominant latent root of $\mathbf{A} \times \mathbf{A}$ is then λ_0^2, and this latent root is of algebraic multiplicity one. The corresponding row and column latent vectors are $\mathbf{y}' \times \mathbf{y}'$ and $\mathbf{x} \times \mathbf{x}$ respectively (i.e. the direct products as defined above).

5. The stochastic population model

Let us consider the female population only, at discrete intervals of time $t = 0, 1, 2, \dots$, and consider the $(k+1)$ age groups $0-, 1-, 2-, \dots, k-$, mentioned in connexion with Leslie's theory. Once again changes in the male population structure are assumed to be consistent with the assumption of constant fertility measures $\{F_x\}$. Let us define the following quantities, in addition to those introduced in Leslie's theory.

The number of females in the age group $x-$ at time t is a random variable $n_{x,t}$ with expected value $e_{x,t}$ and variance $C_{x,x}^{(t)}$. Let us denote the covariance cov $(n_{x,t}, n_{y,t})$ by $C_{x,y}^{(t)}$.

The number of females in the age group $0-$ at time t whose mothers were aged x at the time of the birth is a random variable $n_{0,t}^{(x)}$ (i.e. their mothers were among the females enumerated by $n_{x,t-1}$). Obviously

$$n_{0,t} = \sum_x n_{0,t}^{(x)}. \tag{16}$$

We now set up a stochastic model as follows. Consider the $n_{x,t}$ females aged x at time t. Each of them has a fixed probability P_x of surviving the unit time interval, and they are assumed independent. Hence $n_{x+1,t+1}$ is a binomial variable $B(n_{x,t}, P_x)$ conditional on $n_{x,t}$. Similarly, each of the $n_{x,t}$ females has a fixed probability F_x of contributing a single daughter in the age group $0-$ at time $t+1$, and they are assumed independent. Hence $n_{0,t+1}^{(x)}$ is a binomial variable $B(n_{x,t}, F_x)$ conditional on $n_{x,t}$. The birth and the death processes are assumed independent.

425

The expectations of the variables $n_{x,t}$ are still given by the Leslie equations (1), viz.

$$
\left.\begin{aligned}
e_{0,t+1} &= \sum_{x=0}^{k} F_x e_{x,t}, \\
e_{1,t+1} &= P_0 e_{0,t}, \\
e_{2,t+1} &= P_1 e_{1,t}, \\
&\ \ \cdot \qquad \cdot \qquad \cdot \\
e_{k,t+1} &= P_{k-1} e_{k-1,t}.
\end{aligned}\right\}
\tag{17}
$$

Using (12), (13) and (14), we obtain the following:

$$
C^{(t+1)}_{x+1,x+1} = p_x^2 C^{(t)}_{x,x} + P_x Q_x e_{x,t} \quad (x \geqslant 0),
\tag{18}
$$

$$
C^{(t+1)}_{x+1,y+1} = P_x P_y C^{(t)}_{x,y} \quad (x,y \geqslant 0,\, x \neq y),
\tag{19}
$$

$$
\mathrm{cov}\,(n^{(x)}_{0,t+1}, n_{x+1,t+1}) = F_x P_x C^{(t)}_{x,x} \quad (x \geqslant 0),
\tag{20}
$$

$$
\mathrm{cov}\,(n^{(x)}_{0,t+1}, n_{y+1,t+1}) = F_x P_y C^{(t)}_{x,y} \quad (x \neq y),
\tag{21}
$$

$$
\mathrm{cov}\,(n^{(x)}_{0,t+1}, n^{(y)}_{0,t+1}) = F_x F_y C^{(t)}_{x,y} \quad (x \neq y),
\tag{22}
$$

$$
\mathrm{var}\,(n^{(x)}_{0,t+1}) = F_x^2 C^{(t)}_{x,x} + F_x G_x e_{x,t} \quad (x \geqslant 0).
\tag{23}
$$

Now by definition,
$$
n_{0,t+1} = \sum_{x=0}^{k} n^{(x)}_{0,t+1}.
$$

Therefore
$$
\begin{aligned}
C^{(t+1)}_{0,0} &= \sum_{x=0}^{k} \mathrm{var}\,(n^{(x)}_{0,t+1}) + \sum\sum_{x \neq y} \mathrm{cov}\,(n^{(x)}_{0,t+1}, n^{(y)}_{0,t+1}) \\
&= \sum_{x=0}^{k} (F_x^2 C^{(t)}_{x,x} + F_x G_x e_{x,t}) + \sum\sum_{x \neq y} F_x F_y C^{(t)}_{x,y},
\end{aligned}
\tag{24}
$$

using equations (22) and (23).

Also
$$
\begin{aligned}
C^{(t+1)}_{0,y+1} &= \mathrm{cov}\left(\sum_{x=0}^{k} n^{(x)}_{0,t+1}, n_{y+1,t+1} \right) \\
&= \mathrm{cov}\,(n^{(y)}_{0,t+1}, n_{y+1,t+1}) + \sum_{x \neq y} \mathrm{cov}\,(n^{(x)}_{0,t+1}, n_{y+1,t+1}) \\
&= F_y P_y C^{(t)}_{y,y} + \sum_{x \neq y} F_x P_y C^{(t)}_{x,y},
\end{aligned}
$$

using equations (20) and (21). That is

$$
C^{(t+1)}_{0,y+1} = \sum_{\text{all } x} F_x P_y C^{(t)}_{x,y}.
\tag{25}
$$

Equations (17), (18), (19), (24) and (25) completely define the recurrence relations for the means, variances and covariances. Furthermore, they are linear recurrence relations, and may be written in the form

$$
\begin{bmatrix} \mathbf{e}_{t+1} \\ \mathbf{C}^{(t+1)} \end{bmatrix} = \begin{bmatrix} \mathbf{A} & \mathbf{0} \\ \mathbf{B} & \mathbf{A} \times \mathbf{A} \end{bmatrix} \begin{bmatrix} \mathbf{e}_t \\ \mathbf{C}^{(t)} \end{bmatrix},
\tag{26}
$$

where $\mathbf{A}$ is the $(k+1) \times (k+1)$ Leslie matrix. The vector $\mathbf{C}^{(t)}$ has as its elements the variances and covariances $C^{(t)}_{ij}$ and these are listed in dictionary order according to the subscripts i and j (for $i \neq j$, $C^{(t)}_{ij}$ and $C^{(t)}_{ji}$ are both listed).

It is of interest to note that, by successive application of recurrence relation (26), we may readily obtain:

$$
\begin{bmatrix} \mathbf{e}_n \\ \mathbf{C}^{(n)} \end{bmatrix} = \begin{bmatrix} \mathbf{A} & \mathbf{0} \\ \mathbf{B} & \mathbf{A} \times \mathbf{A} \end{bmatrix}^n \begin{bmatrix} \mathbf{e}_0 \\ \mathbf{C}^{(0)} \end{bmatrix} = \begin{bmatrix} \mathbf{e}_n \\ (\mathbf{A} \times \mathbf{A})^n \mathbf{C}^{(0)} + \sum_{j=1}^{n} (\mathbf{A} \times \mathbf{A})^{n-j} \mathbf{B} \mathbf{e}_{j-1} \end{bmatrix}.
$$

That is, we have the following relation for the variances and covariances

$$\mathbf{C}^{(n)} = (\mathbf{A} \times \mathbf{A})^n \mathbf{C}^{(0)} + \sum_{j=1}^{n} (\mathbf{A} \times \mathbf{A})^{n-j} \mathbf{B} \mathbf{e}_{j-1}. \tag{27}$$

Let us now write $\mathbf{C}^{[n]}$ for the vector of non-central quadratic product-moments at time n. Obviously

$$\mathbf{C}^{[n]} = \mathbf{C}^{(n)} + (\mathbf{e}_n \times \mathbf{e}_n) = \mathbf{C}^{(n)} + (\mathbf{A} \times \mathbf{A})^n (\mathbf{e}_0 \times \mathbf{e}_0).$$

Hence we have

$$\mathbf{C}^{[n]} = (\mathbf{A} \times \mathbf{A})^n \mathbf{C}^{[0]} + \sum_{j=1}^{n} (\mathbf{A} \times \mathbf{A})^{n-j} \mathbf{B} \mathbf{e}_{j-1}. \tag{28}$$

This is equation (4·3) on page 37 of Harris (1963), but written in a slightly different form. We have derived it for our special cases of the Galton–Watson process.

6. The asymptotic behaviour of the means and variances

From Leslie's theory, we already know the asymptotic behaviour of the means. Equation (8) gives us these results. It is easy to write down the eigenvectors $\mathbf{x}$ and $\mathbf{y}'$. To determine the asymptotic behaviour of the variances and covariances, we need the dominant latent root of the matrix in equation (26). This matrix has characteristic equation

$$\begin{vmatrix} \mathbf{A} - \lambda \mathbf{I}_{k+1} & \mathbf{0} \\ \mathbf{B} & (\mathbf{A} \times \mathbf{A}) - \lambda \mathbf{I}_{(k+1)^2} \end{vmatrix} = 0,$$

which is equivalent to $\quad |\mathbf{A} - \lambda \mathbf{I}_{k+1}| |(\mathbf{A} \times \mathbf{A}) - \lambda \mathbf{I}_{(k+1)^2}| = 0. \tag{29}$

Hence the matrix has as its latent roots all the latent roots of $\mathbf{A}$ together with all the latent roots of $\mathbf{A} \times \mathbf{A}$. From §4, we know that the dominant latent root of $\mathbf{A} \times \mathbf{A}$ is λ_0^2, and we recall that it has algebraic multiplicity one. Therefore we have three cases to consider:

(i) $\lambda_0 > 1$. In this case $\mathbf{A} \times \mathbf{A}$ contributes the algebraically simple dominant latent root λ_0^2, with latent vector $\begin{bmatrix} \mathbf{0} \\ \mathbf{x} \times \mathbf{x} \end{bmatrix}$.

(ii) $\lambda_0 = 1$. In this case there is a pair of dominant latent roots and each is equal to unity.

(iii) $\lambda_0 < 1$. In this case $\mathbf{A}$ contributes the algebraically simple dominant latent root λ_0, with latent vector $\begin{bmatrix} \mathbf{x} \\ \mathbf{x}* \end{bmatrix}$, say.

We shall discuss only case (i) in detail. Case (ii) is of no practical significance, and it is difficult to analyse case (iii). ...

Let us now continue with case (i). We know that the dominant latent root is λ_0^2 and that it is of algebraic multiplicity 1. The corresponding right latent vector is

$$\begin{bmatrix} \mathbf{0} \\ \mathbf{x} \times \mathbf{x} \end{bmatrix}.$$

We also note that the recurrence matrix of equation (26) is *not* positive regular. Once again we need the dominant left latent vector, and assume that it is $(\mathbf{u}' \, \mathbf{v}')$. We then have:

$$\{\mathbf{u}'\mathbf{A} + \mathbf{v}'\mathbf{B}, \mathbf{v}'(\mathbf{A} \times \mathbf{A})\} = \lambda_0^2 (\mathbf{u}', \mathbf{v}'). \tag{33}$$

Therefore $\mathbf{v}'(\mathbf{A} \times \mathbf{A}) = \lambda_0^2 \mathbf{v}'$, so that $\mathbf{v}'$ is a multiple of $\mathbf{y}' \times \mathbf{y}'$. In fact we may select $\mathbf{v}' = \mathbf{y}' \times \mathbf{y}'$. From (33), we also have

$$\mathbf{u}'\mathbf{A} + (\mathbf{y}' \times \mathbf{y}')\mathbf{B} = \lambda_0^2 \mathbf{u}'.$$

427

Hence
$$\mathbf{u}' = (\mathbf{y}' \times \mathbf{y}')\,\mathbf{B}(\lambda_0^2 \mathbf{I} - \mathbf{A})^{-1}.$$

Because $(\mathbf{y}' \times \mathbf{y}')(\mathbf{x} \times \mathbf{x}) = 1$, it is clear that

$$(\mathbf{u}'\,\mathbf{v}')\begin{bmatrix}\mathbf{0}\\\mathbf{x}\times\mathbf{x}\end{bmatrix} = 1,$$

where
$$\mathbf{u}' = (\mathbf{y}' \times \mathbf{y}')\,\mathbf{B}(\lambda_0^2 \mathbf{I} - \mathbf{A})^{-1},$$
$$\mathbf{v}' = (\mathbf{y}' \times \mathbf{y}').$$

So using (7) we have

$$\begin{bmatrix}\mathbf{A} & \mathbf{0}\\\mathbf{B} & \mathbf{A}\times\mathbf{A}\end{bmatrix}^n \Big/ \lambda_0^{2n} = \begin{bmatrix}\mathbf{0}\\\mathbf{x}\times\mathbf{x}\end{bmatrix}(\mathbf{u}'\,\mathbf{v}') + o(1) = \begin{bmatrix}\mathbf{0} & \mathbf{0}\\(\mathbf{x}\times\mathbf{x})\mathbf{u}' & \mathbf{x}\mathbf{y}'\times\mathbf{x}\mathbf{y}'\end{bmatrix} + o(1). \tag{34}$$

This gives us the following asymptotic results:

$$\left.\begin{aligned}\mathbf{e}_t/\lambda_0^{2t} &= \mathbf{0} + o(1) \text{ (as it should!)},\\\mathbf{C}^{(t)}/\lambda_0^{2t} &= (\mathbf{u}'\mathbf{e}_0 + (\mathbf{y}'\times\mathbf{y}')\mathbf{C}^{(0)})(\mathbf{x}\times\mathbf{x}) + o(1),\end{aligned}\right\} \tag{35}$$

where
$$\mathbf{u}' = (\mathbf{y}'\times\mathbf{y}')\mathbf{B}(\lambda_0^2\mathbf{I} - \mathbf{A})^{-1}.$$

It is interesting to note that these asymptotic results imply immediately that for large t, when $\lambda_0 > 1$,

$$\text{correlation } (n_{i,t}, n_{j,t}) = 1 - o(1).$$

This fact suggests that the vector random variable $\mathbf{n}_t$ converges in some sense to a random multiple of a fixed vector—the stable age distribution vector. Indeed the following theorem, due to Everett & Ulam (1948) and Harris (1951), describes the behaviour of $\mathbf{n}_t$ precisely.

THEOREM 2. *If* $\mathbf{A}$ *is positive regular and* $\lambda_0 > 1$, *then the vector random variable* $\mathbf{n}_t/\lambda_0^t$ *converges with probability one to a random scalar multiple of the unique stable age distribution* $\mathbf{x}$.

To prove this it is only necessary to apply Theorem 9·2 of Harris (1963, Chapter 2) and to note that his $\mathbf{M}'$ is in fact equal to our $\mathbf{A}$.

REFERENCES

BARTLETT, M. S. (1946). *Stochastic Processes. Notes of a course given at the University of North Carolina in the Fall Quarter 1946*, 39–41.
BERNADELLI, H. (1941). Population waves. *J. Burma Res. Soc.* **31**, 1–18.
EVERETT, C. J. & ULAM, S. (1948). Multiplicative systems in several variables. III. *Los Alamos Scientific Laboratory. LA 707.*
HARRIS, T. E. (1951). Some mathematical models for branching processes. *Second Berkeley Symposium*, pp. 305–28.
HARRIS, T. E. (1963). *The Theory of Branching Processes.* Springer-Verlag.
KENDALL, D. G. (1949). Stochastic processes and population growth. Symposium on Stochastic Processes. *J.R. Statist. Soc.* B **11**, 230–64.
KEYFITZ, N. (1964). The intrinsic rate of natural increase and the dominant latent root of the projection matrix. *Popul. Stud.* **18**, 293–308.
LESLIE, P. H. (1945). On the use of matrices in certain population mathematics. *Biometrika* **33**, 183–212.
LESLIE, P. H. (1948). Some further notes on the use of matrices in population mathematics. *Biometrika* **35**, 213–45.
SHARPE, F. R. & LOTKA, A. J. (1911). A problem in age distribution. *Phil. Mag.* **21**.
TURNBULL, H. W. & AITKEN, A. C. (1951). *An Introduction to the Theory of Canonical Matrices.* Dover.

Cohort and Period, Problem of the Sexes, Sampling

We have deferred to this chapter three problems that much complicate demographic analysis: the distinctions between cohort and period rates, between one-sex and two-sex models, and between population and sample information. These are often ignored in demographic work, either of necessity or for their small effect.

The first two problems are introduced in an article by J. Hajnal (1947), which traces English nuptiality and birth rates between the two World Wars as they were influenced by sex ratios and by the interaction of current with past marriage and fertility patterns. Hajnal's care in separating the various influences and the effort this required will be apparent. In earlier works, G. Udny Yule (1906) anticipated much of his discussion of cohort and period marriage rates but treated less complex shifts, and Robert J. Myers (1941) considered how age distributions and age misreporting might affect male and female net reproduction rates.

Fertility as a two-sex problem was mathematically treated first by Paul Vincent (1946), and, independently and more fully, P. H. Karmel (1947). We include here Karmel's paper, which outlines the limited work done previously and develops the relationships between male and female rates, showing conditions that would hold in stable populations. In subsequent work, Karmel (1948, 1949) suggested using a weighted geometric mean of the male and female growth rates as the joint intrinsic rate in order to account for dominance; i.e., the tendency of fertility rates to follow more closely the numbers of one sex than the other.

A. H. Pollard (1948), whose paper follows Karmel's, extends the analysis to relations between male and female rates in non stable populations. He establishes fertility functions for daughters to fathers and sons to mothers, and so incorporates the interdependence of the sexes. The two-sex intrinsic rate of natural increase becomes, in this model, the arithmetic mean of the male and female intrinsic rates weighted by the respective generation lengths. Luitzen Yntema (1952) pointed out that tracing daughters to fathers and sons to mothers, while giving the same rate of increase for the two sexes, also implies equal dominance of males and females. The model becomes more general if one supposes that out of $k+l$ generations k will be of fathers producing daughters and l will be of mothers producing sons, where the ratio $k:k+l$ expresses relative male dominance.

In stochastic extensions of Karmel's work, David Kendall (1949) and subsequently Leo Goodman (1953) formally reconcile male and female increase rates by setting births as a function of both sexes. Their contributions are included here as papers 52 and 53. The works do not consider age distributions, which much increase the complexity of two-sex models: where there are more boys of age x and girls of age y, not only will more children be born to this combination of ages, but the effect spreads to neighboring ages as well. The fact that children born at any combination of ages is a function of numbers at nearly all ages, and that the functional dependence is non linear, creates very impressive difficulties.

A large literature on two-sex models is now building up, which includes contributions by Henry (1969); Hoem (1969); Fredrickson (1971); Keyfitz (1971); Das Gupta (1972); McFarland (1972): Parlett (1972); Bartlett (1973); Pollard (1973); and Samuelson and Yellin (1974).

Differences of cohort and period measures have been recognized at least since Antoine Deparcieux (1746), who made mortality tables of both types from data on French annuitants, and problems of reconciliation are made clear by Hajnal. The mathematical treatment of the problem is due to Norman Ryder (1964), who showed that cohort fertility could be expressed in terms of the moments of period rates, revealing the extent to which period rates distort cohort behavior. His analysis is presented here in summary form from Keyfitz (1972). As an empirical problem, the difficulty of distinguishing period from cohort behavior is noted in Brass (1974, pp. 556—561. See also the discussion following his paper, pp. 571—583.)

The final problem we will examine is that of sampling, first considered by Major Greenwood (1926) and continued by Edwin B. Wilson (1938) and J. O. Irwin (1949). Setting $\text{Var}(p_i) = p_i q_i / N_i$, where p_i is the probability that a person of exact age i survives to age $i+1$ (hence, $p_i = 1 - q_i = l_i / l_0$) and N_i is sample size, and assuming independence among p_i terms, Greenwood derived the variance of l_x as:

$$\text{Var}(l_x) = \sum_{i=0}^{x-1} [p_i^2 + \text{Var}(p_i)] - \sum_{i=0}^{x-1} p_i^2 \doteq \left(\frac{l_{x-1}}{l_0} \right) \sum_{i=0}^{x-1} \frac{p_i q_i}{N_i}.$$

Wilson, who was unfamiliar with Greenwood's work, developed an expression for $\text{Var}(\mathring{e}_x)$ on the same assumptions and also for the more general case in which q_x terms are derived from the life table death rates m_x. For the simplest case, in which deaths occur linearly in age intervals of width n, his equations are

$$\text{Var}(_nq_x) = \frac{_nP_x\,_nq_x}{N_x} = \left(\frac{_nq_x}{_nm_x} \right)^4 \left[\frac{_nm_x(1 - _nm_x)}{n^2 N_x} \right],$$

$$\text{Var}(\mathring{e}_x) = \sum_{y=x}^{\omega} \left(\frac{l_y}{l_x} \right)^2 \left(\mathring{e}_{y+n} + \frac{n}{2} \right)^2 \text{Var}(_nq_y).$$

More generally, $n/2$ in the latter expression may be replaced by $n - cn$, where cn is "the average life of those who die in the interval." (In modern work $_na_y$ replaces cn.) Wilson's article is presented here as paper 55.

Independent of both authors, Irwin attacked the problem in 1949. His work is quite restricted, but he essentially derives Wilson's $\text{Var}(\mathring{e}_x)$ from Greenwood's $\text{Var}(l_x)$, thereby bringing the series together.

A final development has been the introduction of probability generating functions for terms of the life table by C.L. Chiang (1960b, 1968). An excerpt from his 1968 work closes this chapter.

In reading Wilson's and Chiang's articles, readers should note the strict conditions required for convenient formulation of results, remembering that expressions become much more complex when other than simple random sampling is used, when p_i rates are not independent, and so forth. The number of extensions is large and the formulas given here do no more than establish minimum variances, which most commonly are exceeded. A discussion of these problems and their partial solution will be found in Chiang (1968, pp. 269—296), which is a continuation of the material quoted here.

49. Aspects of Recent Trends in Marriage in England and Wales

JOHN HAJNAL (1947)

From *Population Studies* 1. Excerpts are from pages 80—84, 86—87, 89—92.

We have omitted part of Hajnal's discussion of male and female marriage trends, which reflect year to year differences in age distributions and ages at marriage for the two sexes. Hajnal did not have access to marriage distributions by age of both spouses, which is necessary to a complete analysis. Section 1 of Hajnal's paper, which is compactly summarized in the introduction to Section 2, is also omitted.

2. *Marriage-rates between the wars*

By means of the nuptiality tables shown in § 1 it has been demonstrated that (1) a great increase in marriage-rates has occurred since 1910-12, (2) that this increase has been far greater among women than among men (at the ages where most marriages take place), and (3) that it has been accompanied by a fall in the average age at marriage. To prove that these general features are not the result of comparing the particular years selected, and to show in greater detail when and by what steps these increases occurred, the year to year variations over the period between the two wars may be studied. Table 8 shows the marriage-rates of all unmarried persons over 15 for each year in the period 1921–39.[1]

Table 8. *Marriage-rates 1921–1939. Persons married per 1000 single, widowed and divorced aged 15 and over*

Year	Males	Females	Year	Males	Females
1910–12	50·8	42·5	1930	55·6	42·0
1921	60·4	45·8	1931	53·4	41·6
1922	55·8	42·5	1932	52·5	41·0
1923	53·9	41·1	1933	54·6	42·6
1924	53·6	41·2	1934	59·3	46·1
1925	53·3	40·9	1935	59·8	46·6
1926	50·0	38·3	1936	60·3	46·9
1927	54·8	41·9	1937	60·6	47·4
1928	53·7	40·9	1938	61·2	47·8
1929	55·2	41·9	1939	74·4	58·0

After the post-war boom in marriage-rates had subsided, the marriage-rates of men settled down in the period 1923–30 at a level higher than that obtaining before the war. The rate of women fell to slightly below the pre-war level. In 1926, owing to the general strike, marriage-rates were low. After 1930 the marriage-rate of men showed a slight fall, reflecting the economic depression, but women's rates showed almost no decline during the slump. After 1933 there was a steady rise to 1938, followed by a very sudden marriage boom in 1939, owing to the large number of marriages contracted after the outbreak of war.

Table 9 gives the marriage-rates of 1931–9 by quinquennial age groups.

In 1931 and 1932 marriage-rates were falling slightly, mainly at the younger ages, owing to the depression. The extremely rapid rises after 1933 were steepest at the young ages and about twice as fast among women as among men. The 1938 rates may be seen to follow, in smooth continuation, those of previous years.

The marriage boom which began in 1939 again mainly affected the younger age groups.

---

[1] See, *The Registrar-General's Statistical Review*, 1941, Tables, Part II, Civil, p. 6. This table is the only one in this paper which includes later marriages as well as first marriages. The marriage-rates of bachelors and spinsters by age for each year in the period 1921–30, which are given in the *Registrar-General's Statistical Review* for 1930, Text, p. 110, could not be used. The estimates of population on which they are based must have been subject to very serious error, as the rates for the years just before 1931 are quite incompatible with those later based on the 1931 census.

Table 9. *Marriage-rates 1931–1939*[1]

Age group	1931	1932	1933	1934	1935	1936	1937	1938	1939
	Bachelors								
15–19	3·3	3·4	3·4	3·6	3·2	2·7	2·8	3·3	4·6
20–24	72·3	69·8	70·3	75·0	76·8	80·3	83·3	87·0	120·3
25–29	152·1	147·5	151·9	163·6	167·8	171·1	173·4	176·7	213·1
30–34	111·5	106·3	107·5	117·2	119·1	121·3	123·8	127·4	147·5
35–39	62·6	60·4	60·9	64·2	68·1	69·2	69·3	68·6	77·8
40–44	33·7	33·4	33·1	35·6	36·0	36·1	37·4	37·6	42·7
45–54	16·4	15·4	16·5	16·7	16·5	16·8	17·1	18·5	19·8
	Spinsters								
15–19	17·2	17·8	18·8	20·4	19·2	18·4	19·4	22·6	32·2
20–24	106·8	105·1	109·1	118·9	124·7	132·7	139·5	146·9	195·8
25–29	119·0	116·4	121·3	132·6	138·2	142·4	147·3	153·8	186·1
30–34	57·2	56·1	58·0	63·3	64·1	64·7	65·8	67·1	78·2
35–39	26·9	26·4	26·5	28·97	29·9	31·5	32·5	33·1	37·2
40–44	14·5	14·3	14·8	15·2	15·6	15·9	16·6	16·8	18·6
45–54	7·9	7·5	7·7	7·9	8·1	7·9	8·2	8·6	9·3

3. *The long-term significance of recent trends in marriage-rates*

Will marriage-rates, which rose continuously in the years before 1939, resume their upward trend? Will the high level of recent marriage-rates be maintained? Or was some part of the rise due to temporary influences, such as the economic recovery proceeding in 1932–7?

Some statistical considerations which bear on this question may be set out. As was pointed out at the beginning of this paper. ...   , by the outbreak of war in 1939 it was true, in most age groups, that a greater proportion of the population had been married than at any previous time in this century. The situation as it was at the outbreak of war already implied a considerable change in the marriage habits of the population towards earlier and more frequent marriage. But the full effect of a given level of marriage-rates on the proportions married only works itself out slowly.

To visualize the consequences of the indefinite continuance of a certain set of marriage-rates recourse may again be had to nuptiality tables. Table 10 compares the proportions ever-married which would result, if the marriage-rates of 1930–2 and 1938 continued in operation, with the proportion 'ever-married' (i.e. married, widowed and divorced) as estimated for mid-1938.

The higher marriage-rates of 1938 would, of course, result in higher proportions married than those of 1930–2. At the younger ages the effects of high marriage-rates are soon visible, and the proportions ever-married as estimated for 1938 are close to those which would have resulted from the permanent operation of 1938 nuptiality. At higher ages, however, the proportions ever-married in 1938 were the

[1] The rates for the age group 15–19 fluctuate violently owing to changes in the age distribution within the group. When, in 1935 and 1936, the generations born in 1917–18 were aged 18–19, while the very much more numerous post-war generations were aged 15–16, the marriage-rates in the age group 15–19 fell suddenly. The rates in the table are based on population estimates revised in the light of the National Registration of 1939. They differ slightly, therefore, in some cases from those in the Registrar-General's *Statistical Review*, 1937, Text, p. 198.

Table 10. *Proportions ever-married (%) according to the gross nuptiality of 1930–1932 and 1938 compared with proportions recorded at various dates*

Age group	1930–2 gross nuptiality	1938 gross nuptiality	Actual proportions 1938	Census 1861	Census 1871
			Men		
15–19	(0·3)	(0·3)	0·6	0·5	0·6
20–24	13·9	16·8	16·9	22·5	25·7
25–29	52·8	58·4	54·3	59·8	61·0
30–34	76·5	81·2	77·4	77·4	77·0
35–39	84·6	88·4	84·8	84·7	84·9
40–44	87·9	91·1	88·8	87·0	87·8
45–49	89·4	92·3	90·4	89·2	90·1
50–54	90·4	93·0	90·3	89·8	90·4
			Women		
15–19	1·8	2·9	2·2	3·1	3·2
20–24	26·7	34·9	32·5	33·6	34·8
25–29	60·5	71·0	64·9	63·2	64·4
30–34	74·6	83·4	74·9	76·8	77·6
35–39	79·1	87·0	80·1	82·8	83·2
40–44	81·1	88·5	82·0	85·8	85·9
45–49	82·2	89·2	83·0	87·8	87·6
50–54	82·9	89·6	84·0	88·4	88·3

result of marriage-rates lower than those of 1938. They were rather like the proportions married which would have resulted from the permanent operation of 1930–2 nuptiality.

The maintenance of 1938 marriage-rates at all ages would have involved a considerable rise of the proportions married over the latest level recorded before the war. The rise in marriage proportions would be particularly great in the case of women.

To help assess the plausibility of such a rise it is of interest to compare the proportions ever-married recorded in the last century, before nuptiality fell. Can the recent rise in nuptiality be regarded as a return to earlier conditions?

The last two columns of Table 10 give the proportions 'ever-married' recorded at the Censuses of 1861 and 1871. Two features may be noticed. (1) Comparison of the 1861 and 1871 proportions with those of 1938 shows that the proportions married in 1938 had not in any age group risen above the levels experienced in the last century either for men or for women. (2) The figures for 1938 gross nuptiality show that if the 1938 marriage-rates were to remain in operation, the nineteenth-century proportions would be considerably exceeded both among men and among women at ages over 30.

The maintenance of 1938 marriage-rates would thus result in a situation unprecedented, at any rate in the last hundred years. A somewhat greater proportion of each generation would get married at least once by, say, age 50 than ever before. According to 1938 nuptiality only 7% of men and 10% of women would never have been married. In the nineteenth century the corresponding figures never fell below about 10% for men and 12% for women, as may be gathered from Table 10. That fewer would ultimately remain unmarried according to 1938 nuptiality is due entirely to the higher level of 1938 marriage-rates at the *older* ages. Under 30 years of age the proportions married among men according to 1938 nuptiality were well below those recorded in the last century. There is no sign whatever that the proportion of men marrying under, say, 25 will rise again to anything approaching the nineteenth-century level.

Thus past experience and general impression both suggest that while there is no great scope for a diminution in the proportion of those who never marry, there is ample possibility that the average age at marriage may be lowered further. This belief is supported by the fact that, before the war, marriage-rates were still rising, and rising particularly rapidly at the younger ages.

It may be argued that the year 1938 occurred at the end of a period of economic recovery. We cannot, therefore, be sure that the tendency for marriage-rates to increase will continue if heavy unemployment returns.

It might be plausibly suggested that the high marriage-rates at ages over 30 (in which the novelty of the situation in the 1930's consists) will probably not be maintained after the war. For the generations which, in 1938, experienced high marriage-rates at ages over, say, 30 years, had themselves in their younger years married at far lower rates than those obtaining under 30 in 1938. In a community in which the 1938 marriage-rates had been in force for some time, a smaller proportion of those at older ages would remain unmarried than were unmarried in the 1938 population. Those who would remain unmarried under these conditions would perhaps be less 'marriageable' on average than were those who remained unmarried at the older ages in England and Wales in 1938. On this assumption the maintenance in future of *attitudes* to marriage similar to those of 1938 would bring a gradual decline below the 1938 level in the marriage-rates at higher ages.

This argument is supported by the fact that 1938 was a year occurring at the end of the economic recovery of 1933-7. The effect of a depression, as was illustrated in Table 9, is to decrease marriage-rates at the younger ages, i.e. in a depression people postpone marriage (relatively to the ages at which they would marry in a boom). Correspondingly, it would be expected that the generations who 'postponed' their marriages in the depression years would have higher marriage-rates during the recovery to make up for the marriages they postponed. This influence might push up marriage-rates at the higher ages above the average level which would obtain if full employment were to be maintained indefinitely. ...

4. *The compatibility of male and female nuptiality*

One important aspect of the recent changes in nuptiality—the changing relation between the nuptiality of women and that of men—requires closer analysis. It can, indeed, be shown that the recent rise in women's marriage-rates relative to those of men is likely to continue.

The marriage-rates of men and the marriage-rates of women in any given year are compatible with one another because of the relation between the number of men and the number of women which exists in that year. If the ratio of women to men in the marriageable ages falls, the chances of marriage for women increase and vice versa. The recent increase in the nuptiality of women relative to that of men has been due to a change in the sex constitution of the population at marriage-

437

able ages. The ratio of women to men has been falling. The ratio of women to men found at various censuses, for the ages at which most marriages take place, is shown in Table 12.

Before 1901 the fluctuations are irregular, but between 1901 and 1911 the excess of females diminished slightly in almost all age groups. This fact was implied in the changes of female relative to male nuptiality occurring at that period, which were described in § 1. Between 1911 and 1921, the effects of the war supervened, and the figures for 1931 and 1939 show abnormal excesses of females in the groups affected by war losses. By 1939, however, the youngest group of men of whom the war took a substantial toll was over 40 years of age. Under 40 the sex proportions were in all age groups more favourable to women than they had ever been since 1871.

Table 12. *Females per 1000 males by age*

Ages	1871	1881	1891	1901	1911	1921	1931	1939
15–19	1010	1008	1014	1019	1016	1027	1009	997
20–24	1106	1093	1122	1119	1113	1176	1056	1009
25–29	1111	1087	1115	1126	1115	1209	1061	1032
30–34	1090	1077	1073	1100	1091	1186	1132	1036
35–39	1093	1069	1059	1074	1072	1156	1185	1048
40–44	1084	1079	1075	1062	1077	1127	1167	1177

It should be realized, however, that the proportion of women to men, though far lower now than in the past, is still above that which would obtain if the distribution of the population by sex were determined only by the proportions of male to female births and current mortality conditions. In fact, in the past the sex ratio has never been near what would be expected from the mortality conditions. ...

It is clear, therefore, that women's marriage-rates must rise further in relation to those of men. In view of the level of female nuptiality implied by the male nuptiality of 1938, the marriage-rates of men can hardly rise any further in the long run, in the absence of further disturbances of the sex balance through war and migration. Any further rise in male marriage-rates would be arrested by a 'shortage' of women.

5. *Paternal and maternal reproduction rates*

It now remains to discuss the relation of the developments in nuptiality described, to the prospects of population growth. In this connexion the differences that have been shown between the marriage-rates of men and those of women are of particular interest. What would be the result if the prospects of population growth were considered, and population projections constructed, on the basis of the marriage and fertility rates of men and not, as has been universal in the past, on the basis of women? This problem is conveniently discussed by means of paternal reproduction rates, i.e. reproduction rates based on male births and male populations in the same way as ordinary reproduction rates are based on female births and female populations.

It is desirable, as a preliminary step, to distinguish certain factors which affect the comparability of paternal and maternal reproduction rates. In the first place

438

paternal net reproduction rates differ from maternal net reproduction rates not only because the fertility rates of men differ from the fertility rates of women, but also because (1) more boys than girls are born, (2) male mortality is heavier than female mortality.

Secondly, for a stable population the paternal and maternal net reproduction rates would not be equal as the mean length of a generation is greater for fathers than for mothers. In a stable population the 'true rate of natural increase' (which concept is due to Dr Lotka) must, however, be the same whether measured on the male or on the female population. To test the compatibility of the male and female reproduction rates of a given year in this sense, i.e. to test whether a stable population with the age-specific fertility and mortality rates of females could at the same time have the fertility and mortality rates of males, it is necessary to calculate the 'true rate of natural increase'.

These points are illustrated in Table 14 which shows how various indices of fertility and replacement differ when calculated on the basis of men and not women.

Table 14. *Paternal and maternal fertility indices* 1938

	Paternal	Maternal
Total fertility rate	1·982	1·840
Gross reproduction rate	1·016	0·897
Net reproduction rate	0·881	0·808
True rate of natural increase	−0·0038	−0·0073

The total fertility rate (i.e. the sum of the age-specific fertility rates based on all live births, male and female) is slightly higher when based on men than when based on women. The paternal gross reproduction rate shows a greater excess over the maternal rate, because of the sex ratio at birth. Paternal and maternal net reproduction rates again approximate more closely than gross reproduction rates because the excess of the mortality of males over that of females narrows the difference.

What is important about a reproduction rate is, however, not its absolute value, but the extent to which it differs from unity. By computing true rates of natural increase, the paternal and maternal reproduction rates are shown to imply strikingly different annual rates of increase. A stable population with the fertility and mortality rates of women in 1938 would decrease almost twice as fast as a stable population with the mortality and fertility rates of men in 1938.

As the average age of fathers is slightly higher than the average age of mothers the maternal net reproduction rate compatible with a given (negative) paternal rate would be slightly higher and not, as in 1938, lower. It would measure the effect of a given annual rate of decrease over a shorter number of years than the paternal rate.

With these considerations in mind the course of the paternal and maternal reproduction rates between 1911 and 1938 may now be compared. The data are given in Table 15.

Even in 1938, the prospects of population growth appeared very different according to whether paternal or maternal reproduction rates are considered. But in earlier years the differences were greater still. Thus the paternal reproduction rate

439

of 1910–12 related to a stable population growing more than twice as fast as that implied in the maternal rate for the same year, while the paternal rate of 1931 related to a stable population decreasing less than one-third as slowly as that implied in the maternal rate for that year. Between 1920–2 and 1938 the gap between paternal and maternal reproduction narrowed appreciably.

Table 15. *Reproduction rates, 1911–1938*

Date	Paternal rates		Maternal rates		Ratio of paternal net reproduction rate to maternal net reproduction rate
	Gross reproduction rate	Net reproduction rate	Gross reproduction rate	Net reproduction rate	
1910–12	1·725	1·274	1·444	1·129	1·128
1920–2	1·712	1·346	1·345	1·111	1·212
1931	1·119	0·928	0·929	0·805	1·153
1938	1·016	0·881	0·897	0·808	1·090

The reason for the growing approximation of the two rates to one another is to be found in the changes of the proportions married among women relative to those of men which have been described in this paper. ...

Do these conclusions (that the paternal and maternal reproduction rates are not compatible and that the difference between them depends on the proportions married) affect the validity of the usual procedure by which the maternal net reproduction rate is used as an indication of the prospects of long-run population growth? The significance of a reproduction rate derives from the fact that, in the absence of migration, a population permanently subject to the fertility and mortality rates on which the reproduction rate is based will, in the long run, decrease (or increase) at the rate indicated by the reproduction rate. To pay attention to a reproduction rate is, therefore, to contemplate the consequences of the maintenance, in the long run, of the fertility and mortality rates on which the reproduction rate is based.

50. The Relations between Male and Female Reproduction Rates

P. H. KARMEL (1947)

From *Population Studies* 1. Excerpts are from pages 249—260.

We omit part of Karmel's discussion of nuptuality, and an example illustrating interactions between the male and female rates.

I. THE PROBLEM

Although reproduction rates are usually calculated in terms of maternities, it is well established that they can also be calculated in terms of paternities. Hence it is possible to calculate for a given region at a given time both male and female gross and net reproduction rates. Frequently, such male and female rates give contradictory results, in that they do not lie on the same side of unity. More often they give results, which, although not so obviously paradoxical, are nevertheless inconsistent in that they yield very different stable rates of growth for the two sexes.

In spite of the almost exclusive preoccupation of demographers with the female part of the population, some writers have paid a little attention to male reproduction rates. Thus R. R. Kuczynski, in his book *Fertility and Reproduction*, calculated male and female net reproduction rates for France 1920–3, and found the male rate to be 1·194 and the female rate 0·977. He explained that the reason for this lay in deficiencies in the male population due to war casualties, concluded that 'it may be useful in connexion with certain studies in differential fertility to include males but for any general study of the balance of births and deaths it seems preferable to relate births and deaths to the female population only',[1] and left it at that. A. J. Lotka pointed out that the net reproduction rate 'is influenced by the mean interval between generations, and for this reason the net reproduction rate calculated from the male part of the population is higher than that which the female part yields'. Here he is clearly referring to what obtains in the theoretically stable population. But when he says that 'although in the long run the male population and the female population ought to have equal rates of growth...during a limited period it is quite possible for there to be, and in general there will be a certain amount of disagreement between the two rates of growth', he is referring to the actual course of events.[1'] C. Tietze has discussed the use of the total paternity rate (i.e. the male counterpart to the total fertility rate) in analyses of differential reproduction, but gives no precise formulation of the relation between the male and female rates. He says, 'It is obvious that total paternity and fertility rates will be identical if there is no excess of either sex in the adult population', but this would seem to be true only if the numbers in each adult age group of each sex were equal, which could only be so in very special circumstances.[2] R. J. Myers has drawn attention to the causes of possible discrepancies between male and female net reproduction rates, but gives no clear indication of what he believes the relation between the two rates to be, other than that they should be equal in a stationary population.[3] But the most satisfactory discussion of this matter so far is that of Paul Vincent.[4] M. Vincent reveals some of the relations between the male and

---

[1] R. R. Kuczynski, *Fertility and Reproduction* (1932), pp. 36–8.

[1'] A. J. Lotka, 'Analyse démographique avec application particulière à l'espèce humaine', *Actualités Sci. Industr.* 780 (1939), pp. 102 and 134 (writer's translations).

[2] C. Tietze, 'Measurement of differential reproduction by paternity rates', *Eugen. Rev.* 1938, no. 2, pp. 101–7; 'Differential reproduction in England', *Milbank Mem. Fd Quart. Bull.* 1939, no. 3, pp. 288–93. The quotation is from the latter paper.

[3] R. J. Myers, 'The validity and significance of male net reproduction rates', *J. Amer. Statist. Ass.*, June 1941, pp. 275–82.

[4] Paul Vincent, 'De la mesure du taux intrinsèque d'accroissement naturel dans les populations monogames', *Population*, 1946, no. 4, pp. 699–712. The present study was substantially completed before the writer saw M. Vincent's paper, and it agrees with the general conclusions of the latter. However, M. Vincent concerned himself with net reproduction rates only, and the present study gives a more general formulation of the relations between the various rates.

female net reproduction rates and works out some of the implications of these relations.

The objects of this paper are, first, to examine the relations which obtain between male and female net and between male and female gross reproduction rates in the theoretically stable population, and, secondly, to stress the implications of these relations in so far as they affect the logical internal consistency of the concepts of the net and gross reproduction rates and of those concepts associated with the stable population.

2. A CONSTANT DIFFERENCE BETWEEN AGE OF MOTHERS AND AGE OF FATHERS AT BIRTH OF CHILDREN ASSUMED

We shall first examine the problem by considering the simplified case where for each birth the age of the father differs by a constant number of years from the age of the mother. Let there be calculated from an actual population in which this assumption holds the following functions:

$l_F(x)$, the probability of a female surviving to age x.

$l_M(y)$, the probability of a male surviving to age y.

$b_F(x)$, the number of female births per head of females aged x.

$b_M(y)$, the number of male births per head of males aged y.

Let m be the masculinity of births, where m is constant. Then we have

$$R_{0M} = \int_0^\infty l_M(y) \, b_M(y) \, dy, \tag{1}$$

$$R_{0F} = \int_0^\infty l_F(x) \, b_F(x) \, dx, \tag{2}$$

$$R_M = \int_0^\infty b_M(y) \, dy, \tag{3}$$

$$R_F = \int_0^\infty b_F(x) \, dx, \tag{4}$$

where R_{0M}, R_{0F}, R_M, R_F are the male and female net and the male and female gross reproduction rates, respectively.

Lotka and others have shown that with given female mortality and fertility conditions, as defined by $l_F(x)$ and $b_F(x)$, a female population, subject to these conditions constantly, will ultimately have a stable age distribution and will increase or decrease at a certain rate per generation—the net reproduction rate—and at a certain rate per annum—the stable rate of growth—such rates being known functions of $l_F(x)$ and $b_F(x)$. Similarly, with given male mortality and fertility conditions, as defined by $l_M(y)$ and $b_M(y)$, a stable male population can be deduced and such a population will increase or decrease at certain rates per generation and per annum. It is clear that in ultimate stability the two sexes must increase or decrease at the same rate per annum, but in practice the stable rate of growth calculated on the basis of male mortality and fertility conditions may give a very different answer from that calculated on the basis of female conditions. This may be put by saying that, although $l_M(y)$ and $l_F(x)$ may be taken as independent functions, this is not the case with $b_M(y)$ and $b_F(x)$, for, if we take $b_M(y)$ as given, then the values of $b_F(x)$ in any actual population will depend on the sex ratio between males aged y and females aged $(y-d)$, where d is the constant excess of age of father over age of mother at the birth of their child, and vice versa. Since the sex ratios in the population at any particular point of time are the result of past demographic history and are changing from time to time, the question arises whether, if we hold $l_F(x)$, $l_M(y)$ and $b_M(y)$ continuously constant, we can determine the function $b_F(x)$ such that the whole set of fertility and mortality conditions are consistent with an ultimately stable male and female population.

In other words, given $l_F(x)$, $l_M(y)$ and $b_M(y)$ (and hence R_{0M} and R_M), is there a set of rates $\bar{b}_F(x)$ (and hence female net and gross reproduction rates, $\bar{R}_{0F}$ and $\bar{R}_F$) consistent with ultimate stability? Such rates will be termed 'consistent' rates. Furthermore, if such rates exist, what are the relations between the male rates and their consistent female counterparts?

Since, in stability, the male and female populations must be growing at the same stable rate, r, the male stable population at the age of y is given by[1]

$$me^{-ry}l_M(y),$$

and the female stable population at age x is given by

$$e^{-rx}l_F(x).$$

Male births to the male stable population at age y will be

$$me^{-ry}l_M(y)\,b_M(y).$$

But these male births are associated with female births born to females aged $(y-d)$, hence

$$\bar{b}_F(x) = \bar{b}_F(y-d) = e^{-rd}\frac{l_M(y)}{l_F(y-d)}b_M(y) \tag{5}$$

and

$$\bar{R}_{0F} = \int_0^\infty l_F(y-d)\bar{b}_F(y-d)\,dy = e^{-rd}R_{0M} \tag{6}$$

and

$$\bar{R}_F = e^{-rd}\int_0^\infty \frac{l_M(y)}{l_F(y-d)}b_M(y)\,dy. \tag{7}$$

It is not very difficult to prove that the stable rates of growth derived from (1) and (6) are the same and hence that both male and female stable populations are growing at the same rate.

Let r_M be the stable rate of growth of the male population and r_F be the stable rate of growth of the female population. Then r_M is to be derived from the integral equation

$$\int_0^\infty e^{-r_My}l_M(y)\,b_M(y)\,dy = 1 \tag{8}$$

and r_F from

$$\int_0^\infty e^{-r_Fx}l_F(x)\,b_F(x)\,dx = 1, \tag{9}$$

---

[1] In this paper, for the sake of convenience, the stable population at age y is taken to be the number at the exact age y instead of the number in the age group y and under $(y+1)$, i.e. at average age $(y+\frac{1}{2})$, as is more usual.

where $b_F(x) = \bar{b}_F(x)$ and $x = y - d$, i.e. r_F is to be derived from

$$e^{-d(r_M - r_F)} \int_0^\infty e^{-r_F y} l_M(y) b_M(y) \, dy = 1. \tag{10}$$

Now Lotka has shown[1] that equation (8) expressed in terms of cumulants becomes

$$S_1 r_M - S_2 \frac{r_M^2}{2!} + S_3 \frac{r_M^3}{3!} - \ldots - \log_e R_{0M} = 0, \tag{11}$$

where $\kappa_1, \kappa_2, \ldots$ are, in this case, the cumulants of the distribution of ages of fathers at the birth of their children in the stationary population. Similarly equation (10) can be expressed as

$$\kappa_1 r_F - \kappa_2 \frac{r_F^2}{2!} + \kappa_3 \frac{r_F^3}{3!} - \ldots - \log_e R_{0M} = dr_F - dr_M. \tag{12}$$

We now have to solve for r_F given (11) and (12). Substituting the value of $\log_e R_{0M}$, given by (11), in (12), we get

$$\kappa_1 r_F - \kappa_2 \frac{r_F^2}{2!} + \kappa_3 \frac{r_F^3}{3!} - \ldots - dr_F = \kappa_1 r_M - \kappa_2 \frac{r_M^2}{2!} + \kappa_3 \frac{r_M^3}{3!} - \ldots - dr_M, \tag{13}$$

from which it is clear[1'] that $r_F = r_M$.

The implications of the above analysis are very far-reaching. In the first instance, let us concentrate on the net reproduction rate. It is clear that, unless the actual population has the age and sex composition of the stable population, $b_F(x)$ will not equal $\bar{b}_F(x)$. This being so, we cannot hold constant both $b_M(y)$ and $b_F(x)$. We have assumed that $b_M(y)$ is held constant and hence that $b_F(x)$ changes as the actual population develops towards the stable one; but there is no more reason to assume that $b_M(y)$ can be held constant than to assume the same about $b_F(x)$. In fact, it is most likely that both will change. If this is so, r cannot be exactly located, nor can the stable population be precisely defined. This is a difficulty which is already well recognized by some demographers,[2] but the blow which this deals to the concepts of both the net reproduction rates and the stable population has not been sufficiently emphasized. The stable rate of growth derived from the net reproduction rate measures the ultimate rate of growth of a population, based on the hypothesis that current fertility and mortality conditions will hold continuously. But, if it is logically impossible to maintain this hypothesis, then the concept of stable rate of growth, and the net reproduction rate, which goes with it, loses much meaning. They become internally inconsistent. In general, we may say that it may well be a logical impossibility to hold male and female fertility conditions, as defined above, constant. One or the other or both must change. If we hold $b_M(y)$ constant, then as the actual population develops towards the stable one, $b_F(x)$ must approach $\bar{b}_F(x)$, R_{0F} must approach $\bar{R}_{0F}$ and R_F must approach $\bar{R}_F$.

---

[1] Lotka, op. cit. p. 69.

[1'] Of course, there are other roots for r_F in equation (13), but as Lotka has pointed out (op. cit. p. 70) there is only one root which corresponds to the conditions of the problem and that root is in the neighbourhood of the first order approximation to the solution. In our notation that root must be in the neighbourhood of $r_M = \log_e R_{0M}/S_1$ in equation (11) and in the neighbourhood of $r_F = r_M$ for equation (13). Since in fact one root is $r_F = r_M$, this one fulfils the conditions of the problem.

[2] Cf. Vincent, op. cit. pp. 709–10.

If we assume that $b_M(y)$ is held constant, then from (6) above, it is seen that in the stable population the female net reproduction rate is related to the male net reproduction rate by the simple formula

$$\bar{R}_{0F} = e^{-rd} R_{0M}.$$

Hence, the male and female net reproduction rates are equal only when r equals o (i.e. in the case of a stationary population) or where d equals o (i.e. where males and females pair off at exactly the same age); in other cases, given the male fertility and mortality, $\bar{R}_{0F}$ is uniquely determined quite independently of female fertility and mortality as defined by $l_F(x)$ and $b_F(x)$. This latter conclusion no longer holds where d is not taken as a constant (see §3 below).

Turning to the question of gross reproduction rates, we see that, given $b_M(y)$, i.e. male fertility, the consistent female gross reproduction rate depends not only on $b_M(y)$ but also on $l_M(y)$ and $l_F(x)$. This means that $\bar{R}_F$ is not independent of the level of male and female mortality. Since in practice both $b_M(y)$ and $b_F(x)$, i.e. both male and female fertility rates, are likely to change as the actual population develops towards the stable one, this means that the ultimate values for both male and female gross reproduction rates will depend on the level of male and female mortality. This rather upsets the notion that the gross reproduction rate is a pure measure of fertility.

It can be seen from (6) above that so long as $d > 0$ where $r > 0$, $R_{0M} > \bar{R}_{0F}$, and where $r < 0$, $R_{0M} < \bar{R}_{0F}$, i.e. so long as the mothers are younger than the fathers, the male net reproduction rate is further from unity than its consistent female counterpart. But the same sort of reasoning cannot be applied to the gross reproduction rates. For example:

Let $r < 0$, then $\qquad\qquad\qquad \bar{R}_{0F} > R_{0M},$

but $\qquad\qquad\qquad \bar{R}_F = e^{-rd} \int_0^\infty \frac{l_M(y)}{l_F(y-d)} b_M(y)\, dy.$

Now suppose that $l_M(y)/l_F(y-d)$ is constant Q for all ages, $Q < 1$, then

$$\bar{R}_F = e^{-rd} Q R_M,$$

and if $\qquad\qquad\qquad e^{-rd} Q < 1,$

then $\qquad\qquad\qquad \bar{R}_F < R_M.$

Hence the relation between the consistent gross reproduction rates is not as clear-cut as that between the consistent net reproduction rates. For whilst $\bar{R}_{0F}$ is given when R_{0M} is given, $\bar{R}_F$ depends not only on the function $b_M(y)$ which enters into R_M, but also on $l_M(y)$ which enters into R_{0M} and $l_F(x)$ which does not enter into the calculation of any of the male rates. Thus, given R_{0M}, there is only one value of R_{0F} consistent with it, but given both R_{0M} and R_M, there is an indefinite number of R_F consistent with them.

3. THIS ASSUMPTION DROPPED

The above analysis has been restricted to the case where the difference between ages of mothers and fathers is constant. If we drop this assumption, the analysis becomes more complicated.

Suppose we again assume that we hold $l_M(y)$, $l_F(x)$ and $b_M(y)$ constant. Then, as before, the number of male births to males aged y in the stable population is given by

$$me^{-ry} l_M(y) b_M(y).$$

446

But the way in which the female counterparts to these births are distributed over the female stable population is unknown. So we may write

$$\bar{b}_F(x) = \frac{B(x)}{e^{-rx} l_F(x)}, \tag{14}$$

where $B(x)$ represents female births to females aged x in the stable population and

$$\int_0^\infty B(x)\, dx = \int_0^\infty e^{-ry} l_M(y)\, b_M(y)\, dy.$$

This means that we have no unique set of $\bar{b}_F(x)$ rates, as we had before, hence there is no unique consistent female net reproduction rate $\bar{R}_{0F}$. But it is possible to show how at least one plausible $\bar{R}_{0F}$ can be calculated.

We can calculate the stable rate of growth from the formula

$$r = \frac{\dfrac{R_{1M}}{R_{0M}} - \sqrt{\left\{\left(\dfrac{R_{1M}}{R_{0M}}\right)^2 - 2\left[\dfrac{R_{2M}}{R_{0M}} - \left(\dfrac{R_{1M}}{R_{0M}}\right)^2\right] \log_e R_{0M}\right\}}}{\dfrac{R_{2M}}{R_{0M}} - \left(\dfrac{R_{1M}}{R_{0M}}\right)^2}, \tag{15}$$

where R_{1M}/R_{0M} and R_{2M}/R_{0M} are the first and second moments about zero origin of the age of fathers in the stationary population, respectively.

Now the stable rate of growth calculated from the consistent female net reproduction rate is given by

$$r = \frac{\dfrac{\bar{R}_{1F}}{\bar{R}_{0F}} - \sqrt{\left\{\left(\dfrac{\bar{R}_{1F}}{\bar{R}_{0F}}\right)^2 - 2\left[\dfrac{\bar{R}_{2F}}{\bar{R}_{0F}} - \left(\dfrac{\bar{R}_{1F}}{\bar{R}_{0F}}\right)^2\right] \log_e \bar{R}_{0F}\right\}}}{\dfrac{\bar{R}_{2F}}{\bar{R}_{0F}} - \left(\dfrac{\bar{R}_{1F}}{\bar{R}_{0F}}\right)^2}. \tag{16}$$

Given r, $\bar{R}_{0F}$ can only be determined if $\bar{R}_{1F}/\bar{R}_{0F}$ and $\bar{R}_{2F}/\bar{R}_{0F}$ are also known. This is so because there are an indefinite number of values of $\bar{R}_{1F}$ and $\bar{R}_{2F}$ which could be associated with the same $\bar{R}_{0F}$. However, if we assume that $\bar{b}_F(x) = C b_F(x)$, where C is constant for all x, i.e. $\bar{R}_{0F} = C R_{0F}$, then the terms $\bar{R}_{1F}/\bar{R}_{0F}$ and $\bar{R}_{2F}/\bar{R}_{0F}$ in (16) can be replaced by R_{1F}/R_{0F} and R_{2F}/R_{0F} respectively, and $\bar{R}_{0F}$ can be calculated from the following formula:

$$\log_e \bar{R}_{0F} = \frac{\left(\dfrac{R_{1F}}{R_{0F}}\right)^2 - \left\{r\left[\dfrac{R_{2F}}{R_{0F}} - \left(\dfrac{R_{1F}}{R_{0F}}\right)^2\right] - \dfrac{R_{1F}}{R_{0F}}\right\}^2}{2\left[\dfrac{R_{2F}}{R_{0F}} - \left(\dfrac{R_{1F}}{R_{0F}}\right)^2\right]}. \tag{17}$$

Hence, a value for $\bar{R}_{0F}$ can be calculated. This value depends on the shape of the $b_F(x)$ distribution, but not on the general level of magnitude of the distribution.

Thus when we relax the assumption about the constant difference between the age of mother and father, the relation between the consistent net reproduction rates which would exist in the ultimately stable population become very vague and so, *a fortiori*, does the relation between the gross reproduction rates.

In the above analysis, the formulae are taken correct to the second order. The approximate relation between R_{0M} and $\bar{R}_{0F}$ may be seen more clearly if the formulae are expressed to the first order only. Thus we have

$$\sqrt[T_M]{R_{0M}} \simeq \sqrt[T_F]{\bar{R}_{0F}}, \tag{18}$$

where, here, T_M and T_F are respectively the average ages of the males and females at the birth of their children in the stationary population. And hence

$$\log \bar{R}_{0F} \simeq (\log R_{0M}) \frac{T_F}{T_M}. \tag{19}$$

The consistent rate $\bar{R}_{0F}$ is thus seen to depend not only on male mortality and fertility as expressed in R_{0M} and T_M, but also on the value of T_F, of which there is no unique value. However, we can say that, so long as T_M is greater than T_F, R_{0M} will be further away from unity than $\bar{R}_{0F}$.

If the values of R_{0F}, R_{0M}, T_F and T_M are calculated from an actual population and relation (19) is found not to hold, then if we take R_{0M} and T_M as fixed either R_{0F} or T_F or both must change. In any case, this involves a change in the rates $b_F(x)$, as the actual population develops towards the stable one and hence we may say that the rates $b_M(y)$ and $b_F(x)$ are inconsistent, but the important thing to note is that from relation (19) no unique value for the consistent female net reproduction rate $\bar{R}_{0F}$ can be calculated.

4. Nuptiality introduced

From the above analysis, it is clear that actual male and female fertility conditions, as expressed in male and female age-specific fertility rates respectively, may be mutually inconsistent with ultimate stability and that the relation between the male and female specific fertility rates and hence between male and female net reproduction rates is clearly definable only in the case where the difference between age of fathers and age of mothers at the birth of their children is the same for all births. Furthermore, even in this case there is no clear-cut relation between the gross reproduction rates.

Now this inconsistency between the male and female age-specific fertility rates can be overcome in a population where monogamous marriage is the rule and there are no illegitimate births. In such a population it is clear that all fathers and mothers are paired off by marriage and hence the separate age-specific fertility rates for males and females can be replaced by a rate showing the number of births per head of the number of married couples of husband aged y and wife aged x. Instead of the two sets of rates $b_F(x)$ and $b_M(y)$, we have one set of rates $b(x, y)$, where $b(x, y)$ means the number of female births per married couple of husband aged y and wife aged x and $m b(x, y)$ the corresponding number of male births (m being the masculinity at birth). There can be no inconsistency, in the above sense, in such a set of fertility rates, since the same set of rates applies equally to males and females.

Now let us assume that there is neither illegitimacy nor divorce nor remarriage, and that we have given the probabilities of never-married males surviving a year at each age, the probabilities of married couples surviving a year at each pair of ages, the probabilities of never-married males marrying at each age and the proportions in which they marry brides of each age. It is then a simple enough task to calculate what may be called, a net 'joint-nuptiality' table, showing not only married male survivors at each age, as in the case of a simple net nuptiality table, but also the distribution of such survivors as between the ages of their wives. If we write $M(x, y)$ for the number[1] of married couples of husband aged y and wife aged x in the net joint-nuptial population, then clearly we can write

$$R_0 = \int_0^\infty \int_0^\infty M(x, y)\, b(x, y)\, dx\, dy \qquad (20)$$

for the net reproduction rate, and this rate is the same whether we calculate it on the basis of males or females, because the radix of the table must contain males and females in the proportion of $m : 1$, just as the resulting births must contain males and females in the same proportion.

Hence this eliminates the inconsistency in fertility rates, but in fact it does so only at the expense of introducing another inconsistency. For whilst it is possible to build up a table of $M(x, y)$ on the basis of the probabilities of never-married males marrying at each age and the proportions in which they marry brides of each age, so it is possible to build up a similar table on the basis of the probabilities of never-married females marrying at each age and the proportions in which they marry bridegrooms of each age. It is extremely unlikely that the two joint-nuptiality tables would give consistent answers, for they would only do so if the age and sex composition of the actual population were the same as that in the theoretically stable population. Hence the inconsistency in fertility rates has been replaced by an inconsistency in marriage rates, and two mutually inconsistent values for the net reproduction rate can be calculated, one based on male and the other on female nuptiality conditions.[2] This means that whilst it is possible to hold mortality and fertility conditions as defined above constant, it is logically impossible to hold both male and female nuptiality conditions constant; one or the other or both of these sets of conditions must change if the actual population is developing towards a stable one. And if both change, which seems the most likely, the stable population can no longer be defined because there is no way of telling (other than by informed guesswork) just how the male and female nuptiality conditions will change.

Now suppose that the given fertility conditions as defined by $b(x, y)$, and the given mortality conditions as defined by probabilities of never-married males and married couples surviving, and the given male nuptiality conditions as defined by probabilities of never-married males marrying at each age and the proportions in which they marry brides of each age, are held constant. Then the value of R_0 as defined in (20) can

[1] This 'number' is determined only when the radix of the table is determined, and it will be used with this qualification in the rest of the paper. The radix must, of course, contain males and females in the proportion of $m : 1$, where m is the masculinity at birth.

[2] Since writing the above paper the first issue of *Population Studies* (June 1947) has appeared, and in it articles by Carl-Erik Quensel, 'Population movements in Sweden in recent years', pp. 29–43, and J. Hajnal, 'Aspects of recent trends in marriage in England and Wales', pp. 72–98. Both articles contain much material relevant to the present discussion and a comment on Hajnal's paper appears in note 1, p. 261 below. Quensel (p. 37) says that 'as a criterion of the reliability of the method of calculation (of the reproduction rate) the following test might be of use: applied to the male population, the method should give the same result as when applied to the female population', and he suggests as an improved method a measure very similar to the net reproduction rate for married women as defined by Lim (see note 1, p. 262 below) except that age at marriage and duration of marriage are used as controls instead of just age at birth of children, and illegitimacy is allowed for. However, this does not remove the fundamental difficulty, for on his own showing (pp. 38–9) he gets different answers according as male and female nuptiality is used as base.

be calculated. Is it possible to deduce $\bar{R}_0$, based on female nuptiality conditions, such that the whole system of conditions is internally consistent with ultimate stability?

Since R_0 is based on male nuptiality conditions, we derive the value for r, the stable rate of growth, from the integral equation

$$\int_0^\infty e^{-ry} \int_0^\infty M(x,y)\,b(x,y)\,dx\,dy = 1, \tag{21}$$

and then the number of married couples of each age group in the stable population is given by

$$e^{-ry}\,M(x,y).$$

Now $M(x,y)$ represents the number of married couples of each age group in the net joint-nuptial population, based on male nuptiality. Let $\bar{M}(x,y)$ be the analogous number in the consistent net joint-nuptial population, based on female nuptiality. Then

$$e^{-rx}\,\bar{M}(x,y) = e^{-ry}\,M(x,y),$$

i.e.

$$\bar{M}(x,y) \quad\; = e^{-r(y-x)}\,M(x,y). \tag{22}$$

Hence

$$\bar{R}_0 = \int_0^\infty \int_0^\infty e^{-r(y-x)}\,M(x,y)\,b(x,y)\,dx\,dy. \tag{23}$$

As before, it is not very difficult to prove that the stable rates of growth derived from (20) and (23) are the same. Again, letting r_M and r_F be the stable rates of growth of the male and female populations respectively, we have r_M derived from equation (21) and r_F from

$$\int_0^\infty e^{-r_M y} \int_0^\infty e^{-(r_F - r_M)x}\,M(x,y)\,b(x,y)\,dx\,dy = 1, \tag{24}$$

and obviously $r_F = r_M$ is a solution.

Thus we see that there is a unique value of $\bar{R}_0$ consistent with R_0. So by shifting the inconsistency from fertility to nuptiality conditions we eliminate the indefinite relation between the male and female net reproduction rates which was introduced by relaxing the assumption that the difference of age between fathers and mothers at the birth of their children was constant. The relation between R_0 and $\bar{R}_0$, although complex, is contained in the three formulae (20), (21) and (23). From (21) and (23) we see that $\bar{R}_0 = R_0$ for $r = 0$ or for $y - x = 0$ and, by expanding (23) to the first order, we see that, in general, R_0 will be further from unity than $\bar{R}_0$, if on the average the age of fathers is greater than the age of mothers at the birth of their children in the net joint-nuptial population.

Since there is a unique value $\bar{R}_0$, based on female nuptiality conditions, consistent with R_0, based on male nuptiality conditions, it is necessary to see whether the consistent female nuptiality conditions can be deduced and whether there is a unique set of them. Let us write $u(x,y)$ for the number of marriages of bridegrooms aged y to brides aged x in the net joint-nuptial population, based on male nuptiality conditions. Then the analogous number in the consistent net joint-nuptial population, based on female nuptiality conditions, will be given by

$$\bar{u}(x,y) = e^{-r(y-x)}\,u(x,y) \tag{25}$$

and

$$\bar{u}(x) = e^{rx} \int_0^\infty e^{-ry}\,u(x,y)\,dy, \tag{26}$$

where $\bar{u}(x)$ is the number of marriages of never-married females aged x in the consistent net joint-nuptial population, based on female nuptiality.

It is possible to deduce by means of (26) the proportions in which brides of each age should marry bridegrooms of each age to be consistent with male nuptiality conditions. These proportions can be uniquely determined and can be represented by

$$\frac{e^{-ry}\,u(x,y)\,dy}{\displaystyle\int_0^\infty e^{-ry}\,u(x,y)\,dy} \qquad \text{for each } x. \qquad (27)$$

There is also sufficient information to deduce the consistent probabilities of marrying for never-married females at each age. For we know the number of marriages $\bar{u}(x)$ which must take place in the consistent net joint-nuptial population at each age, and if we assume the probabilities of survival for never-married females we can deduce the probabilities of their marrying at each age. But here two qualifications must be noted. First, the probabilities of marrying will depend on the assumed probabilities of survival for never-married females, and hence there will be no unique consistent set of rates. This means that there is no unique relation in the ultimately stable population between the gross reproduction rates based on male and those based on consistent female nuptiality conditions, as there is between the net reproduction rates. This is so because a gross reproduction rate based on the fertility conditions $b(x,y)$ and the male nuptiality conditions can be calculated through the medium of a gross joint-nuptial population which is not dependent on mortality conditions. Similarly, a gross reproduction rate based on the same fertility conditions and the consistent female nuptiality conditions can be calculated. But the latter depend on the mortality conditions of never-married females, which conditions do not enter into the calculation of either R_0 or $\bar{R}_0$ or the gross reproduction rate based on male nuptiality. Hence there are an indefinite number of possible consistent gross reproduction rates based on female nuptiality.

Secondly, if any $\bar{u}(x)$ turns out to be so great that the probability of marrying dependent on it is greater than unity, then the male nuptiality conditions themselves could not possibly be maintained. Although this seems unlikely, it may happen that the female probability of marrying at a particular age turns out to be so high that in practice it could not be maintained. If this is so, neither could the male nuptiality conditions be maintained intact.

51. The Measurement of Reproductivity

A. H. POLLARD (1948)

From *Journal of the Institute of Actuaries* 74. Excerpts are from pages 305—311, 313.

The first part of Pollard's article, which is omitted, summarizes measures of reproductivity in one-sex models and presents models for the effects of changing marriage rates on births. The final section of the paper is an application of reproductivity formulas to Australian data, also omitted here.

4. *Male v. Female Reproductivity Formulae*

One of the most serious objections which can be levelled against all the formulae of section 2 has not been mentioned so far. It will be considered in some detail in the following paragraphs.

The formulae of section 2 are based on a determination of the rate at which a given sex is replacing itself. For various reasons (shorter reproductive period makes for shorter calculation; the required data are more often available for the female sex; illegitimate births are easily referable to the mother, etc.) the female sex is commonly used. There is no reason, of course, why the male sex should not be used as a basis. It is here, however, that the anomaly arises. In practice, for reasons discussed below, two quite different values are obtained for a given measure of reproductivity using the two sexes. If these two values of ρ for males and for females were to continue until stable conditions eventually emerged one sex would in course of time swamp the other. The conception of two stable populations, one for males and one for females, with different values of ρ is therefore untenable. The population as a whole, and both sexes in particular, must therefore ultimately increase at the same rate, which presumably would lie somewhere between the values obtained for the separate sexes. Unfortunately this frequently leaves a large range (for example, see section 6) anywhere within which the required value might fall and seriously detracts from the value of the method. Translated into other terms, the anomaly means that it is an impossible hypothesis to assume that the rates of fertility obtained for each sex can continue indefinitely in the future.

To assist us, in a particular case, in deciding where, within the range bounded by the values of ρ for the separate sexes, the true rate of increase lies, we will consider briefly some possible reasons for the difference between the male and female rates. R.J. Myers (10) elaborates these reasons with some actual figures.

(I) If there is a temporary excess of females (e.g. as a result of the ravages of war) at the reproductive ages, age specific fertility rates and the net reproduction rate for women will be relatively low as compared with those for men.

(II) Excess female over male immigration at the reproductive ages, if females are already well represented, is unlikely appreciably to increase the number of births. This would lower the female rate relative to the male.

(III) The tendency for females over 30 to underestimate, and females under 20 to overestimate, their age tends to lower the computed female net reproduction rate.

(IV) Because husbands are, on the average, about 5 years older than their wives the supply of husbands will tend to fall the more rapidly the population is increasing, and hence the female rate will become smaller relative to the male rate the more rapid the population increase.

Male rates have generally been found to be appreciably higher than female and hence, in view of the discussion above, it might well be that some of the pessimistic discussion in demographic literature, caused by the fall of the female net reproduction rate below unity, may not be well founded.

The serious theoretical difficulty of the male *v.* female rates discussed in this section, and the serious practical difficulty of having an inherent rate of increase only known to lie within two (perhaps) widely separated limits, amply justify considerable further investigation whether a unique index of reproductivity can be found. An index which can easily be calculated, for which data are readily available, which is theoretically unique and which lies between the male and female rates will now be discussed.

5. *Joint Rate of Increase*

5.1. *Aim*

The aim of this section is to outline the properties of an index of reproductivity which has all the advantages of those previously discussed and yet which does not suffer from their main weakness—the anomaly of section 4.

5.2. *Basic data*

The basic data required for the determination of this index consist of the probability at birth that a male will have a female child between ages x and $x+dx$ (written $\phi(x)dx$) and the probability at birth that a female will give birth to a male child between ages y and $y+dy$ (written $\xi(y)dy$). For almost all countries, the annual male and female births are published according to age of mother, and, for most countries, according to age of father also. If these data are not published they are of such a simple nature that they can readily be obtained from the birth records. Combining these fertility data with mortality gives the net fertility functions $\phi(x)$ and $\xi(y)$.

The theory which, in parts, resembles that applied by Rhodes (6) to R_0 and ρ will be developed in the next few paragraphs and the results obtained summarized in paragraph 5.13. Readers only interested in results could, therefore, turn immediately to that paragraph.

5.3. *Theory*

The female births $F(t)$ and the male births $M(t)$ at time t are given by

$$F(t) = \int_0^\infty M(t-x)\phi(x)dx, \tag{13}$$

and

$$M(t) = \int_0^\infty F(t-y)\xi(y)dy. \tag{14}$$

Hence, we have

$$F(t) = \int_0^\infty \int_0^\infty F(t-x-y)\phi(x)\xi(y)dxdy, \tag{15}$$

$$M(t) = \int_0^\infty \int_0^\infty M(t-x-y)\phi(x)\xi(y)dxdy, \tag{16}$$

455

and thus the total births

$$B(t) = \int\limits_0^\infty \int\limits_0^\infty B(t-x-y)\phi(x)\xi(y)dxdy.\qquad(17)$$

The last three equations are of the same form and hence so also will be their solution. We can see at once that equation (17) will be satisfied by a function of the form

$$B(t) = \sum_n B_n e^{s_n t}.$$

Substituting in (17) we find that that equation is satisfied by this form if the values of s are given by

$$\int\limits_0^\infty \int\limits_0^\infty e^{-(x+y)s}\phi(x)\xi(y)dxdy=1.\qquad(18)$$

This equation (18) is obtained whether we are solving (15), (16) or (17) and hence the values of s obtained apply to male, to female and to total births.

5.4. This equation has only one real solution; for if we suppose s to be real and denote the left-hand side of (18) by f then

$$\frac{df}{ds} = -\int\limits_0^\infty \int\limits_0^\infty (x+y)e^{-(x+y)s}\phi(x)\xi(y)dxdy.$$

Now since $\phi(x), \xi(y), e^{-(x+y)s}$ and $x+y$ are greater than or equal to zero, df/ds must always be negative for all values of s. Hence $f=1$ can have only one real solution σ (say).

5.5. If $\sigma=0$, then

$$\int\limits_0^\infty \int\limits_0^\infty \phi(x)\xi(y)dxdy=1.$$

If $\sigma>0$, then

$$e^{-\sigma(x+y)}<1,$$

and hence, from (18)

$$\int\limits_0^\infty \int\limits_0^\infty \phi(x)\xi(y)dxdy>1.$$

Similarly, if $\sigma < 0$,

$$\int_0^\infty \int_0^\infty \phi(x)\xi(y)dxdy < 1.$$

Therefore $\sigma \gtreqless 0$ according as

$$\int_0^\infty \int_0^\infty \phi(x)\xi(y)dxdy \gtreqless 1.$$

This latter expression which will be denoted by S_0 is analogous to the net reproduction rate and can be used as a measure of reproductivity which is independent of sex. It will be called the 'joint reproduction rate'. It is a rate of increase using as unit of time the total male and female 'generation'. It cannot, therefore, be compared directly with the net reproduction rate or other rates and is not therefore recommended.

5.6. If $s = u + iv$ is a complex root of (18), then, substituting and equating real and imaginary parts

$$\int_0^\infty \int_0^\infty e^{-(x+y)u}\cos\{(x+y)v\}\,\phi(x)\xi(y)dxdy = 1,$$

and

$$\int_0^\infty \int_0^\infty e^{-(x+y)u}\sin\{(x+y)v\}\,\phi(x)\xi(y)dxdy = 0.$$

Therefore $u - iv$ is also a root.

Comparing the former of these equations with the real solution of (18) we have, since $\cos(x+y)v < 1$,

$$e^{-(x+y)u} > e^{-(x+y)\sigma}.$$

Therefore u, the real part of any imaginary root, is less than the real root σ.

Combining conjugate complex roots we may express the solution of (17) as

$$B(t) = B_0 e^{\sigma t} + \sum_n e^{u_n t}(\alpha_n \sin v_n t + \beta_n \cos v_n t),$$

which, since $u_n < \sigma$, tends to $B_0 e^{\sigma t}$ as t becomes large.

We have thus proved that the total births, and may similarly prove that the male and the female births, of a community subject to the net fertility of 5.2 all ultimately increase at an annual rate of σ. σ will be called the 'joint rate of natural increase'.

5.7. σ which from (18) is given by

$$\int_0^\infty e^{-\sigma x}\phi(x)dx \int_0^\infty e^{-\sigma y}\xi(y)dy = 1 \tag{19}$$

may be determined by two methods, corresponding to the method of Lotka (4) and that of Wicksell (5) for determining the true rate of natural increase.

Denoting $\int_0^\infty e^{-\sigma x}\phi(x)dx$ by z, we have $dz/d\sigma = -Cz$, where

$$C = \frac{\int_0^\infty xe^{-\sigma x}\phi(x)dx}{\int_0^\infty e^{-\sigma x}\phi(x)dx} \tag{20}$$

$$= \frac{M_1 - \sigma M_2 + \dfrac{\sigma^2}{2}M_3 - \ldots}{M_0 - \sigma M_1 + \dfrac{\sigma^2}{2}M_2 - \ldots}, \quad \text{where} \quad M_n = \int_0^\infty x^n \phi(x)dx$$

$$= a + b\sigma + c\sigma^2 + \ldots, \quad \text{where} \quad a = \frac{M_1}{M_0}, \quad b = \frac{M_1^2}{M_0^2} - \frac{M_2}{M_0}, \quad \text{etc.}$$

This series has been shown by Lotka to converge very rapidly and only the first two terms need be considered. Hence, substituting in the differential equation, integrating and determining the constant thus introduced, we have

$$z = \int_0^\infty e^{-\sigma x}\phi(x)dx = M_0 e^{-a\sigma - \frac{1}{2}b\sigma^2}. \tag{21}$$

We may obtain a similar expression for $\int_0^\infty e^{-\sigma y}\xi(y)dy$, and, writing N_n, α, β, etc. for the functions of $\xi(y)$ corresponding to M_n, a, b, etc. for $\phi(x)$, we may obtain, by substituting in (19), the following equation for σ:

$$\tfrac{1}{2}(b + \beta)\sigma^2 + (a + \alpha)\sigma - \log_e M_0 N_0 = 0. \tag{22}$$

Alternatively, following Wicksell, we may use a Pearson Type III curve to represent $\phi(x)$, thus

$$\phi(x) = M_0 \frac{t^u}{\Gamma(u)} x^{u-1} e^{-tx},$$

where

$$t = \frac{M_0 M_1}{M_0 M_2 - M_1^2} \quad \text{and} \quad u = \frac{M_1^2}{M_0 M_2 - M_1^2}.$$

Let the Type III curve representing $\xi(y)$ involve constants v and w corresponding to t and u for $\phi(x)$. Then, substituting in (19) and integrating, we have

$$\frac{M_0 \qquad N_0}{\left(1 + \dfrac{\sigma}{t}\right)^u \left(1 + \dfrac{\sigma}{v}\right)^w} = 1. \tag{23}$$

For all practical purposes (23) gives the same results as (22), but here (22) is easier to solve.

5.8. Having determined a value of s, say s_m, from equation (18) or a more accurate form of equation (22) the following method may be used to determine B_m.

Writing l and L for the limits of the reproductive period and substituting for $B(t)$ we have

$$\int_l^{L+l} B(t)e^{-s_m t}\,dt = \sum^{m \neq n} \frac{B_n}{s_n - s_m}\left\{e^{(s_n - s_m)(L+l)} - e^{(s_n - s_m)l}\right\} + B_m L. \qquad (24)$$

Also

$$\int_{2l}^{L+l}\left\{\int_l^{t-l} B(t-x)\phi(x)\,dx\right\}e^{-s_m t}\,dt = \int_l^L\left\{\sum B_n e^{-s_n x}\phi(x)\int_{x+l}^{L+l} e^{(s_n-s_m)t}\,dt\right\}dx$$

$$= \sum^{n \neq m} \frac{B_n}{s_n - s_m}\left\{e^{(s_n-s_m)(L+l)}\int_l^L e^{-s_n x}\phi(x)\,dx - e^{(s_n-s_m)l}\int_l^L e^{-s_m x}\phi(x)\,dx\right\}$$

$$+ B_m L\int_l^L e^{-s_m x}\phi(x)\,dx - B_m\int_l^L xe^{-s_m x}\phi(x)\,dx. \qquad (25)$$

From (24) and (25)

$$\frac{\displaystyle\int_{2l}^{L+l}\left\{\int_l^{t-l} B(t-x)\phi(x)\,dx\right\}e^{-s_m t}\,dt}{\displaystyle\int_l^L e^{-s_m x}\phi(x)\,dx} - \frac{\displaystyle\int_l^{L+l} B(t)e^{-s_m t}\,dt + B_m \dfrac{\displaystyle\int_l^L xe^{-s_m x}\phi(x)\,dx}{\displaystyle\int_l^L e^{-s_m x}\phi(x)\,dx}}{}$$

$$= \sum^{n \neq m} \frac{B_n e^{(s_n-s_m)(L+l)}}{s_n - s_m}\left\{\frac{\displaystyle\int_l^L e^{-s_n x}\phi(x)\,dx}{\displaystyle\int_l^L e^{-s_m x}\phi(x)\,dx} - 1\right\}. \qquad (26)$$

Again using a Pearson Type III curve to represent $\phi(x)$ we may in equation (26) put

$$\frac{\displaystyle\int_l^L e^{-s_n x}\phi(x)\,dx}{\displaystyle\int_l^L e^{-s_m x}\phi(x)\,dx} - 1 = \left(\frac{t+s_n}{t+s_m}\right)^{-u} - 1 = \left(1 + \frac{s_n - s_m}{t+s_m}\right)^{-u} - 1$$

$$= -u\left(\frac{s_n - s_m}{t + s_m}\right) + \frac{u(u+1)}{2}\left(\frac{s_n - s_m}{t+s_m}\right)^2 - \dots. \qquad (27)$$

459

If in the thus modified equation (26) we substitute successively

$$x\phi(x), \quad x^2\phi(x), \quad \ldots,$$

in place of $\phi(x)$ we obtain a series of equations of which all terms on the left-hand side (except B_m) are known and from which the unknown terms on the right-hand side may be eliminated. Writing $I_1 B_m, I_2 B_m$, etc. for the left-hand sides and eliminating unknowns, we have

$$\begin{vmatrix} I_1 B_m, & u, & \dfrac{u(u+1)}{2}, & \cdots \\[2ex] I_2 B_m, & u+1, & \dfrac{(u+1)(u+2)}{2}, & \cdots \\[2ex] I_3 B_m, & u+2, & \dfrac{(u+2)(u+3)}{2}, & \cdots \\[2ex] \vdots & \vdots & \vdots & \end{vmatrix} = 0. \tag{28}$$

Determinant (28) with as many terms as are necessary may be used to determine B_m.

5.9. As $t \to \infty$ the ratio $M(t)/F(t)$ of male to female births tends to

$$M_0 e^{\sigma t} \div F_0 e^{\sigma t} = M_0/F_0$$

which is constant.[1]

If the ratio of male to female births up to time t has been constant and equal to X, then, by expanding the determinant (28) to determine M_0 and F_0, it can be seen by inspection of the form of IB_0 that the ratio M_0/F_0 must be equal to X. If the masculinity ratio has not been constant in the past, M_0/F_0 is a weighted average of past ratios. Hence the ultimate age and sex distribution of the population is determined by the joint rate of natural increase, the male and female mortality and the past sex ratios at birth. This ultimate age and sex distribution is

$$f(x) = l_x^{(f)} e^{-\sigma x} \quad \text{and} \quad m(x) = X l_x^{(m)} e^{-\sigma x}. \tag{29}$$

5.10. If we assume the ratio of male to female births to be constant, independent of age or sex of parent and equal to X, then the true rates of natural increase for males ρ_m and females ρ_f are given by

$$\int_0^\infty e^{-\rho_m x} \phi(x) X \, dx = 1 \quad \text{and} \quad \int_0^\infty e^{-\rho_f y} \xi(y) X^{-1} \, dy = 1.$$

---

[1] Notice that M_0 in section 5.9 is not the same as M_0 in section 5.7.—Eds. *J.I.A.*

Hence, from (19)

$$\int_0^\infty e^{-\sigma x}\phi(x)dx \int_0^\infty e^{-\sigma y}\xi(y)dy = \int_0^\infty e^{-\rho_m x}\phi(x)dx \int_0^\infty e^{-\rho_f y}\xi(y)dy. \qquad (30)$$

Now if

$$\rho_m \geq \sigma, \quad \text{then} \quad e^{-\rho_m x} \leq e^{-\sigma x}.$$

Hence

$$\int_0^\infty e^{-\sigma y}\xi(y)dy \leq \int_0^\infty e^{-\rho_f y}\xi(y)dy.$$

That is,

$$e^{-\sigma y} \leq e^{-\rho_f y} \quad \text{or} \quad \sigma \geq \rho_f.$$

Thus if

$$\rho_m \geq \sigma, \quad \rho_m \geq \sigma \geq \rho_f.$$

Similarly if

$$\rho_f \geq \sigma, \quad \rho_f \geq \sigma \geq \rho_m.$$

Therefore, if the sex ratio at birth is constant for parents of any age or sex, σ must lie between ρ_m and ρ_f.

5.11. If we continue to make the assumption that the sex ratio at birth is constant for parents of any age or sex, we can find approximately the relation between S_0 and the net reproduction rates for males and females R_0^m and R_0^f respectively, and also an approximate relation between σ and ρ_m and ρ_f.
Since

$$S_0 = \int_0^\infty \phi(x)dx \int_0^\infty \xi(y)dy, \quad R_0^m = \int_0^\infty \phi(x)Xdx, \quad R_0^f = \int_0^\infty \xi(y)X^{-1}dy,$$

we have, on the above assumptions,

$$S_0 = R_0^m \cdot R_0^f. \qquad (31)$$

If for the integrals in equation (30) we substitute Lotka's exponential form (21) we obtain, equating powers of e,

$$(a+\alpha)\sigma + \tfrac{1}{2}(b+\beta)\sigma^2 = a\rho_m + \tfrac{1}{2}b\rho_m^2 + \alpha\rho_f + \tfrac{1}{2}\beta\rho_f^2.$$

461

Solving for σ,

$$\sigma = [-(a+\alpha) \pm \{(a+\alpha)^2 + 2(b+\beta)(a\rho_m + \alpha\rho_f + \tfrac{1}{2}b\rho_m^2 + \tfrac{1}{2}\beta\rho_f^2)\}^{\frac{1}{2}}](b+\beta)^{-1}.$$

But, on our assumption, σ must lie between ρ_f and ρ_m. Hence if $\rho_f = \rho_m = 0$, $\sigma = 0$. Substituting these values in the above expression for σ shows that, of the alternatives, we must select the positive sign.

Expanding by the binomial theorem and cancelling, we have

$$\sigma = \frac{a\rho_m + \alpha\rho_f}{a+\alpha} \quad \text{approximately.} \tag{32}$$

5.12. It should be emphasized that the measures S_0 and σ in no way depend on the assumption of constant sex ratio at birth made in the preceding two paragraphs. One of their functions is to allow for variations in this ratio. The usefulness of S_0 and σ in building up a complete theory would be lost if we simply defined S_0 as the right-hand side of equation (31). Relations (31) and (32) while useful practical formulae are limited because of this assumption.

5.13. Summary

Given a population subject to the net fertility functions $\phi(x)$ and $\xi(y)$ we have thus shown, amongst other things, that:

(I) the male, female and total births all ultimately increase at the 'joint rate of natural increase' σ given by equation (19) or approximately by equations (22) and (23);

(II) S_0, the 'joint reproduction rate', defined in paragraph 5.5 is a unique reproduction rate corresponding to σ;

(III) $\sigma \gtreqless 0$ according as $S_0 \gtreqless 1$;

(IV) the ultimate age-sex distribution of the population is given by (29);

(V) if the sex ratio at birth has been constant in the past σ lies between ρ_f and ρ_m;

(VI) on the same assumption, S_0 and σ are related to R_0^m, R_0^f, ρ_f and ρ_m by the approximate relations (31) and (32).

This suggestion could be applied to the Karmel or the Clark-Dyne formulae to avoid, in them, the male v. female anomaly.

References

(1) C.D. Rich (1934). 'The measurement of the rate of population growth.' *J.I.A.* Vol. LXV, p. 38.

(2) A.J. Lotka (1936). 'The geographic distribution of intrinsic natural increase in the United States, and an examination of the relation between several measures of net reproductivity.' *J. Amer. Statist. Ass.* Vol. xxxi, p. 273.

(3) F.R. Sharpe and A.J. Lotka (1911). 'A problem in age distribution.' *Philosophical Magazine,* Vol. xxi, p. 435.

(4) L.I. Dublin and A.J. Lotka (1925). 'On the true rate of natural increase.' *J. Amer. Statist. Ass.* Vol. xx, p. 305.

(5) S.D. Wicksell (1931). 'Nuptiality, fertility and reproductivity.' *Skand. Akt.* Vol. xiv, p. 125.

(6) E.C. Rhodes (1940). 'Population mathematics I, II, and III.' *J.R. Statist. Soc.* Vol. ciii, pp. 61, 218 and 362.

(7) P.H. Karmel (1944). 'Fertility and marriages—Australia 1933—42.' *Econ. Rec.* Vol. xx, No. 38, p. 74.

(8) C. Clark and R.E. Dyne (1946). 'Application and extension of the Karmel formula for reproductivity.' *Econ. Rec.* Vol. xxii, No. 42, p. 23.

(9) L.I. Dublin and A.J. Lotka (1931). 'The true rate of natural increase of the population of the United States. Revision on basis of recent data.' *Metron* Vol. viii, No. 4, p. 107.

(10) R.J. Myers (1941). 'The validity and significance of male net reproduction rates.' *J. Amer. Statist. Ass.* Vol. xxxvi, p. 275.

52. Stochastic Processes and Population Growth

Dᴀᴠɪᴅ G. Kᴇɴᴅᴀʟʟ (1949)

From *Journal of the Royal Statistical Society*, B, 11. Excerpts are from pages 247—249.

The material reproduced here is a continuation of paper 45. Kendall's equation (1), required for this section, is the differential $\dfrac{dN}{dt} = vN$, whose solution is the familiar exponential $N(t) = N(0)e^{vt}$, where $N(t)$ is total population size at time t and v is the intrinsic growth rate.

(ix) *The problem of the two sexes.*—In one of the discussions at the recent Oxford Conference of the Society several speakers pointed out that the possibility of variations in the relative numbers of the two sexes has been too long neglected in population mathematics. The present section is intended to touch on some of the characteristic difficulties of this problem, and to suggest some crude approximations to a solution. A more profound analysis of the question (from the deterministic standpoint) will be found in the papers of P. H. Karmel.

First it is convenient to consider how one could modify the simplest deterministic model, represented in the mathematics by equation (1). If $M(t)$ and $F(t)$ are the numbers of males and females respectively, the most natural way to generalize (1) is to write

$$dM/dt = -\mu M + \tfrac{1}{2}\Lambda(M, F),$$
$$dF/dt = -\mu F + \tfrac{1}{2}\Lambda(M, F), \qquad \qquad \qquad (69)$$

where $\Lambda(M, F)$ is symmetric in M and F and represents the contribution from the birth rate. (It is assumed* that the death rate is the same for the two sexes, and that each birth is equally likely to add a new male or a new female to the population.) Subtraction gives

$$\frac{d}{dt}(M - F) = -\mu(M - F)$$

and so

$$M(t) - F(t) = [M(0) - F(0)]\,e^{-\mu t} \qquad . \qquad . \qquad . \qquad . \qquad (70)$$

i.e. any initial preponderance of one sex over the other disappears in the course of time.

If one keeps to models of this type, further integration of the equations must be preceded by a more detailed assumption about the function Λ. To represent random mating one might set it proportional to MF; this however implies a total number of births per unit time varying as the square of the total population size (when the sex-ratio happens to be constant), and as might be expected the solution takes on an unstable character. To illustrate this, let $M_0 = F_0$, so that $M = F$ for all t, in accordance with (70); M and F then both satisfy an equation of the form

$$dM/dt = -\mu M(1 - M/\alpha) \qquad . \qquad . \qquad . \qquad . \qquad (71)$$

where α is a constant. This can be solved as in section 1 (ii), and it will be found that the nature of the solution is entirely different for different initial values of the population size. Thus M tends to zero when t tends to infinity if $M_0 < \alpha$, while M is constant for all t if $M_0 = \alpha$. If $M_0 > \alpha$, M tends to infinity as t approaches the *finite* value

$$T = -\frac{1}{\mu}\log\left(1 - \frac{\alpha}{M_0}\right).$$

Similar but more complicated phenomena are displayed by the solution in the more general case when $M_0 \neq F_0$.

These difficulties are avoided if Λ is linear in the total population size, and in particular if

$$\Lambda = 2\,\lambda\,\sqrt{(MF)} \qquad . \qquad . \qquad . \qquad . \qquad . \qquad (72)$$

The constant λ is then the birth rate per head per unit of time when M and F happen to be equal. The equations are most easily solved in this case by writing $M = R^2$ and $F = S^2$, so that

$$dR/dt = -\tfrac{1}{2}\mu R + \tfrac{1}{2}\lambda S,$$
$$dS/dt = -\tfrac{1}{2}\mu S + \tfrac{1}{2}\lambda R.$$

Addition and integration now show that M and F tend jointly to infinity when $\lambda > \mu$, and to zero when $\lambda < \mu$, as t tends to infinity. When $\lambda = \mu$, $R + S$ is constant and M and F tend to the same non-zero limit.

A somewhat simpler but less realistic model is obtained if

$$\Lambda = \lambda(M + F), \qquad . \qquad . \qquad . \qquad . \qquad . \qquad (73)$$

so that the birth rate depends on the arithmetic instead of on the geometric mean of M and F. The equations are now

$$dM/dt = -\mu M + \tfrac{1}{2}\lambda(M + F),$$
$$dF/dt = -\mu F + \tfrac{1}{2}\lambda(M + F),$$

---

* These assumptions can, of course, only be admitted as preliminary approximations.

and the solution behaves qualitatively in the same way as before, the critical relation between the constants being still $\lambda = \mu$.

Perhaps the most realistic model is that corresponding to the choice

$$\Lambda = 2\lambda \min(M, F) \qquad . \qquad . \qquad . \qquad . \qquad . \qquad (74)$$

The equations are easily integrable because the algebraic sign of $(M - F)$ is the same for all t, in virtue of (70), and so if there is initially an excess of females,

$$dM/dt = -\mu M + \lambda M,$$
$$dF/dt = -\mu F + \lambda M,$$

and

$$M = M_0 \, e^{(\lambda - \mu)t},$$
$$F = M - (M_0 - F_0) \, e^{-\mu t}.$$

The qualitative behaviour of the solution is the same as for the two preceding models.

Interesting new features arise if one discriminates between married and unmarried persons. Let M, F and N denote at time t the numbers of unmarried males, unmarried females, and married couples, respectively. Then the natural generalization of the preceding model is governed by the equations

$$dM/dt = -\mu M + \lambda N + \mu N - K(M, F),$$
$$dF/dt = -\mu F + \lambda N + \mu N - K(M, F),$$
$$dN/dt = -2\mu N + K(M, F),$$

where μ is the death rate per head per unit of time, λ is the birth rate per married person per unit of time, and $K(M, F)$ is the marriage rate per unit of time. (The age-distribution is, of course, being neglected throughout.) Subtraction of the first and second equations shows that

$$M - F = (M_0 - F_0) \, e^{-\mu t},$$

so that as before any initial excess of males or females disappears in the course of time. On the other hand, if the first and third equations are considered together, it is easy to complete the description of the solution when

$$K(M, F) = 2\nu \min(M, F) \qquad . \qquad . \qquad . \qquad . \qquad . \qquad (75)$$

Suppose, for example, that there is initially an excess of females; then

$$dM/dt = -(\mu + 2\nu) M + (\lambda + \mu)N,$$
$$dN/dt = \quad 2\nu M \quad - \quad 2\mu N,$$

so that both M and N are of the form

$$Ae^{p_1 t} + Be^{p_2 t},$$

where p_1 and p_2 are the roots of the equation

$$p^2 + (3\mu + 2\nu)p + 2\mu^2 + 2\nu(\mu - \lambda) = 0.$$

The roots are always real and distinct, and one of them is always negative, so that it is the greater root which determines the character of the solution. In general it can be seen that the population size tends to infinity, approaches a finite limit, or tends to zero according as $\lambda > \lambda_1$, $\lambda = \lambda_1$, or $\lambda < \lambda_1$ where

$$\lambda_1 = \mu \, (1 + \mu/\nu) \qquad . \qquad . \qquad . \qquad . \qquad . \qquad (76)$$

One can now review briefly the prospects of being able to express these modes of population growth in stochastic form; it is evident that the problem would be a very difficult one. Assumptions such as (72), (74) and (75) seem impossible to discuss with the generating function technique, and one is driven back to that associated with (71), and to (73). The first of these must be dismissed because of the unstable character of the deterministic solution (in any case, the partial differential equation for the generating function would prove to be of the second order, and so would defy solution by the usual methods). The second, and its equivalent

$$K(M, F) = \nu \, (M + F) \qquad . \qquad . \qquad . \qquad . \qquad . \qquad (77)$$

in the final formulation of the problem, while leading to equations for which a solution could certainly be found, still has the disadvantage that it implies a non-vanishing overall marriage rate even when all the unmarried persons are of one sex.

With these somewhat unhelpful comments the topic will now be left.

References

Karmel, P. H. (1947), "The relations between male and female reproduction rates," *Population Studies.* 1, 249–74. (See also two further papers by Karmel, on the same topic, in volumes 1 and 2 of the same journal.)

53. Population Growth of the Sexes

LEO A. GOODMAN (1953)

From *Biometrics* 9. Excerpts are from pages 212—216, 220—224.

We omit a deterministic model introduced by Goodman that distinguishes between married and unmarried persons, and a simple stochastic model for the sex ratio when males and females reproduce independently but may have offspring of either sex.

1. INTRODUCTION

Many authors (see, e.g., references 1–19) have noted that the sex distribution at birth is not equal (more males than females are born) and that the ability of the sexes to withstand the forces of mortality is not the same (the death rates at most ages are higher for males than for females). This phenomenon is not confined to man alone, but it also occurs, though not universally, among a number of other species. The literature contains many discussions of the importance and implications of these two facts.

At a recent meeting of the Royal Statistical Society several speakers pointed out that the possibility of variations in the relative numbers of the two sexes has been too long neglected in population mathematics. D. G. Kendall [22] has discussed some of the characteristic difficulties of this problem and suggested some approximations which we shall extend. Some work on this problem, from a deterministic standpoint, has been carried out (see, e.g., references 1, 13–19). Kendall has mentioned that the problem of expressing the modes of population growth for the two sexes in stochastic form would be a very difficult one. References 20–26 deal with this problem when one sex is considered and their bibliographies give references to much of what has been written in the field.

J. Yerushalmy, in a very interesting paper [1], describes the age-sex composition of the population resulting from natality and mortality conditions. If we are interested in the ultimate stationary population we may determine its over-all sex ratio from Yerushalmy's analysis. In this paper we shall consider the problem of determining the over-all sex ratio for populations which need not be stationary and also the problem of studying the population growth of each sex.

We shall consider various mathematical models of population growth. In order to make possible an analytic treatment of the subject, these models necessarily will be oversimplifications of reality.

¹This paper was prepared in connection with research supported by the Office of Naval Research.

Extending Kendall's [22, p. 247] approach we have the following model: If $M(t)$ and $F(t)$ are the number of males and females respectively at time t, then these quantities satisfy the differential equations

$$\frac{dM}{dt} = -aM + \Lambda(M, F)$$

$$\frac{dF}{dt} = -bF + \Lambda'(M, F)$$

where a is the intrinsic male death rate per male per unit of time, b is the intrinsic female death rate per female per unit of time, $\Lambda(M,F)$ and $\Lambda'(M,F)$ are functions of M and F representing the contributions from the male birth rate and female birth rate, respectively. (Kendall dealt with the case where $a = b$ and $\Lambda = \Lambda'$.)

Consider the case where $\Lambda(M,F) = uF$ and $\Lambda'(M,F) = vF$; i.e., where the birth rates depend on the female population size F (females are marriage dominant, see [17], [18]). We then have

$$\frac{dM}{dt} = -aM + uF$$

$$\frac{dF}{dt} = -bF + vF = (v - b)F.$$

The solution of these equations is

$$M(t) = \frac{uA}{(v - b + a)} e^{(v-b)t} + Be^{-at}$$

$$F(t) = Ae^{(v-b)t},$$

where A and B are determined by the population composition at $t = 0$. We have that the sex ratio is

$$\frac{M(t)}{F(t)} = S(t) = \left\{ 1 + \frac{B(v - b + a)}{uA} e^{-(v-b+a)t} \right\} \bigg/ \left[\frac{v - b + a}{u} \right],$$

and the ultimate sex ratio is $S = S(\infty) = u/[v + a - b]$, when $v + a - b > 0$. We note that M and F tend jointly to infinity when $v - b > 0$, and to zero when $v - b < 0$. When the female death rate equals the female birth rate, $v - b = 0$, M and F tend to nonzero limits and the sex ratio approaches $S = u/a$ the ratio between the male birth rate and the male death rate.

Now consider the case where $\Lambda(M,F) = uM$ and $\Lambda'(M,F) = vM$; i.e., where the birth rates depend on the male population (males are marriage dominant). By applying the methods of the preceding case we may determine $M(t)$, $F(t)$, $S(t)$, and we find

$$S = \frac{u - a + b}{v},$$

when $u - a + b > 0$. Also, M and F behave qualitatively in the same way as before. When $u - a = 0$, M and F tend to nonzero limits, and $S = b/v$.

Let us consider the case where

$$\Lambda(M, F) = u\left(\frac{M + F}{2}\right)$$

and

$$\Lambda'(M, F) = \frac{v(M + F)}{2};$$

i.e., where the contributions from the birth rates depend on the total population size $M + F$ (neither males nor females are marriage dominant). We then have

$$\frac{dM}{dt} = -aM + \frac{u}{2}(M + F) = \left(-a + \frac{u}{2}\right)M + \frac{u}{2}F = fM + cF$$

$$\frac{dF}{dt} = -bF + \frac{v}{2}(M + F) = \left(-b + \frac{v}{2}\right)F + \frac{v}{2}M = gF + kM,$$

where

$$f = \frac{u}{2} - a; \qquad g = \frac{v}{2} - b, \qquad c = \frac{u}{2}, \qquad k = \frac{v}{2}.$$

The solution of these equations is

$$M(t) = Ae^{[f+g+h]t/2} + \left\{\frac{f - g - h}{2k}\right\}Be^{[f+g-h]t/2},$$

$$F(t) = \left\{\frac{-f + g + h}{2c}\right\}Ae^{[f+g+h]t/2} + Be^{[f+g-h]t/2},$$

where $h = +\sqrt{(f - g)^2 + 4ck}$, and A, B are determined by the population composition at $t = 0$. Hence the sex ratio is

$$S(t) = \frac{M(t)}{F(t)} = \frac{1 + \left\{\dfrac{f - g - h}{2k}\right\}\dfrac{B}{A}e^{-ht}}{\left\{\dfrac{-f + g + h}{2c}\right\} + \dfrac{B}{A}e^{-ht}}$$

472

and the ultimate sex ratio is

$$S = S(\infty) = 2c/[g - f + h]$$

$$= u \Big/ \left[\frac{v - u}{2} + a - b + \sqrt{\left(\frac{v - u}{2} + a - b\right)^2 + uv} \right].$$

We could have determined that S was one of two values directly by noting that

$$dS/dt = d\left(\frac{M}{F}\right) \Big/ dt = \left[\frac{F\,dM - M\,dF}{F^2}\right] \Big/ dt$$

$$= \left[\frac{fMF + cF^2 - gFM - kM^2}{F^2}\right]$$

$$= (f - g)S + c - kS^2.$$

The solution of this equation for $dS/dt = 0$ is

$$\frac{g - f \pm \sqrt{(g - f)^2 + 4kc}}{-2k} = \frac{g - f \pm h}{-2k}$$

$$\left(\text{we have } \frac{g - f - h}{-2k} = \frac{2c}{g - f + h} = S\right).$$

We note that M and F tend jointly to infinity when $f + g + h > 0$, and to zero when $f + g + h < 0$. When $f + g + h = 0$, M and F tend to nonzero limits and $S = u/-2f = u/(2a - u) = (2b - v)/v$. The critical relation between the constants is

$$f + g = -h$$

$$(f + g)^2 = (f - g)^2 + 4kc$$

$$fg = kc,$$

or

$$(2a - u)(2b - v) = uv$$

$$a(v - b) + b(u - a) = 0$$

which is a generalization of Kendall's critical relation [22, p. 248].

Let us now consider the case where $\Lambda(M,F) = u\sqrt{MF}$ and $\Lambda'(M,F) = v\sqrt{MF}$, i.e., where the contributions from the birth rates depend on the geometric mean of M and F. We then have

$$\frac{dM}{dt} = -aM + u\sqrt{MF}$$

$$\frac{dF}{dt} = -bF + v\sqrt{MF}.$$

Writing $\sqrt{M} = X$, $\sqrt{F} = Y$,

$$2X \frac{dX}{dt} = -aX^2 + uXY$$

$$\frac{dX}{dt} = fX + cY$$

and

$$2Y \frac{dY}{dt} = -bY^2 + vXY$$

$$\frac{dY}{dt} = gY + kX,$$

where $-a/2 = f$, $-b/2 = g$, $u/2 = c$, $v/2 = k$. Hence using the results for the preceding model, we can compute $X(t)$, $Y(t)$, $X^2(t) = M(t)$, $Y^2(t) = F(t)$, and $S(t)$ directly. We find that

$$S = S(\infty) = \{2c/[g - f + h]\}^2$$

$$= \left\{ u \Big/ \left[\frac{a-b}{2} + \sqrt{\left(\frac{a-b}{2}\right)^2 + uv} \right] \right\}^2.$$

Also M and F behave qualitatively in the same way as before, the critical relation between the constants being

$$fg = kc, \quad \text{or} \quad ab = uv,$$

in which case

$$S = [u/-2f]^2 = [u/a]^2 = [b/v]^2.$$

It is interesting to note that in all the models we have considered heretofore, when the population size tends to a nonzero limit, the sex ratio is what one would expect of a stationary situation; i.e., $S = bu/av$, which equals u/a when $v - b = 0$, and b/v when $u - a = 0$, and $(b/v)^2$ when $ab = uv$ (in the preceding model). ...

4. STOCHASTIC MODELS OF POPULATION GROWTH

We shall now consider a stochastic process which is a probabilistic analogue of the deterministic model describing a population where the females are marriage dominant; i.e., where the birth rates depend on the female population size.[1] Although the analysis which follows pertains to

---

[1]The author wishes to express his indebtedness to David G. Kendall of Magdalen College, Oxford and Princeton University, and Charles Stein of the University of Chicago for their assistance in the analysis of this stochastic process.

populations where the females are marriage dominant, the results obtained may be used in an obvious manner to analyze populations where the males are marriage dominant.

The population under consideration behaves in accordance with the following rules:

(a) the subpopulations generated by two co-existing individuals develop in complete independence of one another;

(b) a female alive at time t has a chance $u\, dt + o(dt)$ of giving birth to a male and a chance $v\, dt + o(dt)$ of giving birth to a female during the following time interval of length dt;

(c) a female at time t has a chance $b\, dt + o(dt)$ of dying and a male alive at time t has a chance $a\, dt + o(dt)$ of dying in the following time interval of length dt.

Let the random variables $M(t)$ and $N(t)$ denote the number of males and females, respectively, alive at time t. We see that the female population follows the rules of a simple birth-and-death process (cf. [22], p. 236), the mean female population size being

$$\overline{N}(t) = e^{(v-b)t}$$

while

$$\text{Var}\,\{N(t)\} = \frac{v+b}{v-b}\, e^{(v-b)t} \{e^{(v-b)t} - 1\},$$

when $N(0) = 1$. Since the subpopulations generated by two coexisting individuals develop in complete independence of one another, we need only multiply $\overline{N}(t)$ and $\text{Var}\{N(t)\}$ by $N(0)$ in the more general case where there are $N(0) \geq 1$ females alive at the initial time $t = 0$. There will be "almost certain" extinction unless the mean female population size is increasing ($v > b$). When $v = b$ and the mean female population size is constant, we still find that there will be "almost certain" extinction.

The behavior of the male population is somewhat more difficult to analyze since male births depend on the female population. The method of analysis will be described in the Appendix. We find that the mean male population size is

$$\overline{M}(t) = \frac{u}{(v - b + a)}\, [e^{(v-b)t} - e^{-at}]$$

while,

$$\mathrm{var}\ \{M(t)\} = \frac{(v+b)u^2}{(v-b+a)^2}\left\{\frac{e^{2(v-b)t}}{(v-b)} - \frac{e^{-2at}}{(v-b+2a)} + \frac{2e^{(v-b-a)t}}{a}\right\}$$

$$+ ue^{(v-b)t}\left\{\frac{1}{(v-b+a)} - \frac{2u(v+b)}{a(v-b)(v-b+2a)}\right\} - \frac{ue^{-at}}{(v-b+a)},$$

when $M(0) = 0$ and $N(0) = 1$. In the more general case where $N(0) \geq 1$ females are alive at the initial time $t = 0$, the formulas $\overline{M}(t)$ and $\mathrm{Var}\{M(t)\}$ are multiplied by $N(0)$, since the subpopulations generate independently,

In the more general case where there are $M(0) \geq 0$ males at the initial time $t = 0$, these $M(0)$ males follow the rules of a simple death process; i.e., the chance that $M_0(t)$ from among the original $M(0)$ males will be alive at time t is

$$C_{M_0(t)}^{M(0)}p(t)^{M_0(t)}q(t)^{M(0)-M_0(t)},$$

where $p(t) = 1 - q(t) = e^{-at}$. Hence,

$$\overline{M}_0(t) = M(0)e^{-at}$$

and

$$\mathrm{Var}\ \{M_0(t)\} = M(0)e^{-at}[1 - e^{-at}].$$

We then add $\overline{M}_0(t)$ and $\mathrm{Var}\{M_0(t)\}$ to $\overline{M}(t)$ and $\mathrm{Var}\{M(t)\}$, respectively, to obtain the mean and variance of the total number of males when there are $M(0)$ males and $N(0)$ females at time $t = 0$. These additions have, of course, no influence on the form of the solution when $t \rightarrow \infty$.

We find that the covariance between $M(t)$ and $N(t)$ is

$$\mathrm{Cov}\ (t) = E\{[M(t) - \overline{M}(t)][N(t) - \overline{N}(t)]\}$$

$$= \frac{u(v+b)}{(v-b+a)}\left\{\frac{e^{2(v-b)t}}{(v-b)} + \frac{e^{(v-b-a)t}}{a}\right\} - \frac{u\ (v+b)}{a\ (v-b)}e^{(v-b)t},$$

when $M(0) = 0$ and $N(0) = 1$. In the more general case where $N(0) \geq 1$, $M(0) \geq 0$, then the covariance is multiplied by $N(0)$.

In the special case where the constants are equal ($a = b = u = v = \lambda$), we have

$$\overline{N}(t) = N(0), \qquad \overline{M}(t) = N(0)[1 - e^{-\lambda t}] + M(0)e^{-\lambda t},$$

$$\mathrm{Var}\ \{N(t)\} = N(0)2\lambda t,$$

$$\mathrm{Var}\ \{M(t)\} = N(0)[2\lambda t - 2 + 3e^{-\lambda t} - e^{-2\lambda t}] + M(0)e^{-\lambda t}[1 - e^{-\lambda t}],$$

$$\mathrm{Cov}\ (t) = N(0)[2\lambda t - 2 + 2e^{-\lambda t}],$$

and the correlation coefficient $\rho(t) = \text{Cov}(t)/\sqrt{\text{Var}\ \{M(t)\}\ \text{Var}\ \{N(t)\}}$ between the number of males and the number of females in the population is seen to approach one as t becomes large.

The methods which we have used to analyze the stochastic process representing a marriage dominant (female or male) population, can be extended in order to analyze a stochastic process where the contributions from the birth rates depend on the total population size (neither male nor female is marriage dominant). That is, the population under consideration behaves in accordance with the following rules:

(a) the subpopulation generated by two coexisting individuals develops in complete independence of one another;

(b) an individual alive at time t has a chance $u\ dt/2 + o(dt)$ of reproducing a male and a chance $v\ dt/2 + o(dt)$ of reproducing a female during the following time interval of length dt;

(c) a female alive at time t has a chance $b\ dt + o(dt)$ of dying and a male alive at time t has a chance $a\ dt + o(dt)$ of dying in the following time interval of length dt.

Using the second method described in the Appendix we may obtain the moments and cross moments of $M(t)$ and $N(t)$, the number of males and females, respectively, alive at time t. The means, variances, and the covariance of $M(t)$ and $N(t)$ satisfy a system of five differential equations with constant coefficients which may be solved to determine these moments explicitly.

Let us consider the special case where the constants are all equal $(a = b = u = v = \lambda)$. Then the probability $P_{m,n}(t)$ that at time t the population contains m males and n females may be determined explicitly when the initial conditions are $P_{0,1}(0) = P_{1,0}(0) = 1/2$. We find that

$$P_{m,n}(t) = C_m^{m+n}\left(\frac{1}{1 + \lambda t}\right)^2\left(\frac{\lambda t}{1 + \lambda t}\right)^{m+n-1}\left(\frac{1}{2}\right)^{m+n}$$

for $m + n \geq 1$, and

$P_{0,0}(t) = \lambda t/(1 + \lambda t)$. Whence we see that there will be "almost certain" extinction. The probability $P_m(t)$ that at time t the population contains m males, and the probability $P'_n(t)$ that at time t the population contains n females may also be determined explicitly. We have

$$P_m(t) = P'_m(t) = \left(\frac{\lambda t}{2 + \lambda t}\right)^{m-1} 2/(2 + \lambda t)^2,$$

for $m \geq 1$ and

$$P_0(t) = P'_0(t) = (1 + \lambda t)/(2 + \lambda t).$$

We find that $\overline{M}(t) = \overline{N}(t) = 1/2$, $\mathrm{Var}\{M(t)\} = \mathrm{Var}\{N(t)\} = (2\lambda t + 1)/4$, $\mathrm{Cov}(t) = (2\lambda t - 1)/4$, and the correlation coefficient $\rho(t) = (2\lambda t - 1)/(2\lambda t + 1)$.

5. APPENDIX

In studying the stochastic process representing a population where the females are marriage dominant, the means, variances, and covariances of the male and female population sizes were given. The following method which was used to obtain the first and second moments could also be used to obtain higher moments:

All moments for the female population size may be obtained directly from its distribution which is a geometric series with a modified zero term (cf. [22] p. 237) or from its moment generating function. All moments for the male population size may be obtained from its moment generating function $\varphi(z,t) = E\{z^{M(t)}\}$. We find that $\varphi(z,t)$ satisfies the differential equation

(1)
$$\frac{\partial \varphi}{\partial t} = (b - v\varphi)(1 - \varphi) + u\varphi(z - 1)e^{-at}$$

and that $\varphi(z,0) = z$, when $M(0) = 0$ and $N(0) = 1$. By differentiating (1) once and setting $z = 1$, we obtain a linear differential equation for the mean $\overline{M}(t)$. By differentiating (1) several times and setting $z = 1$, we obtain linear differential equations for the higher factorial moments of $M(t)$.

The method just described gives the moments of $M(t)$ but does not give the cross moments of $M(t)$ and $N(t)$; e.g., the covariance. Another method might be used to obtain the moments and cross moments of $M(t)$ and $N(t)$ which serves as an independent check on the preceding method. The two methods might be used simultaneously to simplify computation. Let $P_{m,n}(t)$ be the probability that at time t the population contains m males and n females. Then the infinite system of functions $P_{m,n}(t)$, $m = 0, 1, 2, \cdots, n = 0, 1, 2, \cdots$, satisfy a basic system of ordinary differential equations (cf. [23], Chap. 17). By multiplying each differential equation by an appropriate factor and then adding the entire system of differential equations, we obtain ordinary differential equations for the moments; e.g., if we multiply the differential equation

478

containing the term $[\partial P_{m,n}(t)]/\partial t$ by m and then add the entire system of differential equations, we will obtain an equation containing the term

$$\sum_{m,n=0}^{\infty} m \, \frac{\partial P_{m,n}(t)}{\partial t}$$

which is in fact $[\partial \overline{M}(t)]/\partial t$. We then have a differential equation in $\overline{M}(t)$ which may be solved directly.

REFERENCES

1. J. Yerushalmy, "The age-sex composition of the population resulting from natality and mortality conditions," *The Milbank Memorial Fund Quarterly*, Vol. XXI (1943), No. 1, pp. 37–63.
2. Sanford Winston, "The influence of social factors upon the sex ratio at birth," *The American Journal of Sociology*, Vol. 37 (1931), pp. 1–21.
3. Sanford Winston, "Birth control and the sex ratio at birth," *The American Journal of Sociology*, Vol. 38 (1932), pp. 225–231.
4. Christopher Tietze, "A note on the sex ratio of abortions," *Human Biology*, Vol. 20 (1948), pp. 156–160.
5. R. J. Meyers, "Effect of the war on the sex ratio at birth," *American Sociological Review*, Vol. 12 (1947), No. 1, pp. 40–43.
6. Rachel M. Jenss, "An inquiry into methods of studying the sex ratio at birth for the United States during war and postwar years," *Human Biology*, Vol. 15 (1943), No. 3, pp. 255–266.
7. A. S. Parkes, "The physiological factors governing the proportions of the sexes in man," *The Eugenics Review*, Vol. XVII (1926), pp. 275–291.
8. A. S. Parkes, "The mammalian sex ratio," *Biological Review*, Vol. 1 (1926), p. 1.
9. H. Stranskov, "On the variance of human live birth sex ratios," *Human Biology*, Vol. 14 (1941), No. 1, pp. 85–94.
10. A. Ciocco, "Variation in the sex ratios at birth in the United States," *Human Biology*, Vol. 10 (1938), pp. 35–64.
11. W. J. Martin, "A comparison of the trends of male and female mortality," *Journal of the Royal Statistical Society*, Series A, Vol. CXIV (1951), Part 3, pp. 287–298.
12. Hope Tisdale Eldridge and Jacob S. Siegel, "The changing sex ratio in the United States," *The American Journal of Sociology*, Vol. LII (1946), No. 3, pp. 224–234.
13. P. H. Karmel, "An analysis of the sources of magnitudes of inconsistencies between male and female net reproduction rates in actual populations," *Population Studies*, Vol. 2 (1948), Part 2, pp. 240–273.
14. P. H. Karmel, "The relation between male and female reproduction rates," *Population Studies*, Vol. 1 (1947), No. 3, pp. 249–274.
15. P. H. Karmel, "The relation between male and female nuptiality in a stable population," *Population Studies*, Vol. 1 (1948), No. 4, pp. 353–387.
16. J. Hajnal, "Aspects of recent trends in marriage in England and Wales," *Population Studies*, Vol. 1 (1947), No. 1, pp. 72–98.
17. J. Hajnal, "Some comments on Mr. Karmel's paper 'The relation between male and female reproduction rates'," *Population Studies*, Vol. 2 (1948), No. 3, pp. 352–360.

18. P. H. Karmel, "A rejoinder to Mr. Hajnal's comments," *Population Studies*, Vol. 2 (1948), No. 3, pp. 361–372.

19. W. S. Hocking, "The balance of the sexes in Great Britain," *Journal of the Institute of Actuaries*, Vol. 74 (1948), pp. 340–344.

20. G. Udney Yule, "A mathematical theory of evolution based on the conclusions of Dr. J. C. Willis, F.R.S.," Philosophical *Transactions of the Royal Society of London*, Series B, Vol. 213 (1925), pp. 21–87.

21. M. S. Bartlett, "Some evolutionary stochastic processes," *Journal of the Royal Statistical Society*, Series B, Vol. II (1949), pp. 211–229.

22. D. G. Kendall, "Stochastic processes and population growth," *Journal of the Royal Statistical Society*, Series B, Vol. II (1949), pp. 230–264.

23. William Feller, "Diffusion processes in genetics," *Proceedings of the Second Berkeley Symposium of Mathematical Statistics*, University of California Press (1951), pp. 227–246.

24. T. E. Harris, "Some mathematical models for branching processes," *Proceedings of the Second Berkeley Symposium of Mathematical Statistics*, University of California Press (1951), pp. 305–328.

25. P. A. P. Moran, "Estimation methods for evolutive processes," *Journal of the Royal Statistical Society*, Series B, Vol. XIII (1951), No. 1, pp. 141–146.

26. William Feller, *An Introduction to Probability Theory and its Applications*, Vol. 1, John Wiley and Sons, Inc., New York, 1950.

54. On Future Population

NATHAN KEYFITZ (1972)

From *Journal of the American Statistical Association* 67. Excerpts are from pages 351—352.

2.10 Cohorts versus Periods

Everything prior to Section 2.9 is in terms of aggregate births of the several years or other periods, considered as a time series. If each family made its decisions on having or not having children in relation to the conditions of the year, without reference to the number of children it has had in the past, nothing more need be said. But suppose now that couples aim at a certain number of children; good or bad times cause them to defer or to advance their childbearing, but not to change the total number. Then the fluctuations in the time series of births are less consequential; the drop of the birth rate in a depression would be made up in the subsequent business upswing, and the rate of increase of the population would be lower only insofar as older parents imply a greater length of generation. Constancy in the total number of children per family results in constancy in the average number of children per cohort, and the Bureau of the Census and the Scripps Foundation have taken advantage of this constancy in their projections.

The trouble with cohorts is that we do not know their completed size until their childbearing is over. If we project from completed cohorts only we handicap ourselves by disregarding information contained in those cohorts that are still bearing, and this omission would cancel out the advantage over the period method, where the latest information is all routinely put to use. Thus the method of cohorts requires some way of translating period into cohort distributions.

The easy way to effect this translation is in terms of moments. Consider the figure, known as a Lexis diagram, in which the life lines of individual women are plotted.

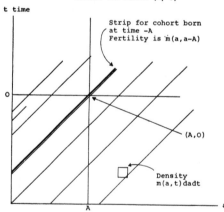

LEXIS DIAGRAM FOR $m(a, t)$ SHOWING SHIFT OF ORIGIN TO POINT $(A, 0)$

Small squares are used to indicate the ages and times when women bear children. Think of women aged a at time t and suppose that $m(a, t)$ is the age-specific rate of childbearing for those women. Then the cohort of women that is aged A at time 0 would be represented by $m(a, a-A)$ as it passed through the several ages a. Its Gross Reproduction Rate (GRR) would be $R_0^*(A) = \int_\alpha^\beta m(a,$

$a - A)da$, the star to identify $R_0^*(A)$ as referring to a cohort. The nth moment around A would be

$$\frac{R_n^*(A)}{R_0^*(A)} = \frac{\int_\alpha^\beta (a - A)^n m(a, a - A)da}{R_0^*(A)} \, .$$

Corresponding parameters of the period distribution at time zero are $R_0 = \int_\alpha^\beta m(a, 0)da$, the ordinary period GRR, and

$$\frac{R_n(A)}{R_0} = \frac{\int_\alpha^\beta (a - A)^n m(a, 0)da}{R_0} \, ,$$

the nth period moment around A. (Note that R_0 does not depend on A.)

Now our problem is to express the cohort $R_n^*(A)$ in terms of the period $R_n(A)$. Using Taylor's theorem to expand $m(a, a - A)$ under the integral sign,

$$R_n^*(A) = \int_\alpha^\beta (a - A)^n m(a, a - A)da$$
$$= \int_\alpha^\beta (a - A)^n [m(a, 0) + (a - A)\dot{m}(a, 0)$$
$$+ ((a - A)^2/2!)\ddot{m}(a, 0) + \cdots]da$$

where $\dot{m}(a, 0)$ is the first derivative of $m(a, 0)$ with respect to time, $\ddot{m}(a, 0)$ the second derivative, $\cdots$. Using the facts that under conditions general enough for our purpose (1) the integral of a sum is the sum of the integrals and (2) the integral of a derivative is the derivative of the integral, we obtain the simple and important result

$$R_n^*(A) = R_n(A) + \dot{R}_{n+1}(A) + \frac{\ddot{R}_{n+2}(A)}{2!} + \cdots \quad (2.11)$$

due to Ryder [48]. This serves to estimate any cohort moment about any age A in terms of the period moments and their changes. It applies equally to net reproduction if the $m(a, t)$ is replaced by $\ell_a m(a, t)$.

The most useful special case is that for $n = 0$,

$$R_0^*(A) = R_0 + \dot{R}_1(A) = R_0\left(1 + \frac{\dot{R}_1(A)}{R_0}\right), \quad (2.12)$$

which we have truncated as though the second derivatives of the second and higher moments are zero. This result has a simple intuitive meaning. Suppose that the period GRR is unchanging, but that the average age of motherhood is increasing through successive periods. Then $\dot{R}_1(A)$ will be positive, and the cohort $R_0^*(A)$ will be greater than the period R_0. The derivative $\dot{R}_1(A)$ gives the amount by which the period moment is a distorted version of the cohort moment.

Since the result (2.11) applies for any A, and since the distribution of the cohort childbearing is implicit in its moments, we can use this in principle to fill out the future childbearing of the incomplete cohorts at time zero.

If we knew that every cohort was aiming at exactly three children (2.11) and (2.12) would not be needed; we would simply deduct the average number of children already recorded from (2.3), and suppose the remainder distributed over time and age in the future in some suitable way. If on the other hand cohorts had nothing to do with the matter we would treat the births to women of given age in successive periods as an ordinary time series and extrapolate. The theory is especially useful for the intermediate case where the R_0^* for cohorts are shifting less rapidly than the R_0 for periods. To take advantage of this we extrapolate $R_0^*(A)$ and one or two further moments, then fill in the cells of the table for future tides and ages, for example by supposing a normal or gamma distribution for childbearing in each cohort. This seems to provide a better estimate of childbearing in each future period than would direct extrapolation of the periods.

[48] Ryder, N. B., "The Process of Demographic Translation," *Demography*, 1 (1964), 74–82.

55. The Standard Deviation of Sampling for Life Expectancy

E. B. WILSON (1938)

Journal of the American Statistical Association 33: 705—708.

Wilson's use of ch "in place of $\frac{1}{2}h$ for the average life of those who die in the interval $[y$ to $y+h]$" is not correct. The fraction of a life added or subtracted by a change in q_y should be $h-ch$, representing the gain for an additional survivor or loss for an additional death, and not ch as his final paragraph suggests. With this correction equation (1) becomes (cf. Keyfitz 1968, p. 342)

$$\delta \mathring{e}_x = -l_y(\mathring{e}_{y+h}+h-ch)\delta q_y/l_x. \tag{1a}$$

Equations (2) and (3) and Wilson's table require similar revision, and we have for equation (7)

$$\sigma^2(\mathring{e}_0)=0.0173+0.0026+ \cdots +0.0117=0.0473. \tag{7a}$$

It has long been customary to take as the standard deviation of sampling for a death rate p per capita the formula $\sigma = \sqrt{pq/n}$ where n is the population on which the rate is based. In comparing rates this standard deviation is used to guide the judgment as to whether the samples are large enough so that the difference in the rates is more than might be expected to be due to random differences between samples. More recently much emphasis has been laid on the expectation of life at birth as an over-all index of mortality in a population and life tables have been computed for relatively small populations, particularly for the purpose of comparing the expectations of life at birth, $\overset{\circ}{e}_0$, in different populations or of comparing the life table death rates $(1000/\overset{\circ}{e}_0)$ per thousand or $1/\overset{\circ}{e}_0$ per capita. Such comparisons are generally made without reference to the corresponding standard deviations.

The expectation of life $\overset{\circ}{e}_x$ at age x is sometimes defined as the average duration of life beyond age x of a cohort l_x of persons starting together at age x. If $\overset{\circ}{e}_x$ were actually computed in accordance with this definition, the sampling error of $\overset{\circ}{e}_x$ would be the standard deviation of the frequency function $d_y, y \geq x$, of deaths at ages from x on divided by the square root of the number l_x of persons involved.[1] But $\overset{\circ}{e}_x$ is never thus computed and the standard deviation of any estimate of a quantity has to be figured from the manner of estimation.

The expectation $\overset{\circ}{e}_x$ at any age x is actually deduced by a mathematical calculation from the age specific death rates; by definition it is the average number of years lived after age x by a group of l_x persons alive at age x, i.e., $l_x \overset{\circ}{e}_x$ is the total number of years lived by the cohort l_x. Now, if in some particular age group y to $y+h$, subsequent to age x, the value of q_y, the chance of dying between those ages, should perchance be increased by δq_y, the number of deaths would be increased by $l_y \delta q_y$ and the years lived would be diminished in that age interval by $\frac{1}{2} h l_y \delta q_y$ and in all ages above $y+h$ by $l_y \overset{\circ}{e}_{y+h} \delta q_y$. Thus the expectation of life $\overset{\circ}{e}_x$ would be changed by

$$\delta \overset{\circ}{e}_x = -l_y(\overset{\circ}{e}_{y+h} + \tfrac{1}{2}h)\delta q_y/l_x. \tag{1}$$

To a first order of approximation, regarding δq as infinitesimal, the total change in $\overset{\circ}{e}_x$ from variations δq in different age groups would be the sum of the individual changes, viz.,

$$\delta \overset{\circ}{e}_x = -\sum_{y \geq x} \left(\frac{l_y}{l_x}\right)(\overset{\circ}{e}_{y+h} + \tfrac{1}{2}h)\delta q_y \tag{2}$$

where the summation extends over all intervals h, not necessarily equal, from age x on to the end of the life table. As the variations δq are supposed to be those of

---

[1] The standard error $\sigma(\overset{\circ}{e}_x)$ may perhaps be roughly estimated as the standard deviation $\sigma(d_y), y \geq x$, of the d_y curve taken from the computed life table divided by the square root of the total number of deaths D above the age x which went into the calculation of the life table, but this estimate has been found somewhat unreliable; we should probably do better to use for D the life table death rate times the population.

random sampling they must be assumed to be uncorrelated and the square of the standard deviation of $\mathring{e}_x$ is therefore

$$\sigma^2(\mathring{e}_x) = \sum_{y \geq x} \left(\frac{l_y}{l_x}\right)^2 (\mathring{e}_{y+h} + \tfrac{1}{2}h)^2 \sigma^2(q_y) \tag{3}$$

In getting the formula it has been assumed that the deaths were spread uniformly over the interval x to $x+h$; for some intervals such as "under 1" or "under 5" this might not be so satisfactory as to assume that the average period lived during the interval was ch instead of $\tfrac{1}{2}h$. Moreover it often happens that deaths or populations at the older ages are available only in a wide group such as "75 and up"; in this case it may perhaps be assumed[2] that

$$\mathring{e}_{75} = \frac{1}{m_{75}}, \quad \delta\mathring{e}_{75} = -\frac{l_{75}}{l_x}\frac{\delta m_{75}}{m_{75}^2}, \quad \sigma^2 m_{75} = \frac{m_{75}(1-m_{75})}{n_{75}} \tag{4}$$

where m_{75} is the rate of mortality per capita for ages 75 and up and n_{75} is the population 75 years of age and older. With respect to σ_q^2 except in the last interval we have[3]

$$q = m \left/ \left(\frac{1}{2}m + \frac{1}{h}\right)\right., \quad \frac{1}{q} = \frac{1}{2} + \frac{1}{hm} \quad \text{or} \quad 1 - c + \frac{1}{hm}, \tag{5}$$

provided we use ch in place of $\tfrac{1}{2}h$ for the average life of those who die in the interval. Then

$$\frac{\delta q}{q^2} = \frac{1}{h}\frac{\delta m}{m^2} \quad \text{and} \quad \sigma^2(q_y) = h^2 \left(\frac{q_y}{hm_y}\right)^4 \frac{m_y(1-m_y)}{n_y}. \tag{6}$$

In the particular group "under 1," q_0 is generally taken as the infant mortality per capita and $\sigma^2(q_0) = q_0(1-q_0)/B$ where B is the number of births on which the infant mortality q_0 is based. In case we compute rates from average deaths for k years, the populations n_y or B are to be multiplied by k.

Thus $\sigma^2(\mathring{e}_x)$ may be computed from the data for the life table and from the entries in it. An example of the computation of $\sigma^2(\mathring{e}_0)$ follows:[4]

---

[2] The final term in (2) would become meaningless, since for the last interval q_y is 1, $\mathring{e}_{y+h}$ is 0 and $\tfrac{1}{2}h$ will not do for the average years lived. The assumption in (4) is that the distribution in the last interval is stationary.

[3] If q were estimated directly as the probability which it is, its standard deviation would be the square root of $q(1-q)/n$ but as q is estimated from m it is necessary to follow its method of estimation.

[4] Of course, a six interval life table is decidedly sketchy; but my colleague Dr. C.R. Doering, who has in preparation a considerable discussion of life tables from which this particular illustration is taken, finds that for many purposes such a method of calculation is adequate. See C.R. Doering and A.L. Forbes, *Proceedings Natl. Acad. Sci.*, 24 (1938), 400—405. On the basis of a thirteen interval life table for this same county, $\mathring{e}_0 = 61.3$ instead of 61.7 and $\sigma(\mathring{e}_0)$ is 0.200 instead of 0.213 as found here. It seems unnecessary for the present illustration to increase the detail by using a larger number of intervals than six.

Onondaga County, N.Y., 1929—1931, Average Deaths

Age	Deaths	Population	m_x^{x+h}	l_x	q_x	$\overset{\circ}{e}_x$
<5	355	23,800	0.01492	1000	0.0701	61.7
5—14	76	51,900	0.001464	930	0.0145	61.3
15—24	104	47,400	0.002194	916	0.0217	52.1
25—44	395	91,500	0.004317	896	0.0828	43.2
45—64	1002	58,400	0.01716	822	0.2930	26.2
65+	1468	18,900	0.07773	581	—	12.9

We take $c = 0.14$ for the first group. Then

Age	l_x/l_0	$\overset{\circ}{e}_{x+h} + ch$	Prod.	$10^{-4}(\text{Prod.})^2$	$10^4 \sigma^2(q)$	Prod.
<5	1.000	62.0	62.0	0.3844	0.0401	0.0154
5—14	0.930	57.1	53.1	0.2822	0.0091	0.0026
15—24	0.916	48.2	44.1	0.1949	0.0148	0.0029
25—44	0.896	36.2	32.4	0.1051	0.0529	0.0056
45—64	0.822	22.9	18.8	0.0353	0.205	0.0072
65+	0.581	—	(96.2)	(0.9259)	(0.0126)	(0.0117)

where the terms in parentheses in the last line are the special values needed for the last term.[5]

$$-\delta \overset{\circ}{e}_0 = 62.0\delta q_0 + 53.1\delta q_5 + 44.1\delta q_{15} + 32.4\delta q_{25} + 18.8\delta q_{45} + 96.2\delta m_{65}$$

$$\sigma^2(\overset{\circ}{e}_0) = 0.0154 + 0.0026 + 0.0029 + 0.0056 + 0.0072 + 0.0117 = 0.0454 \tag{7}$$

Hence

$$\sigma(\overset{\circ}{e}_0) = 0.213, \qquad \overset{\circ}{e}_0 = 61.7 \pm 0.213 = 61.7(1 \pm 0.00345)$$

and

$$1000/\overset{\circ}{e}_0 = 16.21(1 \pm 0.00345) = 16.21 \pm 0.056.$$

Although it is a little to one side of the matter of $\sigma^2(\overset{\circ}{e}_x)$ one may remark that (2), with the proper last term taken from (4) and with the term $\frac{1}{2}h$ replaced by ch for the first interval, gives a value of $\delta(\overset{\circ}{e}_0)$ for any changes in the age specific death rates. The value like that for $\sigma^2(\overset{\circ}{e}_0)$ based upon it must be considered approximate because through the work it has been assumed that δq is infinitesimal and that infinitesimal effects are additive since infinitesimals of higher order are neglected. A test of (2) for the above table may be made from (7) if we assume that the values

---

[5] Namely, $(l_{65}/l_0 m_{65}^2)$, the square of this $\times 10^{-4}$, $10^4 \sigma^2(m_{65})$, and the product of the last two.

of the mortality rates m are all increased or decreased by 10 per cent. From (7) by computing the δq's one finds that $\delta \mathring{e}_0$ is -2.04 or 2.06 respectively on the two assumptions of increase and decrease. A direct calculation of the life tables from the changed rates gives the values -1.95 and 2.23 for $\delta \mathring{e}_0$ to which those obtained from (7) may seem sufficiently near.[6]

---

[6] The analysis of the standard error of a life table applies of course only to the method of direct calculation from the raw data as developed by Dr. Doering and Miss Forbes. Life tables are usually computed only after smoothing the data. How much such processes do toward affecting the values of $\mathring{e}_x$ is an interesting question. For small districts the effect would probably be less than the sampling error, for large districts it might be greater. In any case that method of calculation would have to have its own sampling error properly determined. Finally it may be remarked that only pure chance variation has been admitted in the age specific death rates, whereas it is well known that the actual variations in them which remain unexplained and cannot be corrected for are larger than those due to chance and are not uncorrelated as between the different age groups. The fluctuations which occur in $\mathring{e}_x$ from no as yet assignable causes are probably what the practical statistician might well call chance variations. To study them in a search for their causes it is desirable to have an easy method of calculating life tables and some fairly satisfactory estimate of the sampling error of $\mathring{e}_x$.

56. Probability Distribution of Life Table Functions

C. L. CHIANG (1968)

From *Introduction to Stochastic Processes in Biostatistics*, pp. 218—225. New York: John Wiley & Sons.

1. INTRODUCTION

The concept of the life table originated in longevity studies of man; it was always presented as a subject peculiar to public health, demography, and actuarial science. As a result, its development has not received sufficient attention in the field of statistics. Actually, the problems of mortality studies are similar to those of reliability theory and life testing, and they may be described in terms familiar to the statistically oriented mind. From a statistical point of view, human life is a random experiment and its outcome, survival or death, is subject to chance. The life table systematically records the outcomes of many such experiments for a large number of individuals over a period of time. Thus the quantities in the table are random variables subject to established statistical analysis. The purpose of this chapter is to derive probability distributions of life table functions and to discuss some optimum properties of these functions when they are used as estimates of the corresponding unknown quantities. The presentation will focus on the cohort life table but, whenever necessary, clarification will be made for application to the current life table. A typical abridged life table is reproduced on page 219.
The following symbols are also used in the text:

$$p_{ij} = \Pr\{\text{an individual alive at age } x_i \text{ will survive to age } x_j\},$$

$$i \leq j; \ i, j = 0, 1, \ldots, \quad (1.1)$$

and

$$1 - p_{ij} = \Pr\{\text{an individual alive at age } x_i \text{ will die before age } x_j\},$$

$$i \leq j; \ i, j = 0, 1, \ldots. \quad (1.2)$$

Table 1. Life table

Age Interval (in Years) x_i to x_{i+1}	Number Living at Age x_i l_i	Proportion Dying in Interval (x_i, x_{i+1}) $\hat{q}_i$	Fraction of Last Age Interval of Life a_i	Number Dying in Interval (x_i, x_{i+1}) d_i	Number of Years Lived in Interval (x_i, x_{i+1}) L_i	Total Number of Years Lived Beyond Age x_i T_i	Observed Expectation of Life at Age x_i $\hat{e}_i$
x_0 to x_1	l_0	$\hat{q}_0$	a_0	d_0	L_0	T_0	$\hat{e}_0$
.	.	.	.	.	.	.	.
.	.	.	.	.	.	.	.
.	.	.	.	.	.	.	.
x_w and over	l_w	$\hat{q}_w$		d_w	L_w	T_w	$\hat{e}_w$

492

When $x_j = x_{i+1}$, we drop the second subscript and write p_i for $p_{i,i+1}$. No particular symbol is introduced for the probability $1 - p_{ij}$ except when $x_j = x_{i+1}$, in which case we let $1 - p_i = q_i$. Finally, the symbol e_i is used to denote the true, unknown expectation of life at age x_i, estimated by the "observed expectation of life," $\hat{e}_i$.

All the quantities in the life table, with the exception of l_0 and a_i, are treated as random variables in this chapter. The radix l_0 is conventionally set equal to a convenient number, such as $l_0 = 100,000$, so that the value of l_i clearly indicates the proportion of survivors to age x_i. We adopt the convention and consider l_0 a constant in deriving the probability distributions of other life table functions. The distributions of the quantities in columns L_i and T_i are not discussed because of their limited use. One remark should be made regarding the final age interval (x_w and over): In a conventional table the last interval is usually an open interval, for example 95 and over; statistically speaking, x_w is a random variable and is treated accordingly. This point is discussed in Section 2.1. Throughout this chapter we shall assume a homogeneous population in which all individuals are subjected to the same force of mortality and in which one individual's survival is independent of the survival of any other individual in the group.

1.1. Probability Distribution of the Number of Survivors

The various functions of the life table are usually given for integral ages or for other discrete intervals. In the derivation of the distribution of survivors, however, age is more conveniently treated as a continuous variable with formulas derived for l_x, the number of individuals surviving the age interval $(0, x)$, for all possible values of x.

The probability distribution of l_x depends on the force of mortality, or the intensity of risk of death, $\mu(x)$, defined as follows:

$$\mu(x)\Delta + o(\Delta) = \text{Pr}\{\text{an individual alive at age } x \text{ will die in interval } (x, x + \Delta)\}. \qquad (1.3)$$

It has been shown in the pure death process in Section 5, Chapter 3, that l_x has a binomial distribution with

$$\text{Pr}\{l_x = k \mid l_0\} = \binom{l_0}{k} p_{0x}^k (1 - p_{0x})^{l_0-k} \qquad (1.4)$$

where

$$p_{0x} = \exp\left\{-\int_0^x \mu(\tau) \, d\tau\right\} \qquad (1.5)$$

is the probability that an individual alive at age 0 will survive to age x, and the corresponding p.g.f. is given by

$$G_{l_x|l_0}(s) = (1 - p_{0x} + p_{0x}s)^{l_0}. \tag{1.6}$$

For $x = x_i$, the probability that an individual will survive the age interval $(0, x_i)$ is

$$p_{0i} = \exp\left\{-\int_0^{x_i} \mu(\tau)\, d\tau\right\} \tag{1.7}$$

and the p.g.f. of l_i is

$$G_{l_i|l_0}(s) = (1 - p_{0i} + p_{0i}s)^{l_0}. \tag{1.8}$$

Therefore the expectation and variance of l_i given l_0 are

$$E[l_i \mid l_0] = l_0 p_{0i} \tag{1.9}$$

and

$$\sigma^2_{l_i|l_0} = l_0 p_{0i}(1 - p_{0i}). \tag{1.10}$$

In general, the probability of surviving the age interval (x_i, x_j) is

$$p_{ij} = \exp\left\{-\int_{x_i}^{x_j} \mu(\tau)\, d\tau\right\} \tag{1.11}$$

with the obvious relation

$$p_{\alpha j} = p_{\alpha i} p_{ij} \tag{1.12}$$

for $\alpha \le i \le j$. The p.g.f. for the conditional distribution of l_j given $l_i > 0$ is

$$G_{l_j|l_i}(s_j) = E(s_j^{l_j} \mid l_i) = (1 - p_{ij} + p_{ij}s_j)^{l_i}. \tag{1.13}$$

When $j = i + 1$, (1.13) becomes

$$G_{l_{i+1}|l_i}(s_{i+1}) = E(s_{i+1}^{l_{i+1}} \mid l_i) = (1 - p_i + p_i s_{i+1})^{l_i}. \tag{1.14}$$

Although formula (1.13) holds true whatever $x_i < x_j$ may be, the conditional probabilities of l_j relative to $l_0, l_1, \ldots, l_i$ are the same as those relative to l_i in the sense that for each k

$$\Pr\{l_j = k \mid l_0, \ldots, l_i\} = \Pr\{l_j = k \mid l_i\}. \tag{1.15}$$

In other words, for each u the sequence $l_0, l_1, \ldots, l_u$ is a Markov process. Thus

$$E(l_j \mid l_0, \ldots, l_i) = E(l_j \mid l_i) \tag{1.16}$$

and

$$E(s_j^{l_j} \mid l_0, \ldots, l_i) = E(s_j^{l_j} \mid l_i), \qquad i < j; i, j = 0, 1, \ldots, u. \tag{1.17}$$

494

2. JOINT PROBABILITY DISTRIBUTION OF THE NUMBERS OF SURVIVORS

Let us introduce, for a given u, the p.g.f. of the joint probability distribution of $l_1, \ldots, l_u$,

$$G_{l_1, \ldots, l_u | l_0}(s_1, \ldots, s_u) = E(s_1^{l_1} \cdots s_u^{l_u} \mid l_0) \tag{2.1}$$

where $|s_i| < 1$ for $i = 1, \ldots, u$.

To derive an explicit formula for the p.g.f. (2.1), we use the identity (see Equation (4.22) in Chapter 1):

$$E[s_1^{l_1} \cdots s_{i+1}^{l_{i+1}} \mid l_0] = E[s_1^{l_1} \cdots s_i^{l_i} E\{s_{i+1}^{l_{i+1}} \mid l_0, \ldots, l_i\} \mid l_0] \tag{2.2}$$

and write

$$E[s_1^{l_1} \cdots s_{i+1}^{l_{i+1}} \mid l_0] = E[s_1^{l_1} \cdots s_i^{l_i} E\{s_{i+1}^{l_{i+1}} \mid l_i\} \mid l_0]$$
$$i = 0, 1, \ldots, u-1, \tag{2.3}$$

where the conditional expectation of the quantity inside the braces is the p.g.f. of the conditional distribution of l_{i+1} given l_i, with the explicit function presented in (1.14). Now we use the identity (2.3) for $i = u - 1$ and rewrite the p.g.f. (2.1) as

$$G_{l_1, \ldots, l_u | l_0}(s_1, \ldots, s_u) = E[s_1^{l_1} \cdots s_{u-1}^{l_{u-1}} E\{s_u^{l_u} \mid l_{u-1}\} \mid l_0]. \tag{2.4}$$

In view of (1.14), $E\{s_u^{l_u} \mid l_{u-1}\} = (1 - p_{u-1} + p_{u-1}s_u)^{l_{u-1}}$, hence

$$G_{l_1, \ldots, l_u | l_0}(s_1, \ldots, s_u) = E[s_1^{l_1} \cdots s_{u-1}^{l_{u-1}}(1 - p_{u-1} + p_{u-1}s_u)^{l_{u-1}} \mid l_0]$$
$$= E[s_1^{l_1} \cdots s_{u-2}^{l_{u-2}} z_{u-1}^{l_{u-1}} \mid l_0], \tag{2.5}$$

where

$$z_{u-1} = s_{u-1}(1 - p_{u-1} + p_{u-1}s_u) \tag{2.6}$$

is less than unity in absolute value, and

$$1 - z_{u-1} = (1 - s_{u-1}) + p_{u-1}s_{u-1}(1 - s_u). \tag{2.7}$$

Formula (2.5) contains only $u - 1$ random variables, $l_1, \ldots, l_{u-1}$, or one less than we had originally. Using formulas (2.3) and (1.14) for $i = u - 2$, the random variable l_{u-1} drops out, and the p.g.f. becomes

$$G_{l_1, \ldots, l_u | l_0}(s_1, \ldots, s_u) = E[s_1^{l_1} \cdots s_{u-3}^{l_{u-3}} z_{u-2}^{l_{u-2}} \mid l_0]. \tag{2.8}$$

Here

$$z_{u-2} = s_{u-2} - p_{u-2}s_{u-2}(1 - s_{u-1}) - p_{u-2}p_{u-1}s_{u-2}s_{u-1}(1 - s_u) \tag{2.9}$$

is also less than unity in absolute value, and

$$1 - z_{u-2} = (1 - s_{u-2}) + p_{u-2}s_{u-2}(1 - s_{u-1}) + p_{u-2}p_{u-1}s_{u-2}s_{u-1}(1 - s_u). \tag{2.10}$$

495

By repeated application of the same process, all the random variables are removed from the expression of the p.g.f., and the following formula is finally reached:

$$G_{l_1,\ldots,l_u|l_0}(s_1, \ldots, s_u) = [1 - \{p_{01}(1 - s_1) + p_{02}s_1(1 - s_2)$$

$$+ p_{03}s_1s_2(1 - s_3) + \cdots + p_{0u}s_1s_2\cdots s_{u-1}(1 - s_u)\}]^{l_0}. \quad (2.11)$$

The p.g.f. (2.11) is then used to derive the joint probability function and moments of the random variables $l_1,\ldots, l_u$ by differentiating (2.11) with respect to the arguments. The joint probability function turns out to be

$$\Pr\{l_1 = k_1, \ldots, l_u = k_u \mid l_0\} = \prod_{i=0}^{u-1} \binom{k_i}{k_{i+1}} p_i^{k_{i+1}}(1 - p_i)^{k_i - k_{i+1}}$$

$$k_{i+1} = 0, 1, \ldots, k_i, \quad \text{with} \quad k_0 = l_0. \quad (2.12)$$

The expected values and covariances are

$$E(l_i \mid l_0) = l_0 p_{0i} \quad (2.13)$$

and

$$\sigma_{l_i,l_j|l_0} = l_0 p_{0j}(1 - p_{0i}), \quad i \le j; i, j = 1, \ldots, u. \quad (2.14)$$

When $i = j$, (2.14) reduces to the variance of l_i (Equation (1.10)). These results show that *for a given u, $l_1, \ldots, l_u$ in the life table form a chain of binomial distributions; the p.g.f., the joint probability distribution, the expected values, and covariances are given in (2.11), (2.12), (2.13) and (2.14), respectively.*

▶

2.1. An Urn Scheme

The life table functions can be generated from an entirely different approach. As an example, consider an experiment in which balls are drawn with replacement from an infinite sequence of urns, numbered $0, 1, \ldots$. In the ith urn there is a proportion p_i of white balls and a proportion q_i of black balls with $0 < p_i < 1$ and $p_i + q_i = 1$. Beginning with the 0th urn a number l_0 of balls is drawn of which l_1 are white; then a total of l_1 balls is drawn from the first urn of which l_2 are white; l_2 balls are then drawn from the second urn of which l_3 are white, and so on. In general, the number l_{i+1} of white balls drawn from the ith urn is the number of balls to be drawn from the next or the $(i + 1)$th urn. The experiment terminates as soon as no white balls are drawn. Let the last urn from which balls are drawn be the Wth urn, so that $l_i > 0$ for $i \le W$ and $l_i = 0$ for $i > W$.

The correspondence between the urn scheme and the problem of the life table is evident. For example l_0 is the initial size of the cohort, p_i (or q_i) is the probability of surviving (or dying in) an age interval, and l_i is the number of survivors at age x_i, $i = 0, 1, \ldots$. The number W corresponding to the beginning of the last age interval is also a random variable, which we shall now discuss.

To derive the probability distribution of W, we note that for $W = w$ there must be $l_w = k_w$ drawings from the wth urn for $1 \le k_w \le l_0$ and all k_w balls drawn must be black balls. Therefore we have the probability

$$\Pr\{W = w\} = \sum_{k_w=1}^{l_0} \binom{l_0}{k_w} p_{0w}{}^{k_w}(1 - p_{0w})^{l_0-k_w}(1 - p_w)^{k_w}, \qquad w = 0, 1, \ldots,$$

(2.15)

where, for convenience,

$$p_{0w} = p_0 p_1 \cdots p_{w-1}. \tag{2.16}$$

The expectation of W is more conveniently obtained if the probability in (2.15) is rewritten as

$$\sum_{k_w=1}^{l_0} \binom{l_0}{k_w} p_{0w}{}^{k_w}(1 - p_{0w})^{l_0-k_w}(1 - p_w)^{k_w} = (1 - p_{0,w+1})^{l_0} - (1 - p_{0w})^{l_0}$$

(2.17)

which can be verified by direct computation. The expectation of W is now given by

$$E(W) = \sum_{w=0}^{\infty} w[(1 - p_{0,w+1})^{l_0} - (1 - p_{0w})^{l_0}]. \tag{2.18}$$

For a given v, we write the partial sum

$$\sum_{w=0}^{v} w[(1 - p_{0,w+1})^{l_0} - (1 - p_{0w})^{l_0}] = \sum_{w=1}^{v} [(1 - p_{0,v+1})^{l_0} - (1 - p_{0w})^{l_0}].$$

(2.19)

Letting $v \to \infty$ and $p_{0,v+1} \to 0$, we have from (2.19)

$$E(W) = \sum_{w=1}^{\infty} [1 - (1 - p_{0w})^{l_0}]. \tag{2.20}$$

For $l_0 = 1$,

$$E(W) = p_{01} + p_{02} + p_{03} + \cdots \tag{2.21}$$

which is closely related to the expectation of life e_0

497

If the force of mortality were independent of age with $\mu(\tau) = \mu$ for $0 \leq \tau < \infty$, then the proportion of white balls in each urn would be constant with $p_i = p$. In this case $p_{0i} = p^i$,

$$\Pr\{W = w\} = (1 - p^{w+1})^{l_0} - (1 - p^w)^{l_0}$$

$$= \sum_{k=1}^{l_0} (-1)^{k+1} \binom{l_0}{k} (1 - p^k) p^{wk}, \qquad (2.22)$$

and the expectation, the variance, and the p.g.f. of W all have closed forms. Using (2.22) we compute the p.g.f. of W

$$G_W(s) = \sum_{w=0}^{\infty} s^w \sum_{k=1}^{l_0} (-1)^{k+1} \binom{l_0}{k} (1 - p^k) p^{wk}$$

$$= \sum_{k=1}^{l_0} (-1)^{k+1} \binom{l_0}{k} (1 - p^k)(1 - p^k s)^{-1}, \qquad (2.23)$$

the expectation

$$E(W) = \sum_{k=1}^{l_0} (-1)^{k+1} \binom{l_0}{k} (1 - p^k)^{-1} p^k \qquad (2.24)$$

and the variance

$$\sigma_W^2 = \sum_{k=1}^{l_0} (-1)^{k+1} \binom{l_0}{k} (1 + p^k)(1 - p^k)^{-2} p^k - [E(W)]^2. \qquad (2.25)$$

When $l_0 = 1$, W has the geometric distribution. ...

3. JOINT PROBABILITY DISTRIBUTION OF THE NUMBERS OF DEATHS

In a life table covering the entire life span of each individual in a given population, the sum of the deaths at all ages is equal to the size of the original cohort. Symbolically,

$$d_0 + d_1 + \cdots + d_w = l_0, \qquad (3.1)$$

where d_w is the number of deaths in the age interval (x_w and over). Each individual in the original cohort has a probability $p_{0i} q_i$ of dying in the interval (x_i, x_{i+1}), $i = 0, 1, \ldots, w$. Since an individual dies once and only once in the span covered by the life table,

$$p_{00} q_0 + \cdots + p_{0w} q_w = 1, \qquad (3.2)$$

498

where $p_{00} = 1$ and $q_w = 1$. Thus we have the well-known results: *The numbers of deaths, $d_0, \ldots, d_w$, in a life table have a multinomial distribution with the joint probability distribution*

$$\Pr\{d_0 = \delta_0, \ldots, d_w = \delta_w \mid l_0\} = \frac{l_0!}{\delta_0! \cdots \delta_w!} (p_{00}q_0)^{\delta_0} \cdots (p_{0w}q_w)^{\delta_w}; \quad (3.3)$$

the expectation, variance and covariance are given, respectively, by

$$E(d_i \mid l_0) = l_0 p_{0i} q_i, \quad (3.4)$$

$$\sigma^2_{d_i \mid l_0} = l_0 p_{0i} q_i (1 - p_{0i} q_i), \quad (3.5)$$

and

$$\sigma_{d_i, d_j \mid l_0} = -l_0 p_{0i} q_i p_{0j} q_j \quad for \quad i \neq j; \, i, j = 0, 1, \ldots, w. \quad \blacktriangleright(3.6)$$

In the discussion above, age 0 was chosen only for simplicity. For any given age, say x_α, the probability that an individual alive at age x_α will die in the interval (x_i, x_{i+1}) subsequent to x_α is $p_{\alpha i} q_i$ and the sum

$$\sum_{i=\alpha}^{w} p_{\alpha i} q_i = 1, \quad (3.7)$$

and thus the numbers of deaths in intervals beyond x_α also have a multinomial distribution.

References

Auerbach, F.: Das Gesetz der Bevölkerungskonzentration. Petermanns Mitteilungen **54**, 74—76 (1913).

Bailey, N.T.J.: The Elements of Stochastic Processes with Applications to the Natural Sciences. New York: John Wiley & Sons 1964.

Bartlett, M.S.: A note on Das Gupta's two-sex population model. Theoretical Population Biology **4**, 418—424 (1973).

Basu, D.: A note on the structure of a stochastic model considered by V.M. Dandekar. Sankhya **15**, 251—252 (1955).

Bernardelli, H.: Population waves. Journal of the Burma Research Society **31** (Part 1), 1—18 (1941).

Bernoulli, D.: Essai d'une nouvelle analyse de la mortalité causée par la petite vérole et les avantages de l'inoculation pour la prévenir. Histoire de l'Académie Royale des Sciences, Année 1760, 1—45 (1766). An English translation will be found in L. Bradley: Smallpox Innoculation: An Eighteenth-Century Mathematical Controversy. Nottingham: Adult Education Department of the University of Nottingham 1971. (See also the compact summary of Bernoulli in I. Sutherland's review of Bradley, Population Studies **29**, 317—318 (1975).)

Böckh, R.: Statistisches Jahrbuch der Stadt Berlin. Volume 12. Statistik des Jahres 1884. Berlin 1886.

Bortkiewicz, L.V.: Die Sterbeziffer und der Frauenüberschuß in der stationären und in der progressiven Bevölkerung. Bulletin de l'Institut International de Statistique **19**, 63—138 (1911).

Bowley, A.L.: Births and population in Great Britain. The Economic Journal **34**, 188—192 (1924).

Bowley, A.L.: Estimates of the working population of certain countries in 1931 and 1941. Preparatory Committee Paper CECP 59 (1). International Economic Conference. Geneva 1926.

Brass, W.: Perspectives in population prediction: Illustrated by the statistics of England and Wales. Journal of the Royal Statistical Society A **137**, 532—570 (1974).

Brass, W., Coale, A.J.: Methods of analysis and estimation. In: Brass, W. et.al.: The Demography of Tropical Africa, 88—139. Princeton: Princeton University Press 1968.

Brillinger, D.R.: A justification of some common laws of mortality. Transactions of the Society of Actuaries **13**, 116—119 (1961).

Cannan, E.: The probability of a cessation of the growth of population in England and Wales during the next century. The Economic Journal **5**, 505—515 (1895).

Cardano, G.: Opus Novvm de Proportionibvs Nvmerorvm. Basel 1570.

Chiang, C.L.: A stochastic study of the life table and its applications: I. Probability distributions of the biometric functions. Biometrics **16**, 618—635 (1960a).

Chiang, C.L.: A stochastic study of the life table and its applications: II. Sample variance of the observed expectation of life and other biometric functions. Human Biology **32**, 221—238 (1960b).

Chiang, C.L.: Introduction to Stochastic Processes in Biostatistics. New York: John Wiley & Sons 1968.

Coale, A.J.: How the age distribution of a population is determined. Cold Spring Harbor Symposia on Quantitative Biology **22**, 83—89 (1957a).

Coale, A.J.: A new method for calculating Lotka's r—the intrinsic rate of growth in a stable population. Population Studies **11**, 92—94 (1957b).

Coale, A.J.: Age patterns of marriage. Population Studies **25**, 193—214 (1971).

Coale, A.J.: The Growth and Structure of Human Populations. Princeton: Princeton University Press 1972.

Coale, A.J., Demeny, P.: Regional Model Life Tables and Stable Populations. Princeton: Princeton University Press 1966.

Coale, A.J., McNeil, D.R.: The distribution by age of the frequency of first marriage in a female cohort. Journal of the American Statistical Association **67**, 743—749 (1972).

Coale, A.J., Trussell, T.J.: Model fertility schedules: Variations in the age structure of childbearing in human populations. Population Index **40**, 185—258 (1974).

Cole, L.C.: The population consequences of life history phenomena. Quarterly Review of Biology **29**, 103—137 (1954).

Croxton, F.E., Cowden, D.J., Klein, S.: Applied General Statistics. Third Edition. Englewood Cliffs, New Jersey: Prentice-Hall 1967 (1939).

Das Gupta, P.: On two-sex models leading to stable populations. Theoretical Population Biology **3**, 358–375 (1972).

DeCandolle, A.: Histoire des Sciences et des Savants depuis Deux Siècles. Geneva 1873.

Deevey, E. S., Jr.: Life tables for natural populations of animals. Quarterly Review of Biology **22**, 283—314 (1947).

DeMoivre, A.: Annuities on Lives: Or, the Valuation of Annuities Upon Any Number of Lives; as also, of Reversions. London 1725.

DeMoivre, A.: Miscellanea Analytica de Seriebus et Quadraturis. London 1730. [DeMoivre's treatment of recurrent series is available in English in Demoivre 1738, 193—206.]

DeMoivre, A.: Doctrine of Chances. Second Edition. London 1738.

DeMoivre, A.: Approximatio ad summam terminorum binomii $(a+b)^n$ in seriem expansi. London 1733. Republished with an introduction by R. C. Archibald in Isis **8**, 671—683 (1926). [An English version will be found in DeMoivre 1738, pp. 235—243.]

Deparcieux, A.: Essai sur les Probabilités de la Durée de la Vie Humaine …. Paris 1760 (1746).

DeWit, J.: Waardye van Lyf-renten naer Proportie van Losrenten. The Hague 1671. [An English translation will be found in Hendriks 1852, pp. 232—249.]

Dublin, L. I., Lotka, A. J.: On the true rate of natural increase. Journal of the American Statistical Association **20**, 305—339 (1925).

Dublin, L. I., Lotka, A. J., Spiegelman, M.: Length of Life. New York: Ronald Press 1949 (1936).

Duncan, O. D.: The measurement of population distribution. Population Studies **11**, 27—45 (1957).

Duvillard, É. É.: Analyse et Tableaux de l'Influence de la Petite Vérole sur la Mortalité à Chaque Age …. Paris 1806.

Easterlin, R. A.: Economic-demographic interactions and long swings in economic growth. American Economic Review **56**, 1063—1104 (1966).

Easterlin, R. A.: Population, Labor Force, and Long Swings in Economic Growth: The American Experience. New York: Columbia University Press 1968.

Edmonds, T. R.: On the law of increase of the population of England during the last 100 years. Assurance Magazine **2**, 57—69 (1852).

Elderton, W. P., Johnson, N. L.: Systems of Frequency Curves. Cambridge: Cambridge University Press 1969.

Euler, L.: Introductio in Analysin Infinitorum. Lausanne 1748.

Euler, L.: Recherches générales sur la mortalité et la multiplication du genre humaine. Histoire de l'Académie Royale des Sciences et Belles Lettres **16**, 144—164 (1760). [An English translation will be found in Euler 1970.]

Euler, L.: A general investigation into the mortality and multiplication of the human species. Translated by N. and B. Keyfitz. Theoretical Population Biology **1**, 307—314 (1970).

Farr, W.: English Life Table. London: Longman 1864.

Feeney, G. M.: A model for the age distribution of first marriage. Working Papers of the East-West Population Institute, No. 23 (1972).

Feller, W.: Die Grundlagen der Volterraschen Theorie des Kampfes ums Dasein in Wahrscheinlichkeitstheoretischer Behandlung. Acta Biotheoretica **5**, 11—40 (1939).

Feller, W.: On the logistic law of growth and its empirical verifications in biology. Acta Biotheoretica **5**, 51—66 (1940).

Feller, W.: On the integral equation of renewal theory. Annals of Mathematical Statistics **12**, 243—267 (1941).

Feller, W.: An Introduction to Probability Theory and Its Applications, Volume 1. Third Edition (Revised Printing). New York: John Wiley & Sons 1968 (1950).

Fisher, R. A.: The Genetical Theory of Natural Selection. New York: Dover 1958 (1930).

Fredrickson, A. G.: A mathematical theory of age structure in sexual populations: Random mating and monogamous marriage models. Mathematical Biosciences **10**, 117—143 (1971).

Frobenius, G.: Über Matrizen aus nicht negativen Elementen. Sitzungsberichte der Königlich Preussischen Akademie der Wissenschaften [Berlin], 456—477 (1912).

Furry, W. H.: On fluctuation phenomena in the passage of high energy electrons through lead. Physical Review **52**, 569—581 (1937).

Galton, F., Watson, H. W.: On the probability of extinction of families. Journal of the Anthropological Institute **4**, 138—144 (1874).

Gautier, É., Henry, L.: La Population de Crulai, Paroisse Normande. Paris: Presses Universitaires de France 1958.

Gini, C.: Premières recherches sur la fécondabilité de la femme. Proceedings of the International Mathematics Congress, 889—892. Toronto 1924.

Gompertz, B.: On the nature of the function expressive of the law of mortality. Philosophical Transactions 27, 513—585 (1825).

Goodman, L.A.: Population growth of the sexes. Biometrics 9, 212—225 (1953).

Graunt, J.: Natural and Political Observations Mentioned in a Following Index, and Made upon the Bills of Mortality. London 1662. Republished with an introduction by B. Benjamin in the Journal of the Institute of Actuaries 90, 1—61 (1964).

Greenwood, M.: The natural duration of cancer. Reports on Public Health and Medical Subjects 33, 1—26 (1926).

Greville, T.N.E.: Short methods of constructing abridged life tables. Record of the American Institute of Actuaries 32, 29—42 (1943).

Greville, T.N.E., Keyfitz, N.: Backward population projection by a generalized inverse. Theoretical Population Biology 6, 135—142 (1974).

Gumbel, E.J.: Eine Darstellung statistischer Reihen durch Euler. Jahresbericht der Deutschen Mathematiker-Vereinigung 25, 251—264.

Hadwiger, H.: Eine analytische Reproduktionsfunktion für biologische Gesamtheiten. Skandinavisk Aktuarietidskrift 23, 101—113 (1940).

Hajnal, J.: Aspects of recent trends in marriage in England and Wales. Population Studies 1, 72—92 (1947).

Halley, E.: An estimate of the degrees of the mortality of mankind.... Philosophical Transactions 17, 596—610, 653—656 (1693).

Hendriks, F.: Contributions to the history of insurance, with a restoration of the Grand Pensionary De Wit's treatise on life annuities. Assurance Magazine 2, 121—150, 222—258 (1852); 3, 93—120 (1853).

Henry, L.: Fondements théoriques des mesures de la fécondité naturelle. Revue de l'Institut International de Statistique 21, 135—151 (1953). [An English translation will be found in Henry 1972, pp. 1—26: "Theoretical basis of measures of natural fertility."]

Henry, L.: Schémas de nuptialité: Déséquilibre des sexes et âge au mariage. Population 24, 1067—1122 (1969).

Henry, L.: On the Measurement of Human Fertility. Selected Writings. Translated and edited by M.C. Sheps and E. Lapierre-Adamcyk. New York: Elsevier 1972.

Hernes, G.: The process of entry into first marriage. American Sociological Review 37, 173—182 (1972).

Hoem, J.H.: Concepts of a bisexual theory of marriage formation. Särtryck ur Statistisk Tidskrift 4, 295—300 (1969).

Hogben, L.: Genetic Principles in Medicine and Social Science. London: Williams and Norgate 1931.

Irwin, J.O.: The standard error of an estimate of expectation of life, with special reference to expectation of tumourless life in experiments with mice. Journal of Hygiene 47, 188—189 (1949).

Karmel, P.H.: The relations between male and female reproduction rates. Population Studies 1, 249—274 (1947).

Karmel, P.H.: A rejoinder to Mr. Hajnal's comments. Population Studies 2, 361—372 (1948).

Karmel, P.H.: Le conflit entre les mesures masculine et féminine de la réproduction. Population 4, 471—494 (1949).

Kendall, D.G.: Stochastic processes and population growth. Journal of the Royal Statistical Society B 11, 230—264 (1949).

Kendall, M.G., Stuart, A.: The Advanced Theory of Statistics, Volume 1. Third Edition. New York: Hafner 1969 (1958).

Keyfitz, N.: Introduction to the Mathematics of Population. Reading, Massachusetts: Addison-Wesley 1968.

Keyfitz, N.: The mathematics of sex and marriage. Proceedings of the Sixth Berkeley Symposium on Mathematical Statistics and Probability, Volume IV: Biology and Health, 89—108. Berkeley: University of California Press 1971.

Keyfitz, N.: On future population. Journal of the American Statistical Association 67, 347—363 (1972).

Keyfitz, N., Frauenthal, J.: An improved life table method. Biometrics 31, 889—899 (1975).

King, G.: On the method used by Milne in the construction of the Carlisle table of mortality. Journal of the Institute of Actuaries **24**, 186—204 (1884).

King, G.: Institute of Actuaries Textbook, Part II. Second Edition. London: Charles and Edward Layton 1902.

Kuczynski, R.R.: The Balance of Births and Deaths. Volume 1. New York: Macmillan 1928.

Kuczynski, R.R.: The Balance of Births and Deaths. Volume 2. New York: Macmillan 1931a.

Kuczynski, R.R.: Fertility and Reproduction: Methods of Measuring the Balance of Births and Deaths. New York: Macmillan 1931b.

Kuczynski, R.R.: The Measurement of Population Growth. London: Sidgwick & Jackson 1935.

Lagrange, J.L.: Sur l'intégration d'une équation différentielle à différences finies, qui contient la théorie des suites récurrentes. Miscellanea Taurinensia **1**, 33—42 (1759). Republished in Oeuvres de Lagrange, Volume 1, 23—36. Paris 1867.

Ledermann, S., Breas, J.: Les dimensions de la mortalité. Population **14**, 637—682 (1959).

Legendre, A.M.: Nouvelles méthodes pour la détermination des orbites des comètes. Paris 1805.

Leslie, P.H.: On the use of matrices in certain population mathematics. Biometrika **33**, 183—212 (1945).

Leslie, P.H.: Some further notes on the use of matrices in population mathematics. Biometrika **35**, 213—245 (1948).

Lewis, E.G.: On the generation and growth of a population. Sankhya **6**, 93—96 (1942).

Lexis, W.: Einleitung in die Theorie der Bevölkerungs-Statistik. Strasbourg: K.J. Trübner 1875.

Lopez, A.: Problems in Stable Population Theory. Princeton: Office of Population Research 1961.

Lorimer, F.: The development of demography. In Hauser, P.M. and Duncan, O.D., editors: The Study of Population, 124—179. Chicago: University of Chicago Press 1959.

Lotka, A.J.: Relation between birth rates and death rates. Science N.S. **26**, 21—22 (1907).

Lotka, A.J.: The stability of the normal age distribution. Proceedings of the National Academy of Sciences **8**, 339—345 (1922).

Lotka, A.J.: Industrial replacement. Skandinavisk Aktuarietidskrift, 51—63 (1933).

Lotka, A.J.: Théorie Analytique des Associations Biologiques. Deuxième Partie: Analyse Démographique avec Application Particulière à l'Espèce Humaine. (Actualités Scientifiques et Industrielles, No. 780.) Paris: Hermann & Cie 1939.

Makeham, W.M.: On the law of mortality and the construction of annuity tables. Assurance Magazine **8**, 301—310 (1860).

Makeham, W.M.: On the law of mortality. Journal of the Institute of Actuaries **13**, 325—367 (1867).

Malthus, T.R.: An Essay on the Principle of Population *and* A Summary View of the Principle of Population. Harmondsworth, Middlesex, England: Penguin 1970 (1798, 1830).

McFarland, D.D.: Comparison of alternative marriage models. In Greville, T.N.E., editor: Population Dynamics, 89—106. New York: Academic Press 1972.

Milne, J.: A Treatise on the Valuation of Annuities and Assurances on Lives and Survivorships.... London 1815.

Mitra, S.: The pattern of age-specific fertility rates. Demography **4**, 894—906 (1967).

Myers, R.J.: The validity and significance of male net reproduction rates. Journal of the American Statistical Association **36**, 275—282 (1941).

Nicander, H.: "Om Tabell-Vårkets Tillstånd i Sverige och Finland ifrån 1772 till och med 1795." Kongliga Vetenskaps Academiens nya Handlingar (Stockholm) **20**, 151—162, 239—256 (1799); **21**, 1—6, 75—82, 155—172, 319—328 (1800); **22**, 56—72, 298—338 (1801).

Parlett, B.: Ergodic properties of populations I: The One Sex Model. Theoretical Population Biology **1**, 191—207 (1970).

Parlett, B.: Can there be a marriage function? In Greville, T.N.E., editor: Population Dynamics, 107—136. New York: Academic Press 1972.

Pearl, R.: The Biology of Death. Philadelphia: J.B. Lippincott 1922.

Pearl, R.: Factors in human fertility and their statistical evaluation. Lancet **225**, 607—611 (1933).

Pearl, R., Reed, L.J.: The rate of growth of the population of the United States since 1790 and its mathematical representation. Proceedings of the National Academy of Science **6**, 275—288 (1920).

Pearson, K.: Asymmetrical frequency curves. Nature **48**, 615—616 (1893).

Pearson, K.: Early Statistical Papers. London: Cambridge University Press 1948 (1894—1916).

Perron, O.: Zur Theorie der Matrizen. Mathematische Annalen **64**, 248—263 (1907).

Pollard, A. H.: The measurement of reproductivity. Journal of the Institute of Actuaries **74**, 288—305 (1948).

Pollard, J. H.: On the use of the direct matrix product in analysing certain stochastic population models. Biometrika **53**, 397—415 (1966).

Pollard, J. H.: Mathematical Models for the Growth of Populations. Cambridge: Cambridge University Press 1973.

Potter, R. G., Jr.: Birth intervals: Structure and change. Population Studies **17**, 155—166 (1963).

Potter, R. G., Jr., Parker, M. P.: Predicting the time required to conceive. Population Studies **18**, 99—116 (1964).

Pressat, R.: L'Analyse Démographique. Second Edition. Paris: Presses Universitaires de France 1969 (1961).

Pressat, R.: Demographic Analysis. Translated by J. Matras. New York: Aldine-Atherton 1972.

Pretorius, S. J.: Skew bi-variate frequency surfaces, examined in the light of numerical illustrations. Biometrika **22**, 109—223 (1930).

Price, R.: Observations on Reversionary Payments. London 1771.

Reed, L. J., Merrell, M.: A short method for constructing an abridged life table. American Journal of Hygiene **30**, 33—62 (1939).

Rogers, A.: Matrix Analysis of Interregional Population Growth and Distribution. Berkeley: University of California Press 1968.

Ryder, N. B.: The process of demographic translation. Demography **1**, 74—82 (1964).

Saibante, M.: La concentrazione della popolazione. Metron **7**, 53—99 (1928).

Samuelson, P. A., Yellin, J.: Non-linear population analysis. Unpublished manuscript June 1974.

Schweder, T.: The precision of population projections studied by multiple prediction methods. Demography **8**, 441—450 (1971).

Sharpe, F. R., Lotka, A. J.: A problem in age-distribution. Philosophical Magazine, Ser. 6 **21**, 435—438 (1911).

Sheps, M. C.: On the time required for conception. Population Studies **18**, 85—97 (1964).

Sheps, M. C., Menken, J. A.: Mathematical Models of Conception and Birth. Chicago: University of Chicago Press 1973.

Sheps, M. C., Menken, J. A., Radick, A. P.: Probability models for family building: An analytic review. Demography **6**, 161—183 (1969).

Simpson, T.: The Doctrine of Annuities and Reversions. London 1742.

Steffensen, J. F.: Om Sandsynligheden for at Afkommet uddør. Matematisk Tidsskrift B, 19—23 (1930).

Steffensen, J. F.: Deux problèmes du calcul des probabilités. Ann. Inst. H. Poincaré **3**, 319—344 (1933).

Süssmilch, J. P.: Die göttliche Ordnung. Fourth edition. Three volumes. Berlin 1798. [The materials we quote originally appeared in the second edition (1761, 1762).]

Sutton, W.: On the method used by Dr. Price in the construction of the Northampton mortality table. Journal of the Institute of Actuaries **18**, 107—115 (1874).

Sutton, W.: On the method used by Milne in the construction of the Carlisle table of mortality. Journal of the Institute of Actuaries **24**, 110—122 (1884).

Sykes, Z. M.: Some stochastic versions of the matrix model for population dynamics. Journal of the American Statistical Association **64**, 111—130 (1969).

Tabah, L.: Représentations matricielles de perspectives de population active. Population **23**, 437—476 (1968).

Thompson, W. R.: On the reproduction of organisms with overlapping generations. Bulletin of Entomological Research **22**, 147—172 (1931).

Thompson, W. S., Whelpton, P. K.: Population Trends in the United States. New York: McGraw-Hill 1933.

Tietze, C.: Pregnancy rates and birth rates. Population Studies **16**, 31—37 (1962).

Trenerry, C. F.: The Origin and Early History of Insurance. London: P. S. King & Son 1926.

United Nations: Age and Sex Patterns of Mortality: Model Life Tables for Underdeveloped Countries. New York: United Nations 1955.

United Nations: United Nations Manual IV: Methods of Estimating Basic Demographic Measures from Incomplete Data. New York: United Nations 1967.

Verhulst, P. F.: Notice sur la loi que la population suit dans son accroissement. Correspondance Mathématique et Physique Publiée par A. Quételet **10**, 113—121 (1838).

Verhulst, P.F.: "Recherches mathématiques sur la loi d'accroissement de la population." Nouveaux Mémoires de l'Académie Royale des Sciences et Belles Lettres (Brussels) **18**, 3—41 (1845); **20** (3), 4—32 (1847).

Vincent, P.: De la mesure du taux intrinsèque d'accroissement naturel dans les populations monogames. Population **1**, 699—712 (1946).

Waugh, W.A.O'N.: An age-dependent birth and death process. Biometrika **42**, 291—306 (1955).

Westergaard, H.: Contributions to the History of Statistics. The Hague: Mouton 1969 (1932).

Whelpton, P.K.: Population of the United States, 1925—1975. American Journal of Sociology **34**, 253—271 (1928).

Whelpton, P.K.: An empirical method for calculating future population. Journal of the American Statistical Association **31**, 457—473 (1936).

Wicksell, S.D.: Nuptuality, fertility, and reproductivity. Skandinavisk Aktuarietidskrift, 125—157 (1931).

Wilson, E.B.: The standard deviation of sampling for life expectancy. Journal of the American Statistical Association **33**, 705—708 (1938).

Wolfenden, H.H.: The Fundamental Principles of Mathematical Statistics. New York: Actuarial Society of America 1942.

Wolfenden, H.H.: Population Statistics and Their Compilation. Chicago: University of Chicago Press 1954 (1925).

Wrigley, E.A.: A simple model of London's importance in changing English society and economy 1650—1750. Past and Present **37**, 44—70 (1967).

Yntema, L.: Mathematical Models of Demographic Analysis. Leiden: J.J. Groen & Zoon 1952.

Yule, G.U.: On the changes in the marriage and birth rates in England and Wales during the past half century with an inquiry as to their probable causes. Journal of the Royal Statistical Society **69**, 88—132 (1906).

Yule, G.U.: A mathematical theory of evolution, based on the conclusions of Dr. J.C. Willis, F.R.S. Philosophical Transactions of the Royal Society B **213**, 21—87 (1924).

Author Index

Subject Index

511

513

A Springer Journal

Journal of

Mathematical Biology

Ecology and Population Biology
Epidemiology
Immunology
Neurobiology
Physiology
Artificial Intelligence
Developmental Biology
Chemical Kinetics

Edited by H.J. Bremermann, Berkeley, CA; F.A. Dodge, Yorktown Heights, NY; K.P. Hadeler, Tübingen; S.A. Levin, Ithaca, NY; D. Varjú, Tübingen.

Advisory Board: M.A. Arbib, Amherst, MA; E. Batschelet, Zurich; W. Bühler, Mainz; B.D. Coleman, Pittsburg, PA; K. Dietz, Tübingen; W. Fleming, Providence, RI; D. Glaser, Berkeley, CA; N.S. Goel, Binghamton, NY; J.N.R. Grainger, Dublin; F. Heinmets, Natick, MA; H. Holzer, Freiburg i. Br.; W. Jäger, Heidelberg; K. Jänich, Regensburg; S. Karlin, Rehovot/Stanford, CA; S. Kauffman, Philadelphia, PA; D.G. Kendall, Cambridge; N. Keyfitz, Cambridge, MA; B. Khodorov, Moscow; E.R. Lewis, Berkeley, CA; D. Ludwig, Vancouver; H. Mel, Berkeley, CA; H. Mohr, Freiburg i. Br.; E.W. Montroll, Rochester, NY; A. Oaten, Santa Barbara, CA; G.M. Odell, Troy, N.Y.; G. Oster, Berkeley, CA; T. Poggio, Tübingen; K.H. Pribram, Stanford, CA; S.I. Rubinow, New York, NY; W.v. Seelen, Mainz; L.A. Segel, Rehovot, Israel; W. Seyffert, Tübingen; H. Spekreijse, Amsterdam; R.B. Stein, Edmonton; R. Thom, Bures-sur-Yvette; Jun-ichi Toyoda, Tokyo; J.J. Tyson, Innsbruck; J. Vandermeer, Ann Arbor, MI.

Springer-Verlag
Berlin
Heidelberg
New York

Journal of Mathematical Biology publishes papers in which mathematics leads to a better understanding of biological phenomena, mathematical papers inspired by biological research and papers which yield new experimental data bearing on mathematical models. The scope is broad, both mathematically and biologically and extends to relevant interfaces with medicine, chemistry, physics and sociology. The editors aim to reach an audience of both mathematicians and biologists.

Lecture Notes in Biomathematics

This series aims to report new developments
in biomathematics research and teaching –
quickly, informally and at a high level.

Springer-Verlag
Berlin
Heidelberg
New York